Encyclopedia of Chemical Engineering

Volume I

Encyclopedia of Chemical Engineering Volume I

Edited by **Lewis Allison**

CLANRYE INTERNATIONAL

New Jersey

Published by Clanrye International,
55 Van Reypen Street,
Jersey City, NJ 07306, USA
www.clanryeinternational.com

Encyclopedia of Chemical Engineering: Volume I
Edited by Lewis Allison

International Standard Book Number: 978-1-63240-176-2 (Hardback)

Printed in the United States of America.

Contents

Preface

Chemical engineering is a comparatively recent branch of engineering. It merges natural sciences, life sciences, mathematics and economics to create, transform and transport chemicals, materials and energy properly.

It is also extremely relevant to fields like nanotechnology, fuel cells and bioengineering, given the field's affinity to trailblazing valuable materials related techniques. Within chemical engineering, there exist broadly two sub-categories which are design, manufacture, and operation of plants and machinery in industrial chemical and related processes, better known as "chemical process engineering" and development of new or adapted substances for products ranging from edibles to cosmetics and cleaners to pharmaceutical ingredients. This is better known as "chemical product engineering".

Though this field of engineering initially appeared in the late 19th century, its development mainly happened towards the latter half of the 20th century. Advancements in computer science found applications, designing and managing plants, simplifying calculations and drawings that previously had to be done manually, consuming a lot more time and effort, thereby hampering productivity. The completion of the Human Genome Project is considered to be a benchmark, both for chemical engineering and genetic engineering and genomics as well. Chemical engineering principles find applications in the production of DNA sequences in large quantities. This is often looked upon as a new trend in biotechnology.

There are various fields of chemical engineering like chemical reaction engineering, plant design, process design and transport phenomena. This book takes a look at the collective field of chemical engineering, bringing forth the latest advancements and developments. It should be of great use to students and professionals.

I would personally like to thank all the contributors who have put in their hard work and efforts in these researchers. I would also like to thank the publishing team who have been extremely supportive and my family for their support at every step.

Editor

A *Priori* Direct Numerical Simulation Modelling of the Curvature Term of the Flame Surface Density Transport Equation for Nonunity Lewis Number Flames in the Context of Large Eddy Simulations

Mohit Katragadda and Nilanjan Chakraborty

School of Mechanical and Systems Engineering, Newcastle University, Claremont Road, Newcastle-upon-Tyne NE1 7RU, UK

Correspondence should be addressed to Nilanjan Chakraborty, nilanjan.chakraborty@newcastle.ac.uk

Academic Editor: Mahesh T. Dhotre

A Direct Numerical Simulation (DNS) database of freely propagating statistically planar turbulent premixed flames with Lewis numbers Le ranging from 0.34 to 1.2 has been used to analyse the statistical behaviours of the curvature term of the generalised Flame surface Density (FSD) transport equation, in the context of the Large Eddy Simulation (LES). Lewis number is shown to have significant influences on the statistical behaviours of the resolved and sub-grid parts of the FSD curvature term. It has been found that the existing models for the sub-grid curvature term C_{sg} do not capture the qualitative behaviour of this term extracted from the DNS database for flames with Le $\ll$ 1. The existing models of C_{sg} only predict negative values, whereas the sub-grid curvature term is shown to assume positive values within the flame brush for the Le = 0.34 and 0.6 flames. Here the sub-grid curvature terms arising from combined reaction and normal diffusion and tangential diffusion components of displacement speed are individually modelled, and the new model of the sub-grid curvature term has been found to capture C_{sg} extracted from DNS data satisfactorily for all the different Lewis number flames considered here for a wide range of filter widths.

1. Introduction

Flame Surface Density (FSD) based reaction rate closure is well established in the context of Reynolds Averaged Navier-Stokes (RANS) simulations of turbulent premixed flames [1, 2]. The increased affordability of high performance computing has made Large Eddy Simulation (LES) an alternative simulation tool, where the large-scale physical processes are resolved, but modelling is still required for the subgrid quantities. The FSD-based reaction rate closure has recently been successfully extended for the purpose of LES [3–14]. In LES simulation of premixed combustion, a Favre-filtered reaction progress variable transport equation is solved alongside other filtered conservation equations. The reaction progress variable is defined as $c = (Y_{R0} - Y_R)/(Y_{R0} - Y_{R\infty})$, where Y_R is the mass fraction of a suitable reactant and the subscripts 0 and ∞ denote the values in the fully unburned and burned gases, respectively. The generalised

FSD is defined as $\Sigma_{gen} = \overline{|\nabla c|}$ [3–14], where the overbar indicates the LES filtering process. The Favre-filtered reaction progress variable transport equation takes the following form:

$$\overline{\rho}\frac{\partial \tilde{c}}{\partial t} + \overline{\rho}\tilde{u}_j\frac{\partial \tilde{c}}{\partial x_j} = \overline{\frac{\partial}{\partial x_j}\left(\rho D \frac{\partial c}{\partial x_j}\right)} + \overline{\dot{w}} - \frac{\partial}{\partial x_j}\left(\overline{\rho}\left(\widetilde{u_j c} - \tilde{u}_j\tilde{c}\right)\right), \quad (1)$$

where $\tilde{Q} = \overline{\rho Q}/\overline{\rho}$ indicates the Favre filtered value of a general variable Q, u_j is the velocity component in the jth direction, ρ is the density, D is the molecular diffusivity, and $\overline{\dot{w}}$ is the filtered reaction rate. The first two terms on right hand side of (1) denote the filtered molecular diffusion and reaction rates, respectively, and their combined contribution can be modelled using Σ_{gen} in the following manner:

$$\overline{\dot{w} + \nabla \cdot (\rho D \nabla c)} = \overline{(\rho S_d)}_s \Sigma_{gen}, \quad (2)$$

where $\overline{(Q)}_s = \overline{Q|\nabla c|}/\Sigma_{\text{gen}}$ indicates the surface-weighted filtered value of a general quantity Q and $S_d = Dc/Dt/|\nabla c|$ is the displacement speed, which denotes the speed at which a given c isosurface moves normal to itself with respect to an initially coincident material surface. The generalised FSD Σ_{gen} is an unclosed quantity and is closed either by using an algebraic expression or by solving a modelled transport equation alongside other conservation equations. The algebraic closure is valid when the generation rate of flame surface area remains in equilibrium with its destruction rate, but this assumption is rendered invalid under unsteady conditions (e.g., combustion instabilities). Under unsteady conditions, it is often advantageous to solve a modelled transport equation of Σ_{gen}. The exact transport equation for the generalised FSD Σ_{gen} is given as [1, 4–7, 9, 10, 12]:

$$
\frac{\partial \Sigma_{\text{gen}}}{\partial t} + \frac{\partial\left(\tilde{u}_j \Sigma_{\text{gen}}\right)}{\partial x_j}
$$

$$
= -\frac{\partial\left[\left(\overline{(u_i)}_s - \tilde{u}_i\right)\Sigma_{\text{gen}}\right]}{\partial x_i} + \overline{\left(\frac{\left(\delta_{ij} - N_i N_j\right)\partial u_i}{\partial x_j}\right)}_s \Sigma_{\text{gen}}
$$

$$
- \frac{\partial\left[\overline{(S_d N_i)}_s \Sigma_{\text{gen}}\right]}{\partial x_i} + \overline{\left(S_d\left(\frac{\partial N_i}{\partial x_i}\right)\right)}_s \Sigma_{\text{gen}},
$$

(3)

where $N_i = -(\partial c/\partial x_i)/|\nabla c|$ is the ith component of flame normal vector. The terms on the left hand side of (3) denote transient and mean advection effects, respectively. The first three terms on the right hand side of (3) denote the effects of subgrid convection, flame surface area generation due to fluid-dynamic straining, and flame normal propagation, respectively. The last term of (3) describes the production/destruction of Σ_{gen} due to flame curvature $\kappa_m = (\partial N_i/\partial x_i)/2$ and thus referred to as the FSD curvature term [4–7, 9, 10, 12]. It has been found in several previous studies [5–7, 9, 14] that the FSD curvature term remains a leading order contributor to the FSD transport for both unity and nonunity Lewis number turbulent premixed combustion. As the curvature term remains a leading order contributor to the FSD transport, the modelling of $\overline{(S_d \nabla \cdot \vec{N})}_s \Sigma_{\text{gen}}$ is crucial for the transport equation-based FSD closure. The statistical behaviour of $\overline{(S_d \nabla \cdot \vec{N})}_s \Sigma_{\text{gen}}$ is significantly affected by curvature dependence on S_d [9, 10, 12]. Earlier a priori Direct Numerical Simulation (DNS) analyses [9, 10, 12] showed that existing models for the subgrid curvature term C_{sg} do not adequately capture the qualitative behaviour of this term obtained from DNS data. Moreover, the model parameters for the existing subgrid curvature term C_{sg} models are found to be strong functions of the LES filter width Δ [9, 10, 12].

To date, most existing FSD-based models have been proposed for unity Lewis number flames where the differential diffusion of heat and mass has been ignored. The Lewis number is defined as the ratio of thermal diffusivity α_T to mass diffusivity D (i.e., Le $= \alpha_T/D$).

The effects of Le on the statistical behaviour of the FSD curvature term $\overline{(S_d \nabla \cdot \vec{N})}_s \Sigma_{\text{gen}}$ are yet to be analysed in detail, and this paper aims to bridge this gap in the existing literature. It is worth noting that, in a premixed flame, different species have different values of Lewis number. Thus, specifying a global Lewis number Le characterising the whole combustion process is not straightforward. The Lewis number of the deficient reactant is often considered to be the characteristic Le of the combustion process in question [15, 16]. Moreover, several previous studies [16–29] analysed the effects of differential diffusion of heat and mass by modifying the characteristic Lewis number in isolation, and the same procedure has been adopted here. In the present study, a simplified chemistry-based DNS database of statistically planar turbulent premixed flames with global Lewis numbers ranging from 0.34 to 1.2 has been considered to analyse the statistical behaviour of the FSD curvature term $\overline{(S_d \nabla \cdot \vec{N})}_s \Sigma_{\text{gen}}$ in the context of LES. In this context, the main objectives of this study are as follows:

(1) to analyse the statistical behaviours of the subgrid FSD curvature term in the context of LES, for flames with different values of Lewis number;

(2) to propose models for different components of the subgrid FSD curvature terms and assess their performances in comparison to the corresponding quantities extracted from DNS data.

The rest of the paper will be organised as follows. The necessary mathematical background will be provided in the next section. This will be followed by a brief description of the numerical implementation related to the DNS database. Following this, results will be presented and subsequently discussed. The main findings will be summarised, and conclusions will be drawn in the final section.

2. Mathematical Background

The curvature term of the FSD $\overline{(S_d \nabla \cdot \vec{N})}_s \Sigma_{\text{gen}}$ is often decomposed in the following manner [4–7, 9, 10, 12]:

$$
\overline{(S_d \nabla \cdot \vec{N})}_s \Sigma_{\text{gen}} = C_{\text{mean}} + C_{\text{sg}}, \qquad (4)
$$

where C_{mean} and C_{sg} are the resolved and subgrid components of the FSD curvature term, respectively. The resolved curvature term C_{mean} can be expressed in three different manners [5, 9, 10, 12]:

$$
C_{\text{mean}} = \overline{(S_d)}_s \left[\frac{\partial \overline{(N_i)}_s}{\partial x_i}\right] \Sigma_{\text{gen}} \qquad (5a)
$$

$$
C_{\text{mean}} = \overline{(S_d)}_s \left[\frac{\partial M_i}{\partial x_i}\right] \Sigma_{\text{gen}}, \qquad (5b)
$$

$$
C_{\text{mean}} = \left(\delta_{ij} - n_{ij}\right) \frac{\partial\left[\overline{(S_d)}_s \overline{(N_i)}_s\right]}{\partial x_j} \Sigma_{\text{gen}}, \qquad (5c)
$$

where $M_i = -(\partial \bar{c}/\partial x_i)/|\nabla \bar{c}|$ is the ith component of the resolved flame normal vector. It was demonstrated

A Priori Direct Numerical Simulation Modelling of the Curvature Term of the Flame Surface Density Transport Equation
for Nonunity Lewis Number Flames in the Context of Large Eddy Simulations

3

by Chakraborty and Cant [10, 12] that (5a) provides the best option for the resolved curvature term C_{mean}, as it gives rise to the smallest magnitude of C_{sg} among all the possibilities shown in (5a)–(5c). Equation (5a) was found to perform the best among the three possibilities shown in (5a), (5b), (5c) for this database. This is advantageous from the perspective of efficient modelling of the FSD curvature term $\overline{(S_d \nabla \cdot \vec{N})}_s \Sigma_{\text{gen}}$ as most of the modelling uncertainty is associated with C_{sg}. Moreover, (5a) has also been used for the modelling of $\overline{(S_d \nabla \cdot \vec{N})}_s \Sigma_{\text{gen}}$ in previous LES simulations [5–7, 13]. For the present analysis (5a), (i.e., $C_{\text{mean}} = \overline{(S_d)}_s [\partial \overline{(N_i)}_s / \partial x_i] \Sigma_{\text{gen}}$) will be considered for the resolved curvature term C_{mean}.

It is often useful to decompose the flame displacement speed $S_d = (Dc/Dt)/|\nabla c| = [\dot{w} + \nabla \cdot (\rho D \nabla c)]/\rho |\nabla c|$ in the following manner for the purpose of modelling FSD curvature term [9–12, 30, 31]:

$$S_d = S_r + S_n + S_t, \tag{6a}$$

$$S_r = \frac{\dot{w}}{\rho |\nabla c|}, \tag{6b}$$

$$S_n = \frac{\vec{N} \cdot \nabla (\rho D \vec{N} \cdot \nabla c)}{\rho |\nabla c|}, \tag{6c}$$

$$S_t = -D\nabla \cdot \vec{N} = -2D\kappa_m, \tag{6d}$$

where S_r and S_n are the reaction and normal diffusion components of displacement speed and S_t is the tangential diffusion component of displacement speed. The following expression for C_{sg} can be obtained using (6a)–(6d) and (5a) (i.e., $C_{\text{mean}} = \overline{(S_d)}_s [\partial \overline{(N_i)}_s / \partial x_i] \Sigma_{\text{gen}}$):

$$C_{\text{sg}} = C_{\text{sg1}} + C_{\text{sg2}} = \overline{\left(S_d \frac{\partial N_i}{\partial x_i}\right)_s} \Sigma_{\text{gen}} - \overline{(S_d)}_s \frac{\partial \overline{(N_i)}_s}{\partial x_i} \Sigma_{\text{gen}}, \tag{7}$$

where

$$C_{\text{sg1}} = \left[\overline{\left((S_r + S_n) \frac{\partial N_i}{\partial x_i}\right)_s} \Sigma_{\text{gen}} - \overline{(S_r + S_n)}_s \frac{\partial \overline{(N_i)}_s}{\partial x_i} \Sigma_{\text{gen}} \right], \tag{8a}$$

$$C_{\text{sg2}} = -\left[\overline{\left(D \left(\frac{\partial N_i}{\partial x_i}\right)^2\right)_s} \Sigma_{\text{gen}} - \overline{\left(D \frac{\partial N_i}{\partial x_i}\right)_s} \frac{\partial \overline{(N_i)}_s}{\partial x_i} \Sigma_{\text{gen}} \right]. \tag{8b}$$

Equation (8a) indicates that curvature ($\kappa_m = \nabla \cdot \vec{N}/2$) dependences of $(S_r + S_n)$ and $|\nabla c|$ significantly influence the statistical behaviour of C_{sg1}. Equation (8b) suggests that C_{sg2} is expected to assume negative values throughout the flame brush.

Hawkes and Cant [6, 7] modified a version of the Coherent Flamelet Model (CFM) by Candel et al. [2] for the purpose of LES as:

$$C_{\text{sg}} = -\frac{\alpha_N \beta_1 S_L \Sigma_{\text{gen}}^2}{(1 - \bar{c})}, \tag{9}$$

where $\alpha_N = 1 - \overline{(N_k)}_s \overline{(N_k)}_s$ is a resolution parameter which vanishes when the flow is fully resolved and β_1 is a model parameter. Hawkes [5] discussed a possibility of modifying a RANS model proposed by Cant et al. [1] for the purpose of LES as:

$$C_{\text{sg}} = -\frac{C_H S_L \Sigma_{\text{gen}}^2}{(1 - \bar{c})}, \tag{10}$$

where $C_H = \alpha_N \beta_2 (1 - (1/3)[1 - \exp(-10(1 - \bar{c})\sqrt{\tilde{k}}/\Sigma_{\text{gen}} S_L \Delta)]$, $A = 10.0$, $u'_\Delta = \sqrt{2\tilde{k}/3}$ is the subgrid turbulent velocity fluctuation, $\tilde{k} = (\overline{\rho u_i u_i} - \bar{\rho} \tilde{u}_i \tilde{u}_i)/2\bar{\rho}$ is the subgrid kinetic energy, and β_2 is a model parameter. Another model of C_{sg} was proposed by Charlette et al. [4]:

$$C_{\text{sg}} = -\frac{\beta_3 S_L (\Sigma_{\text{gen}} - |\nabla \bar{c}|) \Sigma_{\text{gen}}}{\bar{c}(1 - \bar{c})}, \tag{11}$$

where β_3 is a model parameter. The models given by (9)–(11) (henceforth will be referred to as CSGCFM, CSGCPB, and CSGCHAR, resp.) ensure that C_{sg} vanishes when the flow is fully resolved (i.e., $\overline{(N_k)}_s \overline{(N_k)}_s = 1.0$ and $\Sigma_{\text{gen}} = |\nabla \bar{c}|$). A priori DNS assessment of the CSGCFM, CSGCPB, and CSGCHAR models and the modelling of C_{sg1} and C_{sg2} will be addressed in Section 4 of this paper.

3. Numerical Implementation

In principle combustion, DNS should account for both three dimensionality of turbulence and detailed chemical mechanism. However, until recently, most combustion DNS studies were carried out either in two dimensions with detailed chemistry or in three dimensions with simplified chemistry due to the limitation of computer storage capacity. Although it is now possible to carry out three-dimensional DNS with detailed chemistry, they remain extremely expensive (e.g., millions of CPU hours and thousands of processors [32]) and the cost of an extensive parametric analysis based on three-dimensional detailed chemistry-based DNS often becomes prohibitive. As the present analysis concentrates on an extensive parametric variation in terms of Lewis number, the chemical mechanism is simplified here by an Arrhenius-type irreversible single-step chemical reaction (i.e., Reactants → Products) following several previous studies [1–12, 14]. It has been found that the strain rate and curvature dependences of S_d and $|\nabla c|$ obtained from three-dimensional simplified chemistry DNS [25–27, 33, 34] are found to be qualitatively similar to the corresponding behaviours obtained from detailed chemistry-based DNS simulations [16, 30, 31, 35]. As the statistical behaviours of the FSD curvature term $\overline{(S_d \nabla \cdot \vec{N})}_s \Sigma_{\text{gen}}$ are strongly dependent on the curvature dependences of S_d and $|\nabla c|$, the results for this analysis are expected to be valid even for detailed chemistry based simulations at least in a qualitative sense without much loss of generality. Several studies [3–7, 9–12] have concentrated on a priori DNS modelling of FSD based on simplified chemistry in the past and the same approach has been adopted here.

A compressible three-dimensional DNS code SENGA [36] was used for the simulations where the conservation equations of mass, momentum, energy, and species are solved in nondimensional form. A cubic domain of each side equal to $24.1\delta_{th}$ is considered for the present DNS database where δ_{th} is the thermal flame thickness, which is defined as $\delta_{th} = (T_{ad} - T_0)/\text{Max}|\nabla \hat{T}|_L$, and the subscript L refers to quantities in an unstrained planar laminar flame with T_{ad}, T_0, and $\hat{T}$ being the adiabatic flame, unburned gas, and instantaneous gas temperatures, respectively. The computational domain was discretised using a Cartesian grid of $230 \times 230 \times 230$ with equal grid spacing in each direction. The grid spacing Δx is determined based on the flame resolution, and about 10 grid points are kept within the thermal flame thickness δ_{th} for all the cases considered here. This grid spacing Δx corresponds to 0.73η, where η is the Kolmogorov length scale. The boundaries in the mean flame propagation were taken to be partially nonreflecting and were implemented using the Navier-Stokes Characteristic Boundary Conditions (NSCBC) technique [37]. The boundary conditions in the transverse direction were taken to be periodic. The spatial derivatives for the internal grid points were evaluated using a tenth-order central differencing scheme, and the order of differentiation gradually decreases to a one-sided 2nd order scheme at the partially nonreflecting boundaries. The time advancement was carried out using an explicit low storage third-order Runge-Kutta scheme [38].

For the current DNS database, the turbulent velocity field was initialised using a pseudospectral method [39] following the Batchelor-Townsend turbulent kinetic energy spectrum [40]. The flame is initialised using a steady planar unstrained laminar flame solution. The initial values of normalised root mean square (rms) turbulent velocity fluctuation u'/S_L, integral length scale to thermal flame thickness ratio l/δ_{th}, heat release parameter $\tau = (T_{ad} - T_0)/T_0$, Damköhler number $Da = lS_L/u'\delta_{th}$, and Karlovitz number $Ka = (u'/S_L)^{3/2}(l/\delta_{th})^{-1/2}$ are listed in Table 1. According to Peters [41], all the cases considered here can be taken to represent the thin reaction zone regime combustion, as Ka remains greater that unity. Standard values are considered for Prandtl number Pr and the Zel'dovich number β (i.e., Pr = 0.7 and $\beta = T_{ac}(T_{ad} - T_0)T_{ad}^2 = 6.0$, where T_{ac} is the activation temperature).

Under decaying turbulence, DNS simulations should be carried out for a simulation time $t_{sim} \geq \text{max}(t_f, t_c)$ [42], where $t_f = l/u'$ is the initial eddy turn over time and $t_c = \delta_{th}/S_L$ is the chemical time scale. For this database, the statistics were extracted after about three eddy turn over times (i.e., $3t_f = 3l/u'$), which corresponded to one chemical time scale (i.e., $t_c = \delta_{th}/S_L$). This simulation time remains small but comparable to several studies [3, 24, 28, 43–47] which contributed significantly to the FSD-based modelling in the past. The statistics presented in this paper did not change significantly since halfway through the simulation (i.e., $1.5t_f = 1.5l/u'$). The value of u'/S_L in the unburned gas ahead of the flame had decayed by 50% of its initial value when the statistics were extracted. By this time, the

TABLE 1: Initial values of the simulation parameters and non-dimensional numbers relevant to the DNS database.

Case	Le	u'/S_L	l/δ_L	l/δ_{th}	δ_{th}/η	τ	Re_t	Da	Ka
A	0.34	7.5	1.13	2.45	7.32	4.5	47.0	0.33	34.3
B	0.6	7.5	1.76	2.45	7.32	4.5	47.0	0.33	19.40
C	0.8	7.5	2.13	2.45	7.32	4.5	47.0	0.33	14.70
D	1.0	7.5	2.45	2.45	7.32	4.5	47.0	0.33	11.75
E	1.2	7.5	2.72	2.45	7.32	4.5	47.0	0.33	9.80

normalised integral value l/δ_{th} had increased to around 1.7 times of its initial value. The values of u'/S_L and l/δ_{th} at the time statistics were extracted are also representative of the thin reaction zones regime combustion [41]. This DNS database was used extensively earlier for the purpose of RANS modelling [27, 28, 48, 49], and the interested readers are referred to these papers for further details.

The DNS data was explicitly LES filtered using a Gaussian filter kernel in physical space for the purpose of a priori analysis. The filtered quantity $\overline{Q}(\vec{x}, t)$, is given by

$$\overline{Q}(\vec{x}, t) = \int Q(\vec{x} - \vec{r})G(\vec{r})d\vec{r}, \qquad (12)$$

where $G(\vec{r})$ is the Gaussian filter kernel, which is defined in the following manner:

$$G(\vec{r}) = \left(\frac{6}{\pi\Delta^2}\right)^{3/2} \exp\left(\frac{-6\vec{r}\cdot\vec{r}}{\Delta^2}\right). \qquad (13)$$

The filtered quantities of interest were extracted for filter widths Δ ranging from $0.4\delta_{th}$ to $2.4\delta_{th}$ in steps of $0.4\delta_{th}$. These filter sizes are comparable to the range of Δ used in several previous studies [3, 4, 9–12, 14] for a priori DNS analysis and span a useful range of length scales (i.e., from $\Delta \approx 0.4\delta_{th}$, where the flame is partially resolved, up to $2.4\delta_{th}$, where the flame becomes fully unresolved and Δ is comparable to the integral length scale).

4. Results and Discussion

The instantaneous isosurfaces of c ranging from 0.01 to 0.99 at $t_c = \delta_{th}/S_L$ are shown in Figure 1, which indicates that the flame wrinkling increases with decreasing Lewis number and this tendency is particularly prevalent for the Le $\ll$ 1 flames due to thermodiffusive instabilities [17–29]. The unburned reactants diffuse into the reaction zone at a faster rate than the rate at which heat diffuses out in the Le < 1 flames. This gives rise to simultaneous presence of high temperature and reactant concentration in the reaction zone for the Le < 1 flames, which in turn leads to greater burning rate and flame surface area generation in comparison to the unity Lewis number flame. By contrast, heat diffuses faster than the diffusion rate of reactants into the reaction zone in the case of Le > 1, which reduces the burning rate and the rate of flame area generation in comparison to the unity Lewis number flame. The increase in burning rate and flame area generation with decreasing Lewis number can be substantiated by the values of normalised turbulent

A Priori Direct Numerical Simulation Modelling of the Curvature Term of the Flame Surface Density Transport Equation for Nonunity Lewis Number Flames in the Context of Large Eddy Simulations

5

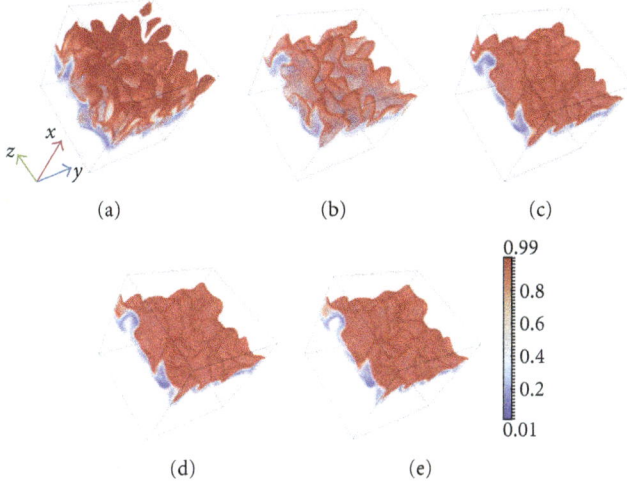

FIGURE 1: Instantaneous isosurfaces of c ranging from 0.01 to 0.99 at $t = 3t_f = t_c$ for cases (a) Le = 0.34; (b) Le = 0.6; (c) Le = 0.8; (d) Le = 1.0; (e) Le = 1.2. The domain size is $24.1\delta_{th} \times 24.1\delta_{th} \times 24.1\delta_{th}$.

TABLE 2: The effects of Lewis number on normalised turbulent flame speed S_T/S_L and normalised flame surface area A_T/A_L after 3.0 initial eddy turn over times.

Case	Le	S_T/S_L	A_T/A_L
A	0.34	13.70	3.93
B	0.6	4.58	2.66
C	0.8	2.53	2.11
D	1.0	1.83	1.84
E	1.2	1.50	1.76

flame speed S_T/S_L and normalised flame surface area A_T/A_L which are presented in Table 2. The values of S_T/S_L have been evaluated by volume integrating the reaction rate $\dot{w}$ using the expression $S_T = (1/\rho_0 A_P) \int_V \dot{w} dV$, where A_P is the projected area of the flame in the direction of mean flame propagation, while the values of A_T/A_L have been evaluated by volume integrating $|\nabla c|$ (i.e., $\int_V |\nabla c| dV$) under both turbulent and laminar conditions. Table 2 shows that both S_T/S_L and A_T/A_L increase with decreasing Lewis number, and this effect is particularly prevalent in the flames with Le < 1 due to the presence of thermodiffusive instabilities [17–29]. The increase in flame wrinkling with decreasing Lewis number is also visually evident from the c isosurfaces presented in Figure 1.

The variations of $\overline{(S_d \nabla \cdot \vec{N})_s \Sigma_{gen}}$, C_{mean}, and C_{sg} conditionally averaged in bins of $\tilde{c}$ isosurfaces for cases (a)–(e) are shown in Figure 2 for filter widths $\Delta = 8\Delta_m \approx 0.8\delta_{th}$ and $\Delta = 24\Delta_m \approx 2.4\delta_{th}$, where Δ_m is the DNS grid size. It is evident from Figure 2 that Le significantly affects the statistical behaviours of the curvature terms. The filter widths $\Delta = 8\Delta_m \approx 0.8\delta_{th}$, and $\Delta = 24\Delta_m \approx 2.4\delta_{th}$ span a useful range of length scales (i.e., from $\Delta \approx 0.8\delta_{th}$, where the flame is partially resolved, up to $2.4\delta_{th}$ where the flame becomes fully unresolved and Δ is comparable to the integral length scale). In the Le ≪ 1 flames (e.g., cases (a) and (b)), the FSD curvature term $\overline{(S_d \nabla \cdot \vec{N})_s \Sigma_{gen}}$ behaves as a source term for the major part of the flame brush before assuming negative values towards the burned gas side for $\Delta = 8\Delta_m \approx 0.8\delta_{th}$. For $\Delta = 24\Delta_m \approx 2.4\delta_{th}$, the FSD curvature term $\overline{(S_d \nabla \cdot \vec{N})_s \Sigma_{gen}}$ acts as a source (sink) term towards the unburned (burned) gas side of the flame brush in the Le ≪ 1 flames. In the case of Le ≈ 1.0 flames (i.e., cases (c)–(e)) the curvature term $\overline{(S_d \nabla \cdot \vec{N})_s \Sigma_{gen}}$ behaves as a sink type term throughout the flame brush for all filter widths. It can be seen from Figure 2 that C_{mean} acts as a source (sink) term for cases (a)-(b) ((c)–(e)). The magnitude of C_{mean} (C_{sg}) decreases

(increases) with increasing Δ in all cases, and for large filter widths $\overline{(S_d \nabla \cdot \vec{N})_s \Sigma_{gen}}$ is principally made up of C_{sg}. The LES filtering is a convolution process, and the weighted averaging involved in the filtering process leads to a decrease in the magnitude of C_{mean} with increasing filter width Δ. The flow becomes increasingly unresolved with increasing filter width Δ, and this is reflected in the rise in C_{sg} magnitude with increasing filter width Δ.

The resolved curvature term $C_{mean} = \overline{(S_d)_s} \partial \overline{(N_i)_s}/\partial x_i \Sigma_{gen}$ can be seen to capture the behaviour of the curvature term $\overline{(S_d \nabla \cdot \vec{N})_s \Sigma_{gen}}$, well at small filter widths (i.e., $\Delta \leq \delta_{th}$) for flames with Le ≈ 1.0 (i.e., cases (c)–(e)). However, the magnitude of C_{mean} decreases with increasing Δ and it does not capture the behaviour of the FSD curvature term $\overline{(S_d \nabla \cdot \vec{N})_s \Sigma_{gen}}$ for the Le ≪ 1.0 flames (i.e., cases (a) and (b)). The subgrid curvature term, C_{sg} follows the qualitative behaviour of the FSD curvature term $\overline{(S_d \nabla \cdot \vec{N})_s \Sigma_{gen}}$ for all filter widths. The subgrid curvature term C_{sg} almost entirely makes up the FSD curvature term $\overline{(S_d \nabla \cdot \vec{N})_s \Sigma_{gen}}$ for $\Delta \gg \delta_{th}$, and this is especially true for the Le ≪ 1.0 cases (i.e., cases (a) and (b)). It can further be observed from Figure 2 that C_{sg} assumes positive values towards the unburned gas side of the flame brush in the Le ≪ 1 flames (e.g., cases (a) and (b)), whereas the existing models for C_{sg} allow for only negative values (see (9)–(11)). This suggests that new models for C_{sg} are warranted to account for the influences of nonunity Lewis number (i.e., Le ≠ 1.0).

In order to be able to model the subgrid curvature term C_{sg}, the decomposition prescribed in (8a)-(8b) has been used here. The variations of $\overline{((S_r + S_n)\nabla \cdot \vec{N})_s \Sigma_{gen}}$, $-\overline{(D(\nabla \cdot \vec{N})^2)_s \Sigma_{gen}} = -4\overline{(D\kappa_m^2)_s \Sigma_{gen}}$, C_{sg1} and C_{sg2} conditionally averaged in bins of $\tilde{c}$ isosurfaces for cases (a)–(e) are shown in Figure 3 for filter widths $\Delta = 8\Delta_m \approx 0.8\delta_{th}$ and $\Delta = 24\Delta_m \approx 2.4\delta_{th}$. It is evident from Figure 3 that C_{sg2} remains negative throughout the flame brush for all cases and follows the qualitative behaviour of $(-4\overline{(D\kappa_m^2)_s \Sigma_{gen}})$. A comparison between $\overline{((S_r + S_n)\nabla \cdot \vec{N})_s \Sigma_{gen}}$ and $-4\overline{(D\kappa_m^2)_s \Sigma_{gen}}$ reveals that $-4\overline{(D\kappa_m^2)_s \Sigma_{gen}}$ remains the major contributor to $\overline{(S_d \nabla \cdot \vec{N})_s \Sigma_{gen}}$ for all the flames at all values of Δ, which is consistent with the expected behaviour in the thin reaction zones regime [41]. The contribution of $\overline{((S_r + S_n)\nabla \cdot \vec{N})_s \Sigma_{gen}}$ remains significant for the Le < 1 cases (i.e., cases (a), (b) and (c)), but its contribution remains weak in comparison

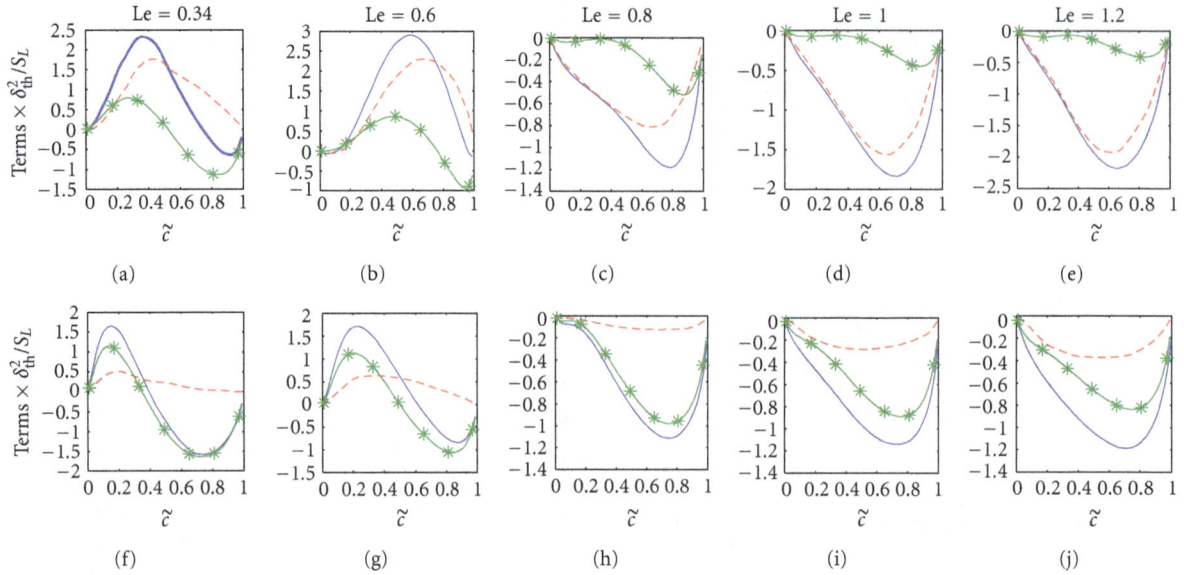

FIGURE 2: Variation of $\overline{(S_d \nabla \cdot \vec{N})_s} \Sigma_{\text{gen}}$ (solid line), $C_{\text{mean}} = \overline{(S_d)_s \partial(N_i)_s / \partial x_i} \Sigma_{\text{gen}}$ (dashed line) and C_{sg} (asterisk line) conditionally averaged in bins of $\tilde{c}$ across the flame brush for filter sizes $\Delta = 8\Delta_m \approx 0.8\delta_{\text{th}}$ (top row): (a) Le = 0.34 (1st column); (b) Le = 0.6 (2nd column); (c) Le = 0.8 (3rd column); (d) Le = 1.0 (4th column); (e) Le = 1.2 (5th column) and for filter size $\Delta = 24\Delta_m \approx 2.4\delta_{\text{th}}$ (bottom row): (f) Le = 0.34 (1st column); (g) Le = 0.6 (2nd column); (h) Le = 0.8 (3rd column); (i) Le = 1.0 (4th column); (j) Le = 1.2 (5th column). All terms are ensemble averaged on $\tilde{c}$ isosurfaces in Figure 2 and subsequent cases. All the curvature terms in this and subsequent figures are normalised by $S_L / \delta_{\text{th}}^2$.

to the magnitude of $-4\overline{(D\kappa_m^2)_s}\Sigma_{\text{gen}}$ in the Le = 1.0 and 1.2 flames (i.e., cases (d) and (e)). Figure 3 demonstrates that C_{sg1} remains close to the magnitude of $\overline{((S_r + S_n)\nabla \cdot \vec{N})_s}\Sigma_{\text{gen}}$ for all Δ for the Le = 1.0 flame (i.e., case (d)), indicating that $\overline{(S_r + S_n)_s \partial(N_i)_s / \partial x_i}\Sigma_{\text{gen}}$ does not play a major role in capturing the behaviour of $\overline{((S_r + S_n)\nabla \cdot \vec{N})_s}\Sigma_{\text{gen}}$. However, there is a significant difference in magnitudes of $\overline{((S_r + S_n)\nabla \cdot \vec{N})_s}\Sigma_{\text{gen}}$ and C_{sg1} for small values of Δ (i.e., $\Delta < \delta_{\text{th}}$) in the nonunity Lewis number flames (i.e., cases (a)–(c) and (e)), which indicates that $\overline{(S_r + S_n)_s \partial(N_i)_s / \partial x_i}\Sigma_{\text{gen}}$ plays a key role for small values of filter width in these flames. For large values of filter width (i.e., $\Delta \gg \delta_{\text{th}}$) C_{sg1} remains the major contributor to $\overline{((S_r + S_n)\nabla \cdot \vec{N})_s}\Sigma_{\text{gen}}$ for all cases considered here, indicating that $\overline{(S_r + S_n)_s \partial(N_i)_s / \partial x_i}\Sigma_{\text{gen}}$ plays progressively less important role for increasing values of Δ.

Figure 3 shows that there is a significant difference between $-4\overline{(D\kappa_m^2)_s}\Sigma_{\text{gen}}$ and C_{sg2} for all cases for small values of Δ, and the difference between these quantities decreases with increasing Δ. As most of the contribution of $-4\overline{(D\kappa_m^2)_s}\Sigma_{\text{gen}}$ remains unresolved for large values of Δ, the subgrid curvature term C_{sg2} remains the major contributor to $-4\overline{(D\kappa_m^2)_s}\Sigma_{\text{gen}}$, indicating that $(-\overline{(D\partial N_i / \partial x_i)_s \partial(N_i)_s / \partial x_i}\Sigma_{\text{gen}})$ plays progressively less important role for increasing values of Δ where the flame is fully unresolved. However, the contribution of $(-\overline{(D\partial N_i / \partial x_i)_s \partial(N_i)_s / \partial x_i}\Sigma_{\text{gen}})$ remains significant for small values of Δ, where the flame is partially resolved. Figure 3 further shows that the order of magnitudes of both C_{sg1} and C_{sg2} remains comparable and thus accurate modelling of C_{sg1} and C_{sg2} is necessary for precise predictions of C_{sg}.

As the range of κ_m values obtained on a flame surface increases with increasing flame wrinkling, the magnitude of $-4\overline{(D\kappa_m^2)_s}$ increases with decreasing Le, which in turn leads to increasing magnitude of $-4\overline{(D\kappa_m^2)_s}\Sigma_{\text{gen}}$ and C_{sg2} (see Figure 3). The positive contribution of C_{sg1} overcomes the negative contribution of C_{sg2} towards the unburned gas side of the flame brush for the Le = 0.34 and 0.6 flames (i.e., cases (a) and (b)) and yields a net positive contribution of C_{sg} towards the reactant side of the flame brush (see Figure 2).

The statistical behaviours of $\overline{((S_r + S_n)\nabla \cdot \vec{N})_s}\Sigma_{\text{gen}}$ and C_{sg1} depend on the nature of the curvature $\kappa_m = \nabla \cdot \vec{N}/2$ dependences of $(S_r + S_n)$ and $|\nabla c|$, and the variation of $\overline{(\kappa_m)_s}$ across the flame brush. The correlation coefficients for $(S_r + S_n) - \kappa_m$ and $|\nabla c| - \kappa_m$ for five different c isosurfaces across the flame brush for all the cases are shown in Figures 4(a) and 4(b), respectively. For all cases, $S_t = -2D\kappa_m$ remains negatively correlated with κ_m with a correlation coefficient close to (-1.0). However, Figures 4(a) and 4(b) demonstrate that Le significantly affects the curvature κ_m dependences of $(S_r + S_n)$ and $|\nabla c|$. It can be seen from Figures 4(a) and 4(b) that $(S_r + S_n)$ and $|\nabla c|$ remain positively (negatively) correlated with κ_m for the Le < 1.0 (Le > 1.0) flames, whereas both $(S_r + S_n)$ and $|\nabla c|$ show weak curvature dependences in the unity Lewis number flame. The positive correlations between $(S_r + S_n)$ and κ_m, and between $|\nabla c|$ and κ_m strengthen with decreasing Le for the Le < 1 flames. The physical explanations for the observed influences of Lewis number on the curvature dependence of $(S_r + S_n)$ and $|\nabla c|$ have been discussed elsewhere [25–27] and thus will not be discussed in this paper.

A Priori Direct Numerical Simulation Modelling of the Curvature Term of the Flame Surface Density Transport Equation
for Nonunity Lewis Number Flames in the Context of Large Eddy Simulations

7

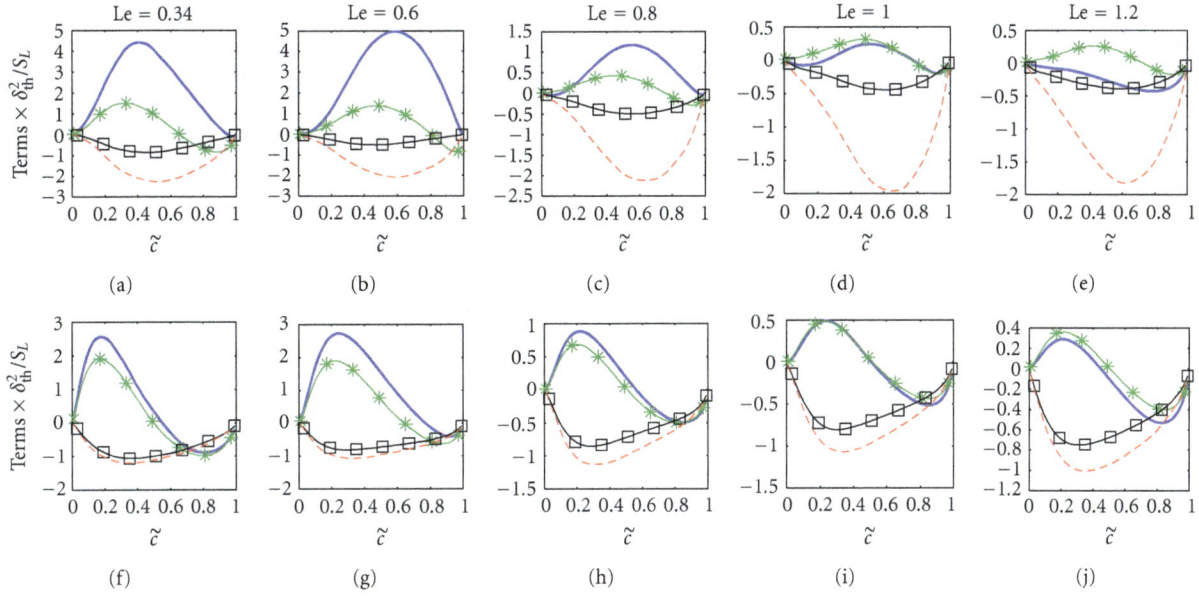

FIGURE 3: Variation of $\overline{((S_r + S_n)\nabla \cdot \vec{N})_s}\Sigma_{gen}$ (solid line), $-4\overline{(D\kappa_m^2)_s}\Sigma_{gen}$ (dashed line), C_{sg1} (asterisk line), and C_{sg2} (squared line) conditionally averaged in bins of $\tilde{c}$ across the flame brush for $\Delta = 8\Delta_m \approx 0.8\delta_{th}$ (top row): (a) Le = 0.34 (1st column); (b) Le = 0.6 (2nd column); (c) Le = 0.8 (3rd column); (d) Le = 1.0 (4th column); (e) Le = 1.2 (5th column) and for filter size $\Delta = 24\Delta_m \approx 2.4\delta_{th}$ (bottom row): (f) Le = 0.34 (1st column); (g) Le = 0.6 (2nd column); (h) Le = 0.8 (3rd column); (i) Le = 1.0 (4th column); (j) Le = 1.2 (5th column).

The variations of $\overline{(\kappa_m)_s}$ conditionally averaged in bins of $\tilde{c}$ isosurfaces for cases A–E are shown in Figures 4(c) and 4(d) for filter widths $\Delta = 8\Delta_m \approx 0.8\delta_{th}$ and $\Delta = 24\Delta_m \approx 2.4\delta_{th}$, respectively. It is evident from Figures 4(c) and 4(d) that $\overline{(\kappa_m)_s}$ assumes positive (negative) values towards the unburned (burned) gas side of the flame brush. For small values of Δ, the surface-weighted filtered value of curvature $\overline{(\kappa_m)_s}$ approaches to κ_m (i.e., $\lim_{\Delta \to 0} \overline{(\kappa_m)_s} = \kappa_m |\nabla c|/|\nabla c| = \kappa_m$) and thus the ensemble averaged value of $\overline{(\kappa_m)_s}$ remains small for small values of filter width as the ensemble averaged value of κ_m remains negligible for statistically planar flames. The difference between the ensemble averaged values of $\overline{(\kappa_m)_s}$ and κ_m increases with increasing filter width Δ, as flame wrinkling increasingly takes place at the subgrid level. For the Le = 1.0 flame (i.e., case D), the combination of positive (negative) value of $\overline{(\kappa_m)_s}$ and weak $(S_r+S_n)-\kappa_m$ and $|\nabla c|-\kappa_m$ correlations gives rise to positive (negative) values of the ensemble averaged values of $\overline{((S_r + S_n)\nabla \cdot \vec{N})_s}\Sigma_{gen}$ and C_{sg1} towards the unburned (burned) gas side of the flame brush for all values of Δ. The predominant positive $(S_r + S_n) - \kappa_m$ and $|\nabla c| - \kappa_m$ correlations give rise to positive values of the ensemble averaged values of $\overline{((S_r + S_n)\nabla.\vec{N})_s}\Sigma_{gen}$ and C_{sg1} throughout the flame brush for small values of Δ in the Le = 0.34, 0.6, and 0.8 flames. By contrast, negative $(S_r + S_n) - \kappa_m$ and $|\nabla c| - \kappa_m$ correlations (see Figures 4(a) and 4(b)) give rise to negative values of the ensemble averaged values of $\overline{((S_r + S_n)\nabla \cdot \vec{N})_s}\Sigma_{gen}$ and C_{sg1} throughout the flame brush for small values of Δ in the Le = 1.2 flame. These local dependences are progressively smeared with increasing Δ because of the convolution operation associated with LES filtering process, and this leads to positive (negative) values

of $\overline{((S_r + S_n)\nabla \cdot \vec{N})_s}\Sigma_{gen}$ and C_{sg1} towards the unburned (burned) gas side of the flame brush for all cases considered here, including the nonunity Lewis number flames where the curvature dependences of $(S_r + S_n)$ and $|\nabla c|$ are particularly strong.

The dependences of $\overline{(S_r + S_n)_s}$ and Σ_{gen} on $0.5 \times \partial\overline{(N_i)_s}/\partial x_i\Sigma_{gen}$ are likely to capture some of κ_m dependences of $(S_r + S_n)$ and $|\nabla c|$ at small values of filter widths Δ (i.e., $\Delta < \delta_{th}$), where the flame is partially resolved. This effect is particularly prevalent in the nonunity Lewis number flames where both $(S_r + S_n)$ and $|\nabla c|$ are strongly correlated with curvature κ_m even though the flames are statistically planar in nature. As a result of this, the contribution of $\overline{(S_r + S_n)_s}\partial\overline{(N_i)_s}/\partial x_i\Sigma_{gen}$ remains close to that of $\overline{((S_r + S_n)\nabla \cdot \vec{N})_s}\Sigma_{gen}$ for small filter widths (i.e., $\Delta < \delta_{th}$) for the non-unity Lewis number flames, which is reflected in the small contribution of C_{sg1} (see $\Delta = 0.8\delta_{th}$ variations in Figures 3(a)–3(c) and Figure 3(e)). The correlation between the resolved quantities (e.g., dependences of $\overline{(S_r + S_n)_s}$ and Σ_{gen} on $0.5 \times \partial\overline{(N_i)_s}/\partial x_i\Sigma_{gen}$) weakens with increasing filter width Δ due to smearing of local information. Moreover, physical processes take place increasingly at the subgrid level for $\Delta \gg \delta_{th}$, and thus $\overline{(S_r + S_n)_s}\partial\overline{(N_i)_s}/\partial x_i\Sigma_{gen}$ does not capture the behaviour of $\overline{((S_r + S_n)\nabla \cdot \vec{N})_s}\Sigma_{gen}$ for large filter widths in all cases considered here, including the nonunity Lewis number flames where the curvature dependences of (S_r+S_n) and $|\nabla c|$ are particularly strong. This leads to $C_{sg1} \approx \overline{((S_r + S_n)\nabla \cdot \vec{N})_s}\Sigma_{gen}$ for $\Delta \gg \delta_{th}$ in all cases considered here (see $\Delta = 2.4\delta_{th}$ variations in Figures 3(f)–3(j)). It can be seen from Figure 3 that the positive contribution of

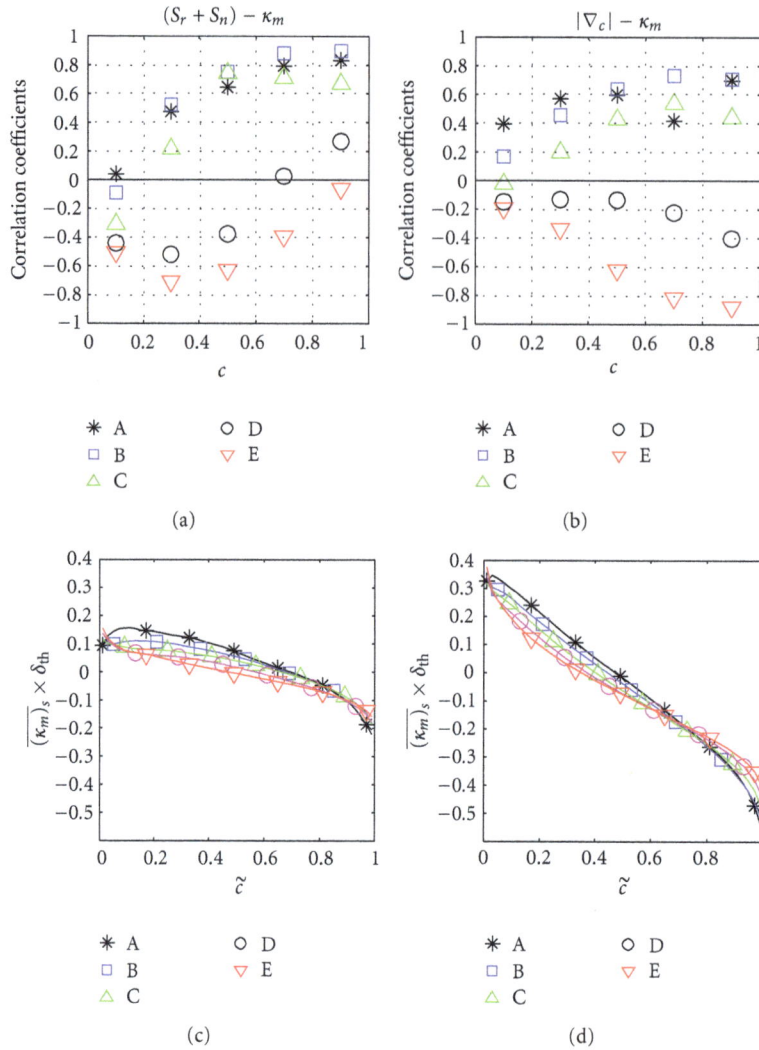

FIGURE 4: Correlation coefficients between (a) $(S_r + S_n)$ and κ_m, and between (b) $|\nabla c|$ and κ_m correlations on $c = 0.1, 0.3, 0.5, 0.7$, and 0.9 isosurfaces for cases A–E. Variation of $\overline{(\kappa_m)}_s \times \delta_{th}$ with $\tilde{c}$ across the flame brush for (c) $\Delta = 8\Delta_m \approx 0.8\delta_{th}$ and (d) $\Delta = 24\Delta_m \approx 2.4\delta_{th}$ for cases A–E.

C_{sg1} overcomes the negative contribution of C_{sg2} towards the unburned gas side of the flame brush in the Le = 0.34 and 0.6 flames, which lead to positive value of $C_{sg} = C_{sg1} + C_{sg2}$ towards the unburned gas side for all values of Δ in these cases (see Figure 2). By contrast, negative values of C_{sg2} overcome the positive contributions of C_{sg1} towards the unburned gas side of the flame brush in the Le = 0.8, 1.0, and 1.2 flames, which lead to negative values of $C_{sg} = C_{sg1} + C_{sg2}$ throughout the flame brush in these cases (see Figure 2).

The subgrid fluctuations of the surface-weighted contributions of $(S_r + S_n)$ and $\nabla \cdot \vec{N}$ are scaled here using S_L and $(\Sigma_{gen} - |\nabla \bar{c}|)$, respectively, to propose the following model for C_{sg1}:

$$C_{sg1} = -\frac{\beta_4 \left(\Sigma_{gen} - |\nabla \bar{c}|\right)(\bar{c} - c^*)S_L\Sigma_{gen}}{\left\{\exp[-a_\Sigma(1 - \bar{c})]\bar{c}(1 - \bar{c})^m\right\}}, \quad (14)$$

where β_4, c^*, a_Σ, and m are the model parameters. The function $(\bar{c} - c^*)/\{\exp[-a_\Sigma(1 - \bar{c})]\bar{c}(1 - \bar{c})^m\}$ in (14)

is used to capture the correct qualitative behaviour of C_{sg1} throughout the flame brush. In a compressible, LES simulation $\tilde{c}$ is readily available and $\bar{c}$ needs to be extracted from $\tilde{c}$. The methodology of extracting $\bar{c}$ from $\tilde{c}$ in the context of LES was discussed elsewhere [9, 10, 12] and will not be discussed in detail in this paper. The model parameter c^* ensures that the transition from positive to negative value of C_{sg1} takes place at the correct location within the flame brush. The quantity $(\Sigma_{gen} - |\nabla \bar{c}|)$ vanishes when the flow is fully resolved (i.e., $\lim_{\Delta \to 0}(\Sigma_{gen} - |\nabla \bar{c}|) = \lim_{\Delta \to 0}(\overline{|\nabla c|} - |\nabla \bar{c}|) = |\nabla c| - |\nabla c| = 0.0$), and thus C_{sg1} becomes exactly equal to zero when the flow is fully resolved (i.e., $\Delta \to 0$) according to (14). It has been found that $m = 1.85$ enables (14) to capture the qualitative behaviour of C_{sg1} when the optimum values of c^* and a_Σ are chosen. The optimum values of $c^*(a_\Sigma)$ tend to increase with decreasing (increasing) Δ. The curvature κ_m dependences of (S_r+S_n) and $|\nabla c|$ are influenced by Le (see Figures 4(a) and 4(b)), and

A Priori Direct Numerical Simulation Modelling of the Curvature Term of the Flame Surface Density Transport Equation for Nonunity Lewis Number Flames in the Context of Large Eddy Simulations

9

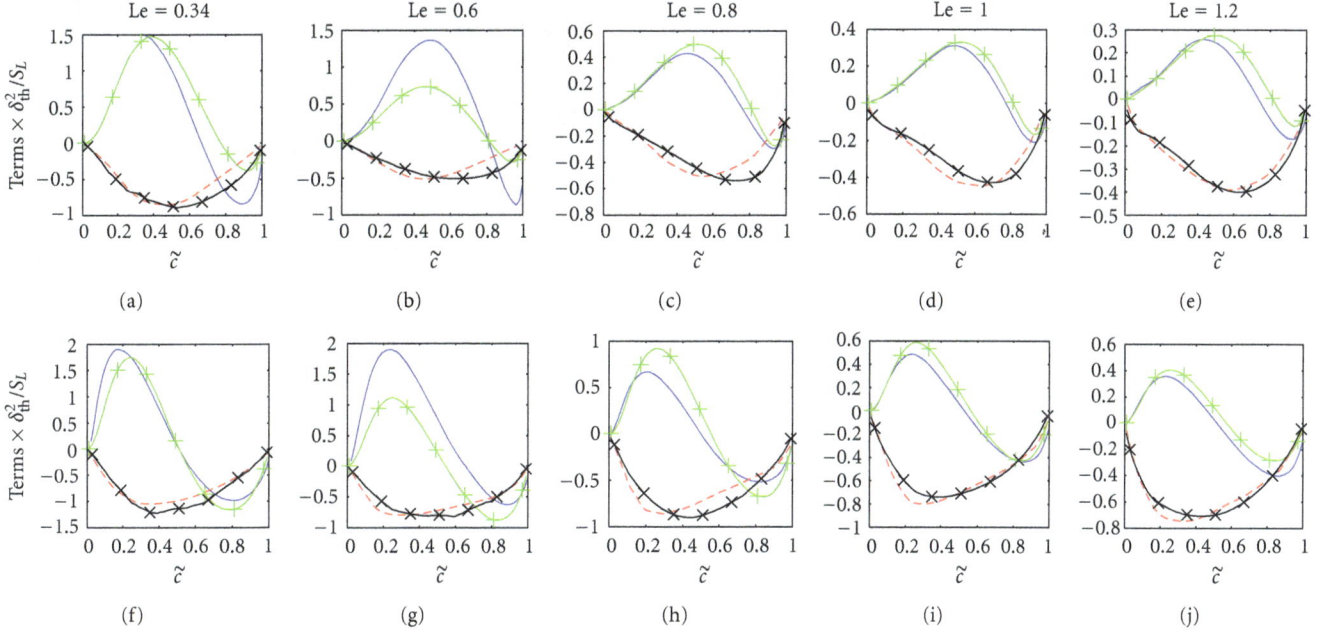

FIGURE 5: Variations of C_{sg1} (solid line) and C_{sg2} (dashed line) conditionally averaged in bins of $\tilde{c}$ across the flame brush along with the predictions of (14) (vertical line) and (16) (crosses line) for $\Delta = 8\Delta_m \approx 0.8\delta_{th}$ (top row): (a) Le = 0.34 (1st column); (b) Le = 0.6 (2nd column); (c) Le = 0.8 (3rd column); (d) Le = 1.0 (4th column); (e) Le = 1.2 (5th column) and for filter size $\Delta = 24\Delta_m \approx 2.4\delta_{th}$ (bottom row): (f) Le = 0.34 (1st column); (g) Le = 0.6 (2nd column); (h) Le = 0.8 (3rd column); (i) Le = 1.0 (4th column); (j) Le = 1.2 (5th column).

these local dependences also appear in the resolved scale but their strength diminishes with increasing Δ due to filtering operation. As the resolved and subgrid curvature terms are closely related [9, 10, 12], the qualitative behaviour of C_{sg1} is also affected by the curvature dependences of displacement speed components and scalar gradient at the resolved scale, which leads to the variation of the optimum values of a_Σ, β_4, and c^* with Le and Δ. The model parameter β_4 needs to be deceased for decreasing values of Σ_{gen} for satisfactory prediction of (14). The prediction of (14) ensemble averaged on $\tilde{c}$ isosurfaces is compared with the ensemble averaged values of C_{sg1} in Figure 5 for all cases considered here for the optimum values of β_4, c^*, and a_Σ for $\Delta = 0.8\delta_{th}$ and $\Delta = 2.4\delta_{th}$ when m is taken to be $m = 1.85$. The optimum values of β_4, c^*, and a_Σ are estimated by calibrating the prediction of (14) with respect to the ensemble averaged values of C_{sg1} obtained from DNS data and the variation of the optimum values of β_4/Σ_{gen}, c^*, and a_Σ with Δ/δ_{th} for all cases are shown in Figure 6. The optimum values of β_4/Σ_{gen}, c^*, and a_Σ are parameterised here as

$$\frac{\beta_4}{\Sigma_{gen}} = 9.81\delta_{th}\left[l_1 + \frac{(l_2 - l_1)}{\{1.0 + \exp[-10.0(Le - 1)]\}^{1/2}}\right],$$

$$(15a)$$

where

$$l_1 = 1.2\frac{\left[\Delta^{2.79} + 1.2(\Delta + \delta_{th})^{2.79}\right]}{(\Delta + \delta_{th})^{2.79}};$$

$$l_2 = 1.34\frac{\left[\Delta^{0.67} + 0.53(\Delta + \delta_{th})^{0.67}\right]}{\left[3.1\Delta^{0.67} + 0.1(\Delta + \delta_{th})^{0.67}\right]},$$

$$(15b)$$

$$c^* = k_1 + \left[\frac{(k_2 - k_1)}{\{1.0 + \exp(-2.0(\Delta/\delta_{th} - 1.5))\}}\right];$$

$$a_\Sigma = \frac{k_4}{(1.0 + \exp(-5.0(\Delta/\delta_{th} - 1.0)))},$$

$$(15c)$$

where

$$k_1 = 0.75 + \frac{0.15}{[1.0 + \exp(-5.0(k_3 - 4.6))]};$$

$$k_2 = 0.65 + \frac{0.05}{[1.0 + \exp(-9.0(k_3 - 4.0))]},$$

$$(15d)$$

$$k_4 = 0.81 - \frac{0.67}{[1.0 + \exp(-5.0(k_3 - 4.6))]};$$

$$k_3 = \frac{\left(Re_\Delta^{0.83} + 0.1\right)}{\left[(\Delta/\delta_{th})^{1.73} + 0.1\right]}; \quad Re_\Delta = \frac{4\rho_0\Delta}{\mu_0}\sqrt{\left(\frac{2\tilde{k}}{3}\right)}.$$

$$(15e)$$

Figure 5 shows that (14) satisfactorily predicts C_{sg1} when m is taken to be $m = 1.85$, and the optimum values of β_4, c^*, and a_Σ are used. According to the parameterisation given by (15a)–(15e), β_4 increases with decreasing Le, as the effects of chemical reaction strengthen with decreasing values of Lewis number (see Table 2). Moreover, β_4/Σ_{gen}, c^*, and a_Σ approach to asymptotic values for large values of Δ and turbulent Reynolds number based on LES filter width $Re_\Delta = 4\rho_0\sqrt{2\tilde{k}/3}\Delta/\mu_0$, where ρ_0 and μ_0 are the unburned gas density and viscosity, respectively.

Here, the contribution of $\overline{(D\kappa_m^2)}_s - \overline{(D\partial N_i/\partial x_i)}_s\partial(N_i)_s/\partial x_i$ is scaled with $(\Xi_\Delta - 1)^n S_L\Sigma_{gen}$ (i.e., $\overline{(D\kappa_m^2)}_s - \overline{(D\partial N_i/\partial x_i)}_s\partial(N_i)_s/\partial x_i \sim (\Xi_\Delta - 1)^n S_L\Sigma_{gen}$) where the

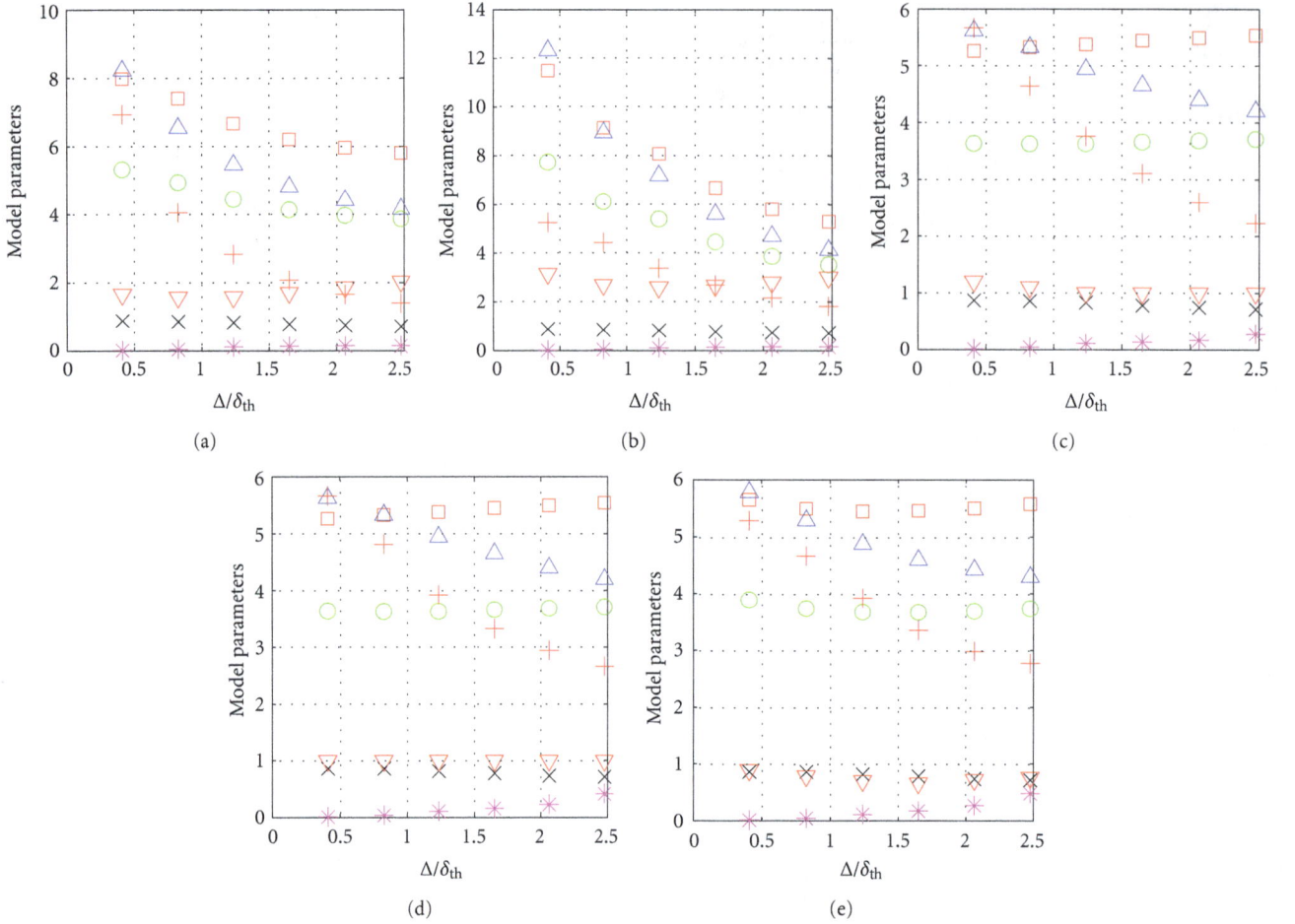

FIGURE 6: Variations of the model parameters β_1 ($\bigcirc$), β_2 ($\square$), β_3 ($\triangle$), $\beta_4/\Sigma_{\text{gen}}$ (down-pointing triangle), β_5 (+), a_Σ ($*$), and c^* ($\times$) with Δ for: (a) Le = 0.34; (b) Le = 0.6; (c) Le = 0.8; (d) Le = 1.0; (e) Le = 1.2.

sub-grid fluctuations of D are taken to scale with S_L/Σ_{gen}. The above relations are utilised here to propose a model for C_{sg2} in the following manner:

$$C_{\text{sg2}} = -\frac{\beta_5 S_L (\Xi_\Delta - 1)^n \Sigma_{\text{gen}}^2}{\overline{c}(1 - \overline{c})}, \qquad (16)$$

where $\Xi_\Delta = \Sigma_{\text{gen}}/|\nabla \overline{c}|$ is the wrinkling factor [8, 11, 43, 50, 51], β_5 and n are the model parameters, and $\overline{c}(1 - \overline{c})$ is used to capture the correct qualitative behaviour of C_{sg2}. The subgrid curvature term C_{sg2} vanishes when the flow is fully resolved according to (16), (i.e., $\lim_{\Delta \to 0} \Xi = \lim_{\Delta \to 0} \Sigma_{\text{gen}}/|\nabla \overline{c}| = \lim_{\Delta \to 0} \overline{|\nabla c|}/|\nabla \overline{c}| = |\nabla c|/|\nabla c| = 1.0$). It has been found that (16) satisfactorily captures the behaviour of C_{sg2} throughout the flame brush for $n = 1.0$ in all cases when a suitable value of β_5 is used. The variation of the global mean optimum values of β_5 with $\Delta/\delta_{\text{th}}$ is shown in Figure 6 for all cases considered here. The optimum values of β_5 have been parameterised here in the following manner:

$$\beta_5 = m(\text{Le}) \left\{ \frac{\text{Re}_\Delta}{(\text{Re}_\Delta + 1.0)} \right\}$$
$$\times \left[r_1 + \left\{ \frac{(r_2 - r_1)}{(1.0 + \exp(-5.0(\text{Re}_\Delta - r_3)))} \right\} \right], \qquad (17a)$$

where

$$r_1 = 1.6 \frac{\left(r_4^{1.23} + 6.24 \right)}{\left(7.17 r_4^{1.23} + 0.26 \right)};$$

$$r_2 = 1.88 \frac{\left(r_4^{2.27} + 5.92 \right)}{\left(8.47 r_4^{2.27} + 0.47 \right)} \qquad (17b)$$

$$r_3 = 35.0 \, \text{erf} \left[\exp\{5.3(r_4 - 1.0)\} \right];$$

$$r_4 = \frac{\Delta}{(\Delta + \delta_{\text{th}})} \qquad (17c)$$

$$m(\text{Le}) = \left(r_5 + \frac{(1.0 - r_5)}{\{1.0 + \exp[-10.0(\text{Le} - 1.0)]\}^{1/4}} \right);$$

$$r_5 = 0.46 \frac{[r_4^{5.22} + 4.53]}{[8.0 r_4^{5.22} + 2.96]}. \qquad (17d)$$

The predictions of (16) ensemble averaged on $\tilde{c}$ isosurfaces are compared with ensemble averaged values of C_{sg2} in Figure 5 for all cases at $\Delta = 0.8\delta_{\text{th}}$ and $\Delta = 2.4\delta_{\text{th}}$, which show that (16) satisfactorily predicts the statistical behaviour of C_{sg2} when n is taken to be $n = 1.0$ and the optimum value of β_5 is used. According to (17a)–(17d), β_5 approaches to asymptotic values for large values of Δ and $\text{Re}_\Delta = 4\rho_0 \sqrt{2\tilde{k}/3} \Delta/\mu_0$.

A Priori Direct Numerical Simulation Modelling of the Curvature Term of the Flame Surface Density Transport Equation
for Nonunity Lewis Number Flames in the Context of Large Eddy Simulations

11

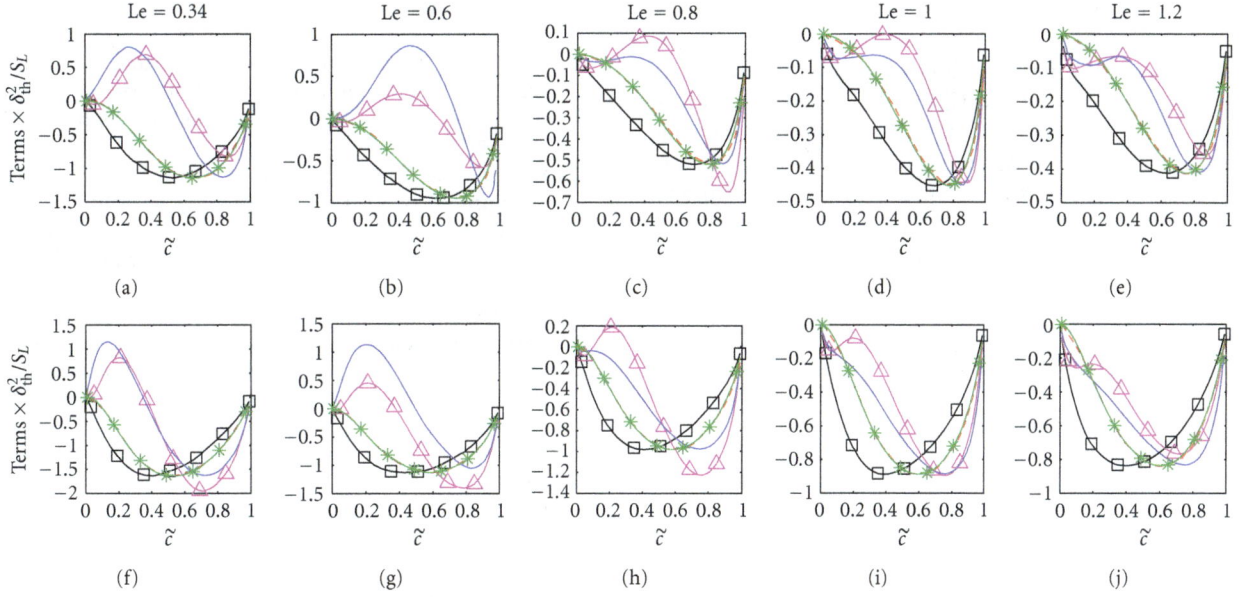

FIGURE 7: Variation of C_{sg} (solid line) conditionally averaged in bins of $\tilde{c}$ across the flame brush along with the predictions of CSGCAND (asterisk line), CSGCANT (dashed line), CSGCHAR (squared line) and CSGNEW (triangle line) for filter sizes $\Delta = 8\Delta_m \approx 0.8\delta_{th}$ (top row): (a) Le = 0.34 (1st column); (b) Le = 0.6 (2nd column); (c) Le = 0.8 (3rd column); (d) Le = 1.0 (4th column); (e) Le = 1.2 (5th column) and for filter size $\Delta = 24\Delta_m \approx 2.4\delta_{th}$ (bottom row): (f) Le = 0.34 (1st column); (g) Le = 0.6 (2nd column); (h) Le = 0.8 (3rd column); (i) Le = 1.0 (4th column); (j) Le = 1.2 (5th column).

Equations (13) and (15a)–(15e) can be combined to propose a model for C_{sg} in the following manner:

$$C_{sg} = \frac{-\beta_4\left(\Sigma_{gen} - |\nabla\bar{c}|\right)(\bar{c} - c^*)S_L\Sigma_{gen}}{\{\exp[-a_\Sigma(1 - \bar{c})]\bar{c}(1 - \bar{c})^m\}} - \frac{\beta_5 S_L(\Xi_\Delta - 1)^n\Sigma_{gen}^2}{\bar{c}(1 - \bar{c})}. \tag{18}$$

The above model will henceforth be referred to CSGNEW model in this paper. Equation (18) allows for a positive contribution of C_{sg} through the contribution of $-\beta_4(\Sigma_{gen} - |\nabla\bar{c}|)(\bar{c} - c^*)S_L\Sigma_{gen}/\{\exp[-a_\Sigma(1 - \bar{c})]\bar{c}(1 - \bar{c})^m\}$, which is absent in the CSGCAND, CSGCANT, and CSGCHAR models. The predictions of the CSGCAND, CSGCANT, CSGCHAR, and CSGNEW models for $\Delta = 0.8\delta_{th}$ and $\Delta = 2.4\delta_{th}$ are compared with C_{sg} obtained from DNS in Figure 7 for the optimum values of β_1, β_2, β_3, and β_5. The optimum values of β_1, β_2, and β_3 are estimated by calibrating the models based on the ensemble averaged value of C_{sg} obtained from DNS data. The variations of the optimum values of β_1, β_2, and β_3 with Δ for all cases are also shown in Figure 6. It is evident from Figure 6 that β_1, β_2, β_3, and β_5 remain greater than unity for all cases. This is found to be consistent with the realisability analysis by Hawkes and Cant [52]. Figure 6 further demonstrates that the optimum values of β_1, β_2, and β_3 change appreciably with increasing Δ, which is consistent with earlier findings [9, 10, 12]. Moreover, optimum values of β_1, β_2, and β_3 for a given Δ are affected by Le (see Figure 6). It is worth noting that parameterisation of the optimum values of β_1, β_2 and β_3 also yields complex relations similar to (15a)–(15e) and (17a)–(17d). However, such parameterisation is

not presented here because the CSGCAND, CSGCANT, and CSGCHAR models do not capture the qualitative behaviour of C_{sg} for the Le = 0.34 and 0.6 flames.

It can further be seen from Figure 7 that the CSGCHAR model tends to overpredict the negative values of C_{sg} towards the unburned gas side in cases C–E (Le = 0.8, 1.0 and 1.2 flames), and this behaviour becomes more prominent with increasing filter size. It is clear from Figure 7 that for $\Delta = 24\Delta_m = 2.4\delta_{th}$, the CSGCHAR model predicts the maximum magnitude of C_{sg} near the middle of the flame whereas the actual maximum magnitude of C_{sg} is attained slightly towards the burned gas side. The CSGCAND and CSGCANT models give comparable performance for optimum values of β_1 and β_2 in cases C–E. However, the CSGCAND and CSGCANT models do not satisfactorily capture the qualitative behaviour of C_{sg} and underpredict (overpredict) the magnitude of C_{sg} towards the burned gas side (middle) of the flame brush in the Le = 0.8, 1.0 and 1.2 flames. Figure 7 demonstrates that the CSGNEW model captures the qualitative behaviour of C_{sg} in a better manner than the CSGCAND and CSGCANT models and the quantitative agreement between C_{sg} and the CSGNEW model remains better than the CSGCAND, CSGCANT, and CSGCHAR models in all cases for all values of Δ when optimum values of β_4, β_5, a_Σ, and c^* are used.

5. Conclusions

The LES modelling of the curvature term $\overline{(S_d\nabla \cdot \vec{N})_s}\Sigma_{gen}$ of the generalised FSD Σ_{gen} transport equation has been addressed here using a simplified chemistry-based DNS database of freely propagating statistically planar turbulent premixed flames with Lewis number Le ranging from 0.34 to 1.2. The statistical behaviours of the subgrid curvature term C_{sg} for a range of different values of Δ have been analysed in terms of its contributions C_{sg1} and C_{sg2} arising from the combined reaction and normal diffusion component and tangential diffusion components of displacement speed (i.e., $(S_r + S_n)$ and $S_t = -2D\kappa_m$), respectively. The Lewis number is shown to have significant influences on the statistical behaviours of the resolved and subgrid components of the FSD curvature term. Detailed physical explanations have been provided for the observed filter size and Lewis number dependences of the different components of $\overline{(S_d \nabla \cdot \vec{N})}_s \Sigma_{gen}$. Models have been identified for individual components of the subgrid curvature term (i.e., C_{sg1} and C_{sg2}), and the performances of these models have been compared to the corresponding quantities extracted from DNS data. It has been found that the new models for C_{sg1} and C_{sg2} satisfactorily capture the statistical behaviours of the corresponding terms extracted from DNS data. It has been found that the existing models for the subgrid curvature term C_{sg} do not satisfactorily capture the qualitative behaviour of the corresponding quantity extracted from DNS data for all the flames considered here. This problem is particularly prevalent for flames with small values of Lewis number (e.g., Le = 0.34 and 0.6) where C_{sg} locally assumes positive values, whereas the existing models can only predict negative values of C_{sg}. The performance of the newly proposed model for C_{sg} has been found to be better than the existing models, and it has been shown to capture positive contributions of C_{sg} for the Le $\ll$ 1 flames. The present analysis has been carried out using a DNS database with moderate value of Re_t in the absence of the effects of detailed chemistry and transport. As simplified chemistry-based DNS qualitatively captures the curvature $\kappa_m = \nabla \cdot \vec{N}/2$ and strain rate dependences of S_d and $|\nabla c|$ obtained from detailed chemistry based simulations, it can be expected that the statistical behaviours of the curvature term $\overline{(S_d \nabla \cdot \vec{N})}_s \Sigma_{gen}$ presented in this paper will be valid at least in a qualitative sense in the context of detailed chemistry. However, the quantitative values of the model parameters (i.e., β_4, β_5, a_Σ, and c^*) may need to be altered in the presence of detailed chemistry. Thus, three-dimensional DNS data with detailed chemistry and experimental data at higher values of Re_t will be necessary for more comprehensive modelling of $\overline{(S_d \nabla \cdot \vec{N})}_s \Sigma_{gen}$ and C_{sg} in the context of LES. Moreover, the newly proposed models need to be implemented in LES simulations for the purpose of *a posteriori* assessments. However, this is kept beyond the scope of this paper. Several previous studies [3–7, 9–12, 43–49] concentrated purely on the model development based on *a priori* analysis of DNS data and the same approach has been adopted here. Implementation of the newly developed models in LES simulations will form the basis of future investigations.

Acknowledgment

The financial support by EPSRC, UK, is gratefully acknowledged.

References

[1] R. S. Cant, S. B. Pope, and K. N. C. Bray, "Modelling of flamelet surface-to-volume ratio in turbulent premixed combustion," *Proceedings of the Combustion Institute*, vol. 23, no. 1, pp. 809–815, 1991.

[2] S. Candel, D. Veynante, F. Lacas, E. Maistret, N. Darabhia, and T. Poinsot, "Coherent flamelet model: applications and recent extensions," in *Recent Advances in Combustion Modelling*, B. E. Larrouturou, Ed., pp. 19–64, World Scientific, Singapore, 1990.

[3] M. Boger, D. Veynante, H. Boughanem, and A. Trouvé, "Direct numerical simulation analysis of flame surface density concept for large eddy simulation of turbulent premixed combustion," *Proceedings of the Combustion Institute*, vol. 27, no. 1, pp. 917–925, 1998.

[4] F. Charlette, A. Trouvé, M. Boger, and D. Veynante, "A flame surface density model for large eddy simulations of turbulent premixed flames," in *Proceedings of the Joint Meeting of the British, German and French Sections of the Combustion Institute*, Nancy, France, 1999.

[5] E. R. Hawkes, *Large eddy simulation of premixed turbulent combustion [Ph.D. thesis]*, Cambridge University Engineering Department, Cambridge, U.K., 2000.

[6] E. R. Hawkes and R. S. Cant, "A flame surface density approach to large-eddy simulation of premixed turbulent combustion," *Proceedings of the Combustion Institute*, vol. 28, no. 1, pp. 51–58, 2000.

[7] E. R. Hawkes and R. S. Cant, "Implications of a flame surface density approach to large eddy simulation of premixed turbulent combustion," *Combustion and Flame*, vol. 126, no. 3, pp. 1617–1629, 2001.

[8] R. Knikker, D. Veynante, and C. Meneveau, "A dynamic flame surface density model for large eddy simulation of turbulent premixed combustion," *Physics of Fluids*, vol. 16, no. 11, pp. L91–L94, 2004.

[9] N. Chakraborty, *Fundamental study of turbulent premixed combustion using direct numerical simulation (DNS) [Ph.D. thesis]*, Cambridge University Engineering Department, Cambridge, U.K., 2005.

[10] N. Chakraborty and R. S. Cant, "*A priori* analysis of the curvature and propagation terms of the flame surface density transport equation for large eddy simulation," *Physics of Fluids*, vol. 19, no. 10, Article ID 105101, 2007.

[11] N. Chakraborty and M. Klein, "*A priori* direct numerical simulation assessment of algebraic flame surface density models for turbulent premixed flames in the context of large eddy simulation," *Physics of Fluids*, vol. 20, no. 8, Article ID 085108, 2008.

[12] N. Chakraborty and R. S. Cant, "Direct numerical simulation analysis of the flame surface density transport equation in the context of large eddy simulation," *Proceedings of the Combustion Institute*, vol. 32, no. 1, pp. 1445–1453, 2009.

[13] F. E. Hernandez-Perez, F. T. C. Yuen, C. P. T. Groth, and O. L. Gulder, "LES of a laboratory-scale turbulent premixed Bunsen flame using FSD, PCM-FPI and thickened flame models," *Proceedings of the Combustion Institute*, vol. 33, no. 1, pp. 1365–1371, 2011.

[14] M. Katragadda and N. Chakraborty, "Effects of Lewis number on flame surface density transport in the context of large eddy simulation," in *Proceedings of the 5th European Combustion Meeting*, Cardiff University, Cardiff, UK, June 2011.

[15] M. Mizomoto, Y. Asaka, S. Ikai, and C. K. Law, "Effects of preferential diffusion on the burning intensity of curved

A Priori Direct Numerical Simulation Modelling of the Curvature Term of the Flame Surface Density Transport Equation
for Nonunity Lewis Number Flames in the Context of Large Eddy Simulations

13

flames," *Proceedings of the Combustion Institute*, vol. 20, no. 1, pp. 1933–1939, 1985.

[16] H. G. Im and J. H. Chen, "Preferential diffusion effects on the burning rate of interacting turbulent premixed hydrogen-air flames," *Combustion and Flame*, vol. 131, no. 3, pp. 246–258, 2002.

[17] G. I. Sivashinsky, "Instabilities, pattern formation, and turbulence in flames," *Annual Review of Fluid Mechanics*, vol. 15, pp. 179–199, 1983.

[18] P. Clavin and F. A. Williams, "Effects of molecular diffusion and thermal expansion on the structure and dynamics of turbulent premixed flames in turbulent flows of large scale and small intensity," *Journal of Fluid Mechanics*, vol. 116, pp. 251–282, 1982.

[19] P. A. Libby, A. Linan, and F. A. Williams, "Strained premix laminar flames with nonunity Lewis numbers," *Combustion Science and Technology*, vol. 34, no. 1–6, pp. 257–293, 1983.

[20] R. G. Abdel-Gayed, D. Bradley, M. N. Hamid, and M. Lawes, "Lewis number effects on turbulent burning velocity," *Proceedings of the Combustion Institute*, vol. 20, no. 1, pp. 505–512, 1985.

[21] W. T. Ashurst, N. Peters, and M. D. Smooke, "Numerical simulation of turbulent flame structure with non-unity Lewis number," *Combustion Science and Technology*, vol. 53, no. 4–6, pp. 339–375, 1987.

[22] D. C. Haworth and T. J. Poinsot, "Numerical simulations of Lewis number effects in turbulent premixed flames," *Journal of Fluid Mechanics*, vol. 244, pp. 405–436, 1992.

[23] C. J. Rutland and A. Trouvé, "Direct simulations of premixed turbulent flames with nonunity Lewis numbers," *Combustion and Flame*, vol. 94, no. 1-2, pp. 41–57, 1993.

[24] A. Trouvé and T. Poinsot, "The evolution equation for the flame surface density in turbulent premixed combustion," *Journal of Fluid Mechanics*, vol. 278, pp. 1–31, 1994.

[25] N. Chakraborty and R. S. Cant, "Influence of Lewis number on curvature effects in turbulent premixed flame propagation in the thin reaction zones regime," *Physics of Fluids*, vol. 17, no. 10, Article ID 105105, 2005.

[26] N. Chakraborty and R. S. Cant, "Effects of Lewis number on turbulent scalar transport and its modelling in turbulent premixed flames," *International Journal of Heat and Mass Transfer*, vol. 49, no. 13-14, pp. 2158–2172, 2006.

[27] N. Chakraborty and M. Klein, "Influence of Lewis number on the surface density function transport in the thin reaction zone regime for turbulent premixed flames," *Physics of Fluids*, vol. 20, no. 6, Article ID 065102, 2008.

[28] N. Chakraborty and R. S. Cant, "Effects of Lewis number on turbulent scalar transport and its modelling in turbulent premixed flames," *Combustion and Flame*, vol. 156, no. 7, pp. 1427–1444, 2009.

[29] N. Chakraborty and R. S. Cant, "Effects of Lewis number on flame surface density transport in turbulent premixed combustion," *Combustion and Flame*, vol. 158, no. 9, pp. 1768–1787, 2011.

[30] N. Peters, P. Terhoeven, J. H. Chen, and T. Echekki, "Statistics of flame displacement speeds from computations of 2-D unsteady methane-air flames," *Proceedings of the Combustion Institute*, vol. 27, no. 1, pp. 833–839, 1998.

[31] T. Echekki and J. H. Chen, "Analysis of the contribution of curvature to premixed flame propagation," *Combustion and Flame*, vol. 118, no. 1-2, pp. 308–311, 1999.

[32] J. H. Chen, A. Choudhary, B. de Supinski et al., "Terascale direct numerical simulations of turbulent combustion using S3D," *Computational Science & Discovery*, vol. 2, no. 1, Article ID 015001, 2009.

[33] N. Chakraborty and S. Cant, "Unsteady effects of strain rate and curvature on turbulent premixed flames in an inflow-outflow configuration," *Combustion and Flame*, vol. 137, no. 1-2, pp. 129–147, 2004.

[34] N. Chakraborty and R. S. Cant, "Effects of strain rate and curvature on surface density function transport in turbulent premixed flames in the thin reaction zones regime," *Physics of Fluids*, vol. 17, no. 6, Article ID 065108, pp. 1–15, 2005.

[35] N. Chakraborty, E. R. Hawkes, J. H. Chen, and R. S. Cant, "The effects of strain rate and curvature on surface density function transport in turbulent premixed methane-air and hydrogen-air flames: a comparative study," *Combustion and Flame*, vol. 154, no. 1-2, pp. 259–280, 2008.

[36] K. W. Jenkins and R. S. Cant, "DNS of turbulent flame kernels," in *Proceedings of the 2nd AFOSR Conference on DNS and LES*, pp. 192–202, Kluwer Academic Publishers, 1999.

[37] T. J. Poinsot and S. K. Lele, "Boundary conditions for direct simulations of compressible viscous flows," *Journal of Computational Physics*, vol. 101, no. 1, pp. 104–129, 1992.

[38] A. A. Wray, "Minimal storage time advancement schemes for spectral methods," Technical Report MS 202 A-1, NASA Ames Research Center, California, Calif, USA, 1990.

[39] R. S. Rogallo, "Numerical experiments in homogeneous turbulence," NASA Technical Memorandum 91416, NASA Ames Research Center, California, Calif, USA, 1981.

[40] G. K. Batchelor and A. A. Townsend, "Decay of turbulence in final period," *Proceedings of the Royal Society A*, vol. 194, no. 1039, pp. 527–543, 1948.

[41] N. Peters, *Turbulent Combustion*, Cambridge University Press, Cambridge, UK, 2000.

[42] D. C. Haworth, R. J. Blint, B. Cuenot, and T. J. Poinsot, "Numerical simulation of turbulent propane-air combustion with nonhomogeneous reactants," *Combustion and Flame*, vol. 121, no. 3, pp. 395–417, 2000.

[43] F. Charlette, C. Meneveau, and D. Veynante, "A power-law flame wrinkling model for LES of premixed turbulent combustion—part I: non-dynamic formulation and initial tests," *Combustion and Flame*, vol. 131, no. 1-2, pp. 159–180, 2002.

[44] N. Swaminathan and K. N. C. Bray, "Effect of dilatation on scalar dissipation in turbulent premixed flames," *Combustion and Flame*, vol. 143, no. 4, pp. 549–565, 2005.

[45] N. Swaminathan and R. W. Grout, "Interaction of turbulence and scalar fields in premixed flames," *Physics of Fluids*, vol. 18, no. 4, Article ID 045102, 2006.

[46] I. Han and K. Y. Huh, "Roles of displacement speed on evolution of flame surface density for different turbulent intensities and Lewis numbers in turbulent premixed combustion," *Combustion and Flame*, vol. 152, no. 1-2, pp. 194–205, 2008.

[47] R. W. Grout, "An age extended progress variable for conditioning reaction rates," *Physics of Fluids*, vol. 19, no. 10, Article ID 105107, 2007.

[48] N. Chakraborty and N. Swaminathan, "Effects of Lewis number on scalar dissipation transport and its modeling in turbulent premixed combustion," *Combustion Science and Technology*, vol. 182, no. 9, pp. 1201–1240, 2010.

[49] N. Chakraborty and N. Swaminathan, "Effects of Lewis number on scalar variance transport in premixed flames," *Flow, Turbulence and Combustion*, vol. 87, no. 2-3, pp. 261–292, 2010.

[50] H. G. Weller, G. Tabor, A. D. Gosman, and C. Fureby, "Application of a flame-wrinkling LES combustion model to a turbulent mixing layer," *Proceedings of the Combustion Institute*, vol. 27, no. 1, pp. 899–907, 1998.

[51] G. Tabor and H. G. Weller, "Large eddy dimulation of premixed turbulent combustion using Ξ flame surface wrinkling model," *Flow, Turbulence and Combustion*, vol. 72, no. 1, pp. 1–27, 2004.

[52] E. R. Hawkes and R. S. Cant, "Physical and numerical realizability requirements for flame surface density approaches to large-eddy and Reynolds averaged simulation of premixed turbulent combustion," *Combustion Theory and Modelling*, vol. 5, no. 4, pp. 699–720, 2001.

Correlation of Power Consumption for Several Kinds of Mixing Impellers

Haruki Furukawa,[1] Yoshihito Kato,[1] Yoshiro Inoue,[2] Tomoho Kato,[1] Yutaka Tada,[1] and Shunsuke Hashimoto[2]

[1] *Department of Life and Materials Engineering, Nagoya Institute of Technology, Gokiso-cho, Showa-ku, Nagoya-shi, Aichi 466-8555, Japan*
[2] *Division of Chemical Engineering, Graduate School of Engineering Science, Osaka University, 1-3 Machikaneyama-cho, Toyonaka-shi, Osaka 560-8531, Japan*

Correspondence should be addressed to Yoshihito Kato, kato.yoshihito@nitech.ac.jp

Academic Editor: See-Jo Kim

The authors reviewed the correlations of power consumption in unbaffled and baffled agitated vessels with several kinds of impellers, which were developed in a wide range of Reynolds numbers from laminar to turbulent flow regions. The power correlations were based on Kamei and Hiraoka's expressions for paddle and pitched paddle impellers. The calculated correlation values agreed well with experimental ones, and the correlations will be developed the other types of impellers.

1. Introduction

Mixing vessels is widely used in chemical, biochemical, food, and other industries. Recently scientific approaches were developed by Inoue and Hashimoto [1, 2]. On the other hand, the power consumption is the most important factor to estimate mixing performance and to design and operate mixing vessels.

To estimate the power consumption, the correlation of Nagata et al. [3] has traditionally been used. However, this correlation was developed for two-blade paddle impellers, which do not always have the same numerical values of power consumption as those of multiblade impellers. Kamei et al. [4, 5] and Hiraoka et al. [6] developed the new correlation of the power consumption of paddle impellers, which is more accurate than Nagata's.

However, the new correlation also cannot reproduce the power consumption for other types of impellers. The propeller- and Pfaudler-type impellers are used for low-viscosity liquid and solid-liquid suspensions, and the propeller type has been widely used in vessels ranging from portable type to large tanks. Kato et al. [7] developed a new correlation of power consumption for propeller- and Pfaudler-type impellers based on the correlations of Kamei and Hiraoka.

The power consumption for an anchor impeller was measured by Kato et al. [8, 9] in a wide range of Reynolds numbers from laminar to turbulent flow regions. In the laminar region, the power number of the anchor was reproduced by the correlations of Nagata and Kamei et al. by considering the anchor as a wide paddle impeller. In the turbulent region, it was reproduced by the correlation of Kamei et al. without the correction of the parameters.

In this paper, the authors reviewed the power correlations developed by authors in unbaffled and baffled mixing vessels with several kinds of impellers.

2. Experimental

Figure 1 shows a photograph of mixing impellers used in this work. Figure 2 shows the geometry of impellers with symbols. The mixing vessel used is shown in Figure 3. The vessels for the measurement of power consumption are flat-bottom cylindrical ones of inner diameter $D = 185$ and 200 mm.

FIGURE 1: Photograph of several kinds of mixing impellers.

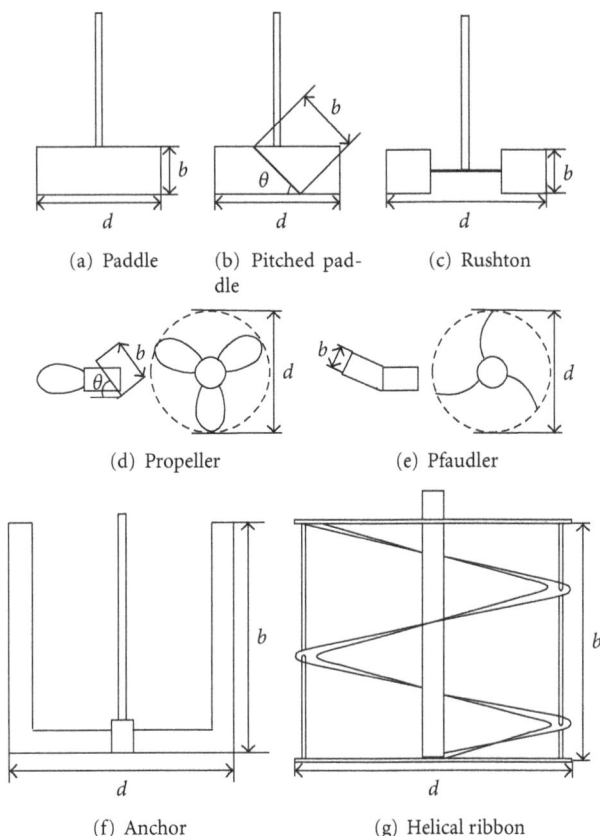

(a) Paddle   (b) Pitched pad-   (c) Rushton
                   dle

(d) Propeller   (e) Pfaudler

(f) Anchor   (g) Helical ribbon

FIGURE 2: Geometry of mixing impellers.

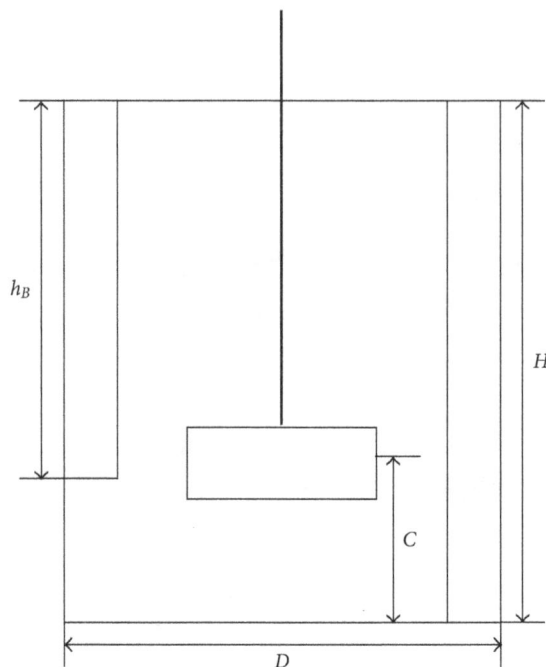

FIGURE 3: Geometry of mixing vessel.

from 60 to 540 rpm to avoid a large vortex at the center of the free surface of the liquid.

3. Results and Discussion

3.1. Paddle Impeller. The correlation equations of Table 1 were developed by Kamei et al. [4, 5] for paddle and Rushton turbine impellers. The comparison between experimental values and calculated values were shown in Figure 4. The correlation equations in Table 1 reproduced the experimental results better than Nagata's correlation equations for the wide range of impeller. The correlation equations for the pitched paddle impeller were shown in Table 2. The correlation equations also reproduced experimental values, which were not shown in a figure here.

3.2. Propeller and Pfaudler Impellers. The blades of the propeller and pfaudler impellers do not have sharp edges. The laminar term C_L in Table 1 can be used without modification. Because the deviation from correlations to measured values in the turbulent region was large, the turbulent terms C_t and m in Table 1 were modified by Kato et al. [7] based on fitting with the modified Reynolds number Re_G and the friction factor f, as follows:

$$C_t = \left[\left(3X^{1.5} \right)^{-7.8} + (0.25)^{-7.8} \right]^{-1/7.8},$$

$$m = \left[\left(0.8X^{0.373} \right)^{-7.8} + (0.333)^{-7.8} \right]^{-1/7.8}. \tag{1}$$

Figures 5 and 6 show the values correlated by the equations in Table 3 and the measured ones. The same equations can be used for the propeller- and the pfaudler-type impellers, regardless of the clearance between the vessel bottom and impeller.

Three kinds of baffled conditions were mainly used: unbaffled, four baffles of $B_W = D/10$ (i.e., the standard baffled condition), and fully baffled. The baffles were plate type. The paddle, pitched paddle, Rushton turbine, propeller, and pfaudler impellers were symmetrically set up at one-half the level of the liquid depth ($C/H = 0.5$). The pfaudler, anchor, and helical-ribbon impellers were set up slightly above the bottom (bottom clearance of 1 mm).

For the measurement of the power consumption, the liquids used were desalted water and varying starch-syrup solutions ($\mu = 0.003$–13 Pa·s). The liquid was filled to the height equal to the vessel diameter ($H = D$).

The power consumption $P(= 2\pi nT)$ was measured with the shaft torque T and rotational speed n by using two types of torque meter (ST-1000 and ST-3000, Satake Chemical Equipment Mfg., Ltd.). The range of rotational speed was

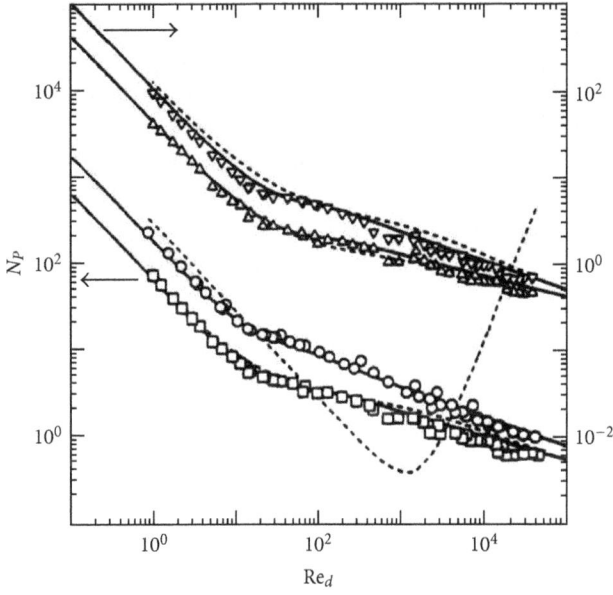

	d/D	b/D	n_p
$\triangledown$	0.585	0.146	8
$\bigcirc$	0.447	0.447	6
$\square$	0.524	0.131	4
$\triangle$	0.585	0.146	2

FIGURE 4: Power diagram of paddle impeller (- - - [3]; — [4, 5]).

TABLE 1: Correlation equations of power number for paddle (a), Rushton turbine (c) and anchor (f).

Unbaffled condition

$$N_{P0} = \{[1.2\pi^4\beta^2]/[8d^3/(D^2H)]\}f$$

$$f = C_L/\text{Re}_G + C_t\{[(C_{tr}/\text{Re}_G) + \text{Re}_G]^{-1} + (f_\curvearrowright/C_t)^{1/m}\}^m$$

$$\text{Re}_d = nd^2\rho/\mu$$

$$\text{Re}_G = \{[\pi\eta\ln(D/d)/(4d/\beta D)]\}\text{Re}_d$$

$$C_L = 0.215\eta n_p(d/H)[1 - (d/D)^2] + 1.83(b\sin\theta/H)(n_p/2\sin\theta)^{1/3}$$

$$C_t = [(1.96X^{1.19})^{-7.8} + (0.25)^{-7.8}]^{-1/7.8}$$

$$m = [(0.71X^{0.373})^{-7.8} + (0.333)^{-7.8}]^{-1/7.8}$$

$$C_{tr} = 23.8(d/D)^{-3.24}(b\sin\theta/D)^{-1.18}X^{-0.74}$$

$$f_\curvearrowright = 0.0151(d/D)C_t^{0.308}$$

$$X = \gamma n_p^{0.7}b\sin^{1.6}\theta/H$$

$$\beta = 2\ln(D/d)/[(D/d) - (d/D)]$$

$$\gamma = [\eta\ln(D/d)/(\beta D/d)^5]^{1/3}$$

$$\eta = 0.711\{0.157 + [n_p\ln(D/d)]^{0.611}\}/\{n_p^{0.52}[1 - (d/D)^2]\}$$

Baffled condition

$$N_P = [(1 + x^{-3})^{-1/3}]N_{P\max}$$

$$x = 4.5(B_w/D)n_B^{0.8}/N_{P\max}^{0.2} + N_{P0}/N_{P\max}$$

Fully baffled condition

$$N_{P\max} = \begin{cases} 10(n_p^{0.7}b/d)^{1.3} & (n_p^{0.7}b/d) \le 0.54 \\ 8.3(n_p^{0.7}b/d) & 0.54 < (n_p^{0.7}b/d) \le 1.6 \\ 10(n_p^{0.7}b/d)^{0.6} & 1.6 < (n_p^{0.7}b/d) \end{cases}$$

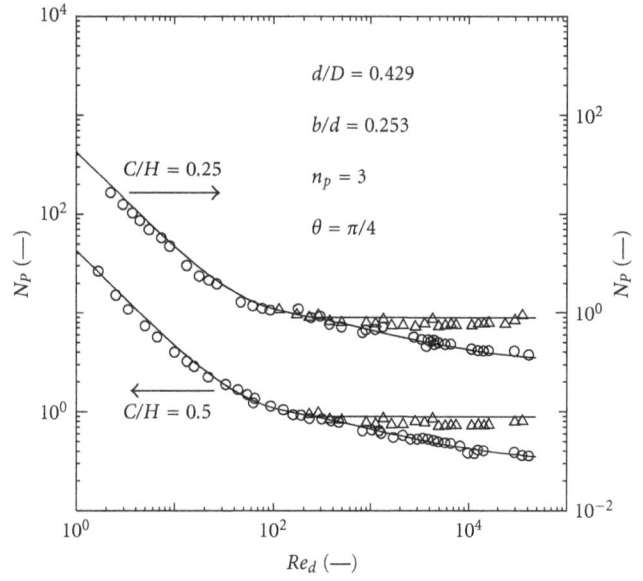

FIGURE 5: Power diagram of propeller impeller (— calc; $\bigcirc$ N_{P0}; $\triangle$ N_P).

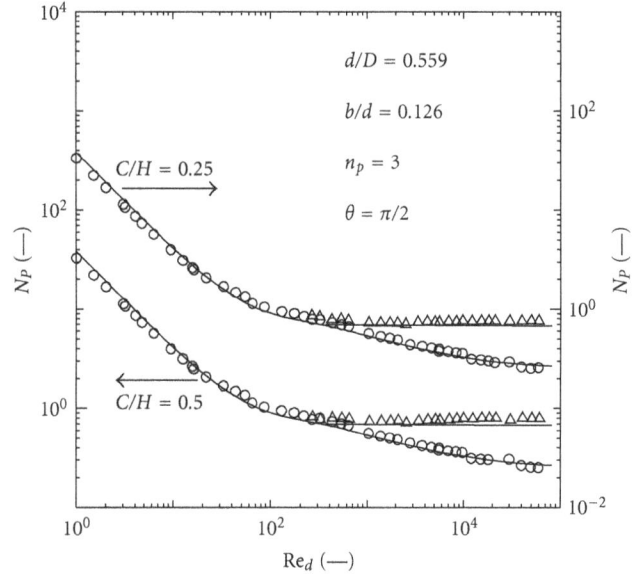

FIGURE 6: Power diagram of pfaudler-type impeller (— calc; $\bigcirc$ N_{P0}; $\triangle$ N_P).

3.3. Anchor Impeller. The anchor impeller was used for the high-viscosity liquid normally. In this work, power consumption for an anchor impeller was measured in a wide range of Reynolds numbers from laminar to turbulent flow regions. In the laminar region, the power number of the anchor was reproduced by the correlations of Nagata and Kamei et al. by considering the anchor as a wide paddle impeller. In the turbulent region, it was reproduced by the correlation of Kamei et al. [4] without the correction of the parameters as shown in Figure 7 [8]. If a large vortex was generated in a turbulent mixing vessel, the experimental values of power number were larger than the calculated ones.

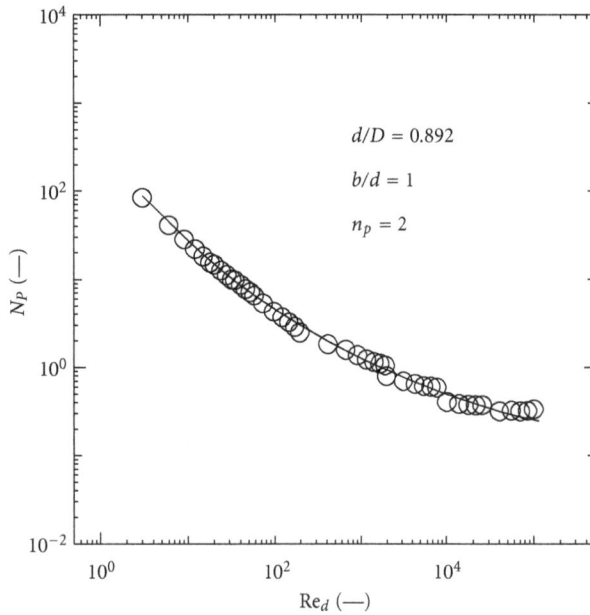

FIGURE 7: Power diagram of anchor impeller (— calc; ○ N_{P0}).

TABLE 2: Correlation equations of power number for pitched paddle (b).

Unbaffled condition

$N_{P0} = \{[1.2\pi^4\beta^2]/[8d^3/(D^2H)]\}f$

$f = C_L/\mathrm{Re}_G + C_t\{[(C_{tr}/\mathrm{Re}_G) + \mathrm{Re}_G]^{-1} + (f_\curvearrowright/C_t)^{1/m}\}^m$

$\qquad \mathrm{Re}_d = nd^2\rho/\mu$

$\qquad \mathrm{Re}_G = \{[\pi\eta\ln(D/d)]/(4d/\beta D)\}\mathrm{Re}_d$

$C_L = 0.215\eta n_p(d/H)[1 - (d/D)^2] + 1.83(b\sin\theta/H)(n_p/2\sin\theta)^{1/3}$

$C_t = [(1.96X^{1.19})^{-7.8} + (0.25)^{-7.8}]^{-1/7.8}$

$m = [(0.71X^{0.373})^{-7.8} + (0.333)^{-7.8}]^{-1/7.8}$

$C_{tr} = 23.8(d/D)^{-3.24}(b\sin\theta/D)^{-1.18}X^{-0.74}$

$f_\curvearrowright = 0.0151(d/D)C_t^{0.308}$

$X = \gamma n_p^{0.7}b\sin^{1.6}\theta/H$

$\beta = 2\ln(D/d)/[(D/d) - (d/D)]$

$\gamma = [\eta\ln(D/d)/(\beta D/d)^5]^{1/3}$

$\eta = 0.711\{0.157 + [n_p\ln(D/d)]^{0.611}\}/\{n_p^{0.52}[1 - (d/D)^2]\}$

Baffled condition

$N_P = [(1 + x^{-3})^{-1/3}]N_{P\max}$

$\qquad x = 4.5(B_W/D)n_B^{0.8}/\{(2\theta/\pi)^{0.72}N_{P\max}^{0.2}\} + N_{P0}/N_{P\max}$

Fully baffled condition

$N_{P\max} = 8.3(2\theta/\pi)^{0.9}(n_p^{0.7}b\sin^{1.6}\theta/d)$

3.4. Helical-Ribbon Impeller.

As an example applied to a helical-ribbon impeller the correlation by Kamei et al. [4] was shown in Figure 8. Since it is the special case, the values of the equations for fitting the experimental values were shown in Table 4. It is the first time to show the power correlation of the helical-ribbon impeller over the wide range of Reynolds numbers.

TABLE 3: Correlation equations of power number for propeller (d) and pfaudler (f).

Unbaffled condition

$N_{P0} = \{[1.2\pi^4\beta^2]/[8d^3/(D^2H)]\}f$

$f = C_L/\mathrm{Re}_G + C_t\{[(C_{tr}/\mathrm{Re}_G) + \mathrm{Re}_G]^{-1} + (f_\curvearrowright/C_t)^{1/m}\}^m$

$\qquad \mathrm{Re}_d = nd^2\rho/\mu$

$\qquad \mathrm{Re}_G = \{[\pi\eta\ln(D/d)]/(4d/\beta D)\}\mathrm{Re}_d$

$C_L = 0.215\eta n_p(d/H)[1 - (d/D)^2] + 1.83(b\sin\theta/H)(n_p/2\sin\theta)^{1/3}$

$C_t = [(3X^{1.5})^{-7.8} + (0.25)^{-7.8}]^{-1/7.8}$

$m = [(0.8X^{0.373})^{-7.8} + (0.333)^{-7.8}]^{-1/7.8}$

$C_{tr} = 23.8(d/D)^{-3.24}(b\sin\theta/D)^{-1.18}X^{-0.74}$

$f_\curvearrowright = 0.0151(d/D)C_t^{0.308}$

$X = \gamma n_p^{0.7}b\sin^{1.6}\theta/H$

$\beta = 2\ln(D/d)/[(D/d) - (d/D)]$

$\gamma = [\eta\ln(D/d)/(\beta D/d)^5]^{1/3}$

$\eta = 0.711\{0.157 + [n_p\ln(D/d)]^{0.611}\}/\{n_p^{0.52}[1 - (d/D)^2]\}$

Baffled condition

$N_P = [(1 + x^{-3})^{-1/3}]N_{P\max}$

$\qquad x = 4.5(B_W/D)n_B^{0.8}/\{(2\theta/\pi)^{0.72}N_{P\max}^{0.2}\} + N_{P0}/N_{P\max}$

Fully baffled condition

$N_{P\max} = 6.5(n_p^{0.7}b\sin^{1.6}\theta/d)^{1.7}$

TABLE 4: Correlation equations of power number for helical ribbon (g).

$N_{P0} = \{[1.2\pi^4\beta^2]/[8d^3/(D^2H)]\}f = 16.0f$

$f = C_L/\mathrm{Re}_G + C_t\{[(C_{tr}/\mathrm{Re}_G) + \mathrm{Re}_G]^{-1} + (f_\curvearrowright/C_t)^{1/m}\}^m$

$\qquad \mathrm{Re}_d = nd^2\rho/\mu$

$\qquad \mathrm{Re}_G = \{[\pi\eta\ln(D/d)]/(4d/\beta D)\}\mathrm{Re}_d = 0.0388\mathrm{Re}_d$

$\qquad C_L = 1.00,\ C_t = 0.100,\ m = 0.333,\ C_{tr} = 2500$

$\qquad f_\curvearrowright = 0.00683$

$\qquad \beta = 2\ln(D/d)/[(D/d) - (d/D)] = 0.999$

$\qquad \eta = 0.538$

3.5. Pitched Paddle and Propeller Impellers with Partial Baffle.

The effects of the baffle length on the power consumption of a mixing vessel with several impellers were studied as shown in Figure 3. The power number was generally correlated with the baffle length, the number of baffles, and the baffle width [9]. The power number of the pitched paddle and the propeller impellers was correlated with only the correction as follows,

$$N_P = \left[\left(1 + x^{-3}\right)^{-1/3}\right]N_{P\max}$$

$$x = \frac{4.5(B_W/D)n_B^{0.8}(h_B/H)}{\left\{(2\theta/\pi)^{0.72}N_{P\max}^{0.2}\right\}} + \frac{N_{P0}}{N_{P\max}}, \tag{2}$$

the results were shown in Figure 9.

3.6. Comparison of Power Number of Propeller Impeller with Rushton's Data.

Figure 10 shows the comparison of power number of a propeller impeller by Rushton et al. [10] with correlation in Table 3. The correlation equations in Table 3

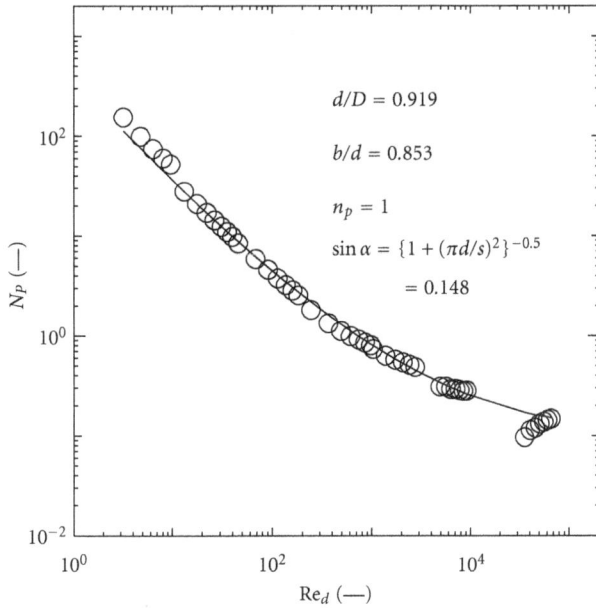

$$d/D = 0.919$$

$$b/d = 0.853$$

$$n_p = 1$$

$$\sin \alpha = \{1 + (\pi d/s)^2\}^{-0.5}$$

$$= 0.148$$

FIGURE 8: Power diagram of helical ribbon-impeller (— calc; $\bigcirc$ N_{P0}).

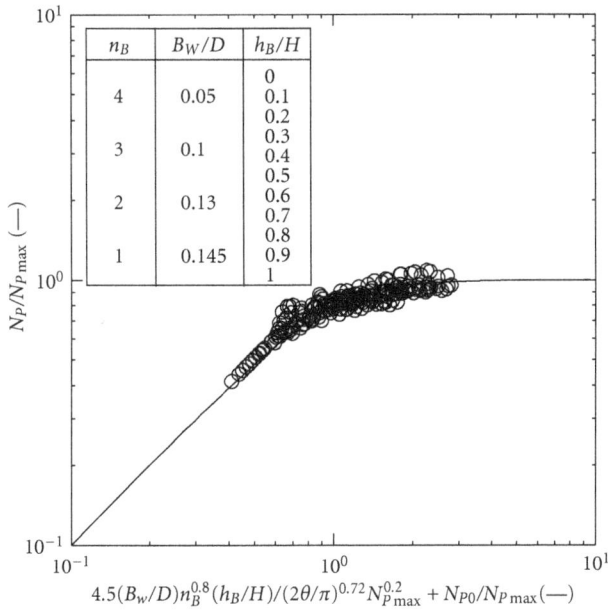

n_B	B_W/D	h_B/H
4	0.05	0
		0.1
		0.2
3	0.1	0.3
		0.4
		0.5
2	0.13	0.6
		0.7
		0.8
1	0.145	0.9
		1

$$4.5(B_w/D)n_B^{0.8}(h_B/H)/(2\theta/\pi)^{0.72}N_{P\,max}^{0.2} + N_{P0}/N_{P\,max}\;(—)$$

FIGURE 9: Correlation of power number for propeller with partial baffle (— calc; $\bigcirc$ exp.).

also reproduced the data which other researchers measured well. As mentioned above, the correlation reproduced in this work has a possibility of correlating the power number of the mixing impellers of all geometry. It is the first time to compare between the data of Rushton's and the power correlation over the wide range of Reynolds number.

4. Conclusions

The authors reviewed the power correlations developed by authors. The new correlation equations of power number,

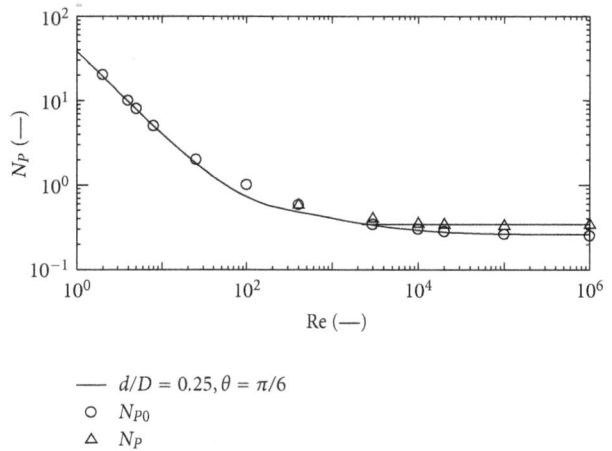

$$— \quad d/D = 0.25, \theta = \pi/6$$
$$\bigcirc \quad N_{P0}$$
$$\triangle \quad N_P$$

FIGURE 10: Comparison of power number for propeller with Rushton's data (— calc; $\bigcirc$, $\triangle$ Rushton's).

based on the correlation of Kamei and Hiraoka, were developed for several kinds of mixing impellers, and it was shown that the calculated values of the power number agreed very closely with the measured ones. In future work, this correlation will be expanded to other impellers.

Nomenclature

b:	Height of impeller blade (m)
C:	Clearance between bottom and impeller (m)
D:	Vessel diameter (m)
d:	Impeller diameter (m)
f:	Friction factor (—)
H:	Liquid depth (m)
h_B:	Baffle length (m)
N_P:	Power number $(=P/\rho n^3 d^5)$ (—)
N_{P0}:	Power number in unbaffled condition (—)
N_{Pmax}:	Power number in fully baffled condition (—)
n:	Impeller rotational speed (—)
n_B:	Number of baffle plates (—)
n_p:	Number of impeller blades (—)
P:	Power consumption (W)
Re_d:	Impeller Reynolds number $(=nd^2\rho/\mu)$ (—)
Re_G:	Modified Reynolds number (—)
T:	Shaft torque (N·m)
θ:	Angle of impeller blade (—)
μ:	Liquid viscosity (Pa·s)
ρ:	Liquid density (kg/m^3).

References

[1] Y. Inoue and S. Hashimoto, "Analysis of mechanism of laminar fluid mixing in 3-D mixing tank based on streakline lobes," *Kagaku Kogaku Ronbunshu*, vol. 36, no. 4, pp. 355–365, 2010.

[2] Y. Inoue, "Invariant structure and variable properties in laminar fluid mixing model based on strealines," *Kagaku Kogaku Ronbunshu*, vol. 37, pp. 211–222, 2011.

[3] S. Nagata, T. Yokoyama, and H. Maeda, "Studies on the power requirement of paddle agitators in cylindrical vessels," *Kagaku Kogaku*, vol. 20, pp. 582–592, 1956.

[4] N. Kamei, S. Hiraoka, Y. Kato et al., "Power correlation for paddle impellers in spherical and sylindrical agitated vessels," *Kagaku Kogaku Ronbunshu*, vol. 21, pp. 41–48, 1995.

[5] N. Kamei, S. Hiraoka, Y. Kato et al., "Effects of impeller and baffle dimensions on power consumption under turbulent flow in an agitated vessel with paddle impeller," *Kagaku Kogaku Ronbunshu*, vol. 22, no. 2, pp. 255–256, 1996.

[6] S. Hiraoka, N. Kamei, and Y. Kato, "Power correlation for pitched blade paddle impeller in agitated vessels with and without baffles," *Kagaku Kogaku Ronbunshu*, vol. 23, no. 6, pp. 974–975, 1997.

[7] Y. Kato, Y. Tada, Y. Takeda, Y. Hirai, and Y. Nagatsu, "Correlation of power consumption for propeller and pfaudler type impellers," *Journal of Chemical Engineering of Japan*, vol. 42, no. 1, pp. 6–9, 2009.

[8] Y. Kato, N. Kamei, Y. Tada et al., "Power consumption of anchor impeller over wide range of reynolds number," *Kagaku Kogaku Ronbunshu*, vol. 37, pp. 19–21, 2011.

[9] Y. Kato, N. Kamei, A. Tada et al., "Effect of baffle length on power consumption in turbulent mixing vessel," *Kagaku Kogaku Ronbunshu*, vol. 37, pp. 377–380, 2011.

[10] J. H. Rushton, E. W. Costich, and H. J. Everett, "Power characteristics of mixing impellers part 1," *Chemical Engineering Progress*, vol. 46, no. 8, pp. 395–404, 1950.

3

Utilization of Agrowaste Polymers in PVC/NBR Alloys: Tensile, Thermal, and Morphological Properties

Ahmad Mousa,[1] **Gert Heinrich,**[2] **Bernd Kretzschmar,**[2] **Udo Wagenknecht,**[2] **and Amit Das**[2]

[1] *Department of Materials Engineering, Faculty of Engineering, Al Balqa Applied University, Salt 19117, Jordan*
[2] *Leibniz-Institut für Polymerforschung Dresden e.V., Hohe Straße 6, 01069 Dresden, Germany*

Correspondence should be addressed to Ahmad Mousa, mousa@rocketmail.com

Academic Editor: Licínio M. Gando-Ferreira

Poly(vinyl chloride)/nitrile butadiene rubber (PVC/NBR) alloys were melt-mixed using a Brabender Plasticorder at 180°C and 50 rpm rotor speed. Alloys obtained by melt mixing from PVC and NBR were formulated with wood-flour- (WF-) based olive residue, a natural byproduct from olive oil extraction industry. WF was progressively increased from 0 to 30 phr. The effects of WF loadings on the tensile properties of the fabricated samples were inspected. The torque rheometry, which is an indirect indication of the melt strength, is reported. The pattern of water uptake for the composites was checked as a function WF loading. The fracture mode and the quality of bonding of the alloy with and without filler are studied using electron scanning microscope (SEM).

1. Introduction

Polymer alloys continue to represent a field of intensive research. One of the most common blends in the modern sense is PVC with NBR [1, 2]. Due to the miscible nature of PVC/NBR blend as evidenced from single glass transition (T_g) the soft blend of PVC/NBR can be categorized as a thermoplastic elastomer (TPE) and more specifically as a melt processable rubber (MPR) [3–5]. Fillers are incorporated mainly to improve service properties or to reduce material cost depending on the source of filler, type of filler, method of preparation, and treatment. Very large quantities of the natural lignocelluloses polymers are produced annually as agrowastes. A very small amount is used as antioxidants or fillers in polymers. The rest is used almost as fuel to generate energy. The field of wood-based agrowastes polymer composites is extensively reviewed in the open literature [5–8]. Recently, we report the effect of virgin olive pomace on the flexural and thermal performance of toughened PVC composites [9]. We found that the virgin olive pomace enhanced the flexural properties to a certain extent, which was due to the hydrogen bond formation, while the thermal stability was improved due to the phenolic hydroxyl group within the lignocellulosic powder. In this work, the effect of wood-flour-based olive residue on the tensile properties, water absorption and morphology of PVC/NBR alloys are reported in the current investigation.

2. Experimental

2.1. Materials and Formulation. Acrylonitrile nitrile rubber with 34% acrylo content was supplied by Bayer AG, Germany. Suspension PVC grade in powder form with a k-value of 67 was supplied by SABIC of Saudi Arabia and stabilized with lead salt. Wood-flour-based agrowastes with particle size equal or less than $45\,\mu m$ were used as received. The WF-based olive mill residue has been fully characterized and reported earlier; the major reactive functional group with its structure was the hydroxyl group from the cellulose and hemicellulose [10, 11]. The samples were formulated according to the following recipe: NBR: 20% PVC: 80%, WF: various in part per hundred-part polymer (php), that is, the filler loading was based on the total amount of resin (PVC) and elastomer (NBR), which is 100 parts.

2.2. Sample Preparation. Mixing was carried out at 180°C and $50\,rev\cdot min^{-1}$ rotor speed using a computerized brabender plasticorder Model PLE 331 for 8 minutes. The NBR was initially loaded into the mixing chamber of Brabender for one minute, followed by PVC and the wood flour.

2.3. Torque Rheometry. Melt rheological properties of the prepared blends were evaluated using a Brabender Plasticorder at the predetermined mixing variables. Mixing was continued until torque and temperature were stabilized to constant values of 8–10 min at 50 rev/min as the optimum mixing shear. The effect of WF loading on the shear heating (ΔT) in a Brabender Plasticorder results in temperature rise given as: ΔT = melt temperature − set temperature. Mixing was performed until constant stabilization torque and temperatures values were recorded.

2.4. Tensile Properties. Tensile properties were carried out according to ASTM D638. The dumbbells specimens were cut from 2 mm thick molded sheets of wood-flour-filled PVC/NBR alloys. Five specimens were tested, and the median value was taken for each formulation.

2.5. Water Absorption. 2 mm thick rectangular samples were weighed in air. The samples were immersed in distilled water for seven days at room temperature. The samples were removed from water, wiped with tissue paper, and reweighed. The % water uptake was calculated according to the following equation:

$$\text{\% Water uptake} = \frac{W_2 - W_1}{W_1} \times 100, \tag{1}$$

where W_1 is the sample weight in air and W_2 is the weight after immersion. The average of three samples was calculated.

2.6. Thermal Analysis. The composites were scanned by a Perkin-Elmer DSC-6 differential scanning calorimeter (DSC) in the range of −40 to 100°C at heating rate of 5°C/min.

2.7. Failure Mode. To check the failure mode and the quality of bonding the surfaces of the tensile fractured samples were viewed under scanning electron microscope (SEM) model (geol Tokyo, Japan). The specimens were sputtered with Au-Pd alloy prior to scanning to avoid electrostatic charges.

3. Results and Discussions

Figure 1 shows the effect of WF loading on tensile modulus of PVC/NBR alloy. It can be seen that tensile modulus increased steadily with filler loading. This trend is in line with earlier work on rigid and toughened PVC [9, 10]. The observed trend recorded in Figure 1 could be attributed to the reduction in free volume between the chains of the PVC/NBR alloy. Reduction in the free volume with filler loading leads to the observed trend shown in Figure 1. Figure 2 shows the influence of WF on the yield tensile strength of the PVC/NBR alloy as a function of filler loading. It is clear that the yield tensile strength decreased with WF loading.

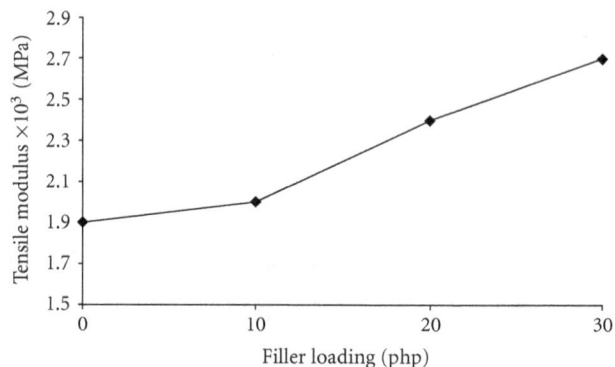

FIGURE 1: The influence of filler loading on tensile modulus of PVC/NBR alloy.

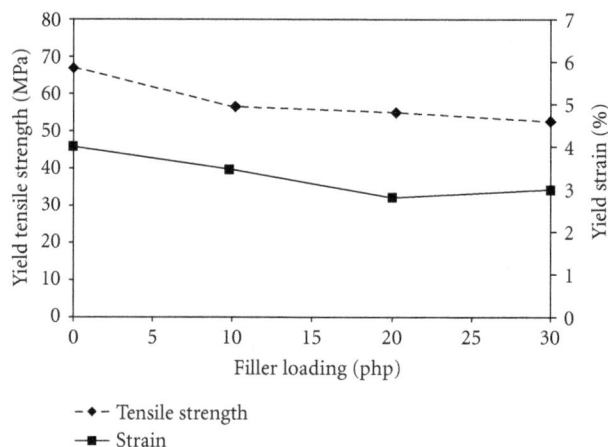

FIGURE 2: The influence of filler loading on the yield tensile strength of PV/NBR alloy.

Such observation might be due to the inability of WF to transfer the stress, possibly due to improper filler dispersion in the matrix and moisture pick-up which lead to insufficient interfacial bonding between the alloy and the filler.

Similar trends were recorded in the case of percentage strain at yield shown in Figure 2. Again this might be due to the rigidity of the PVC/NBR filled with WF due to the rigidity of the filler itself. The progress of the stock temperature as a function of WF loading is represented in Figure 3 for the PVC/NBR alloy. One can see that the stock temperature increased with WF loading over the set processing conditions. However, the extensive shearing causes the stock temperature to rise steeply above the mixing temperature even at the end of the 8th min of the mixing time, until the stock temperature undergoes a steady value. Interestingly, the equilibrium torque displayed in Figure 3, which is an indirect indication of the melt strength, showed a higher value with addition of WF as compared to the control sample. This suggested that the addition of WF has slightly restricted the mobility of the alloy due to the interaction between the matrix and the filler and hence increased the

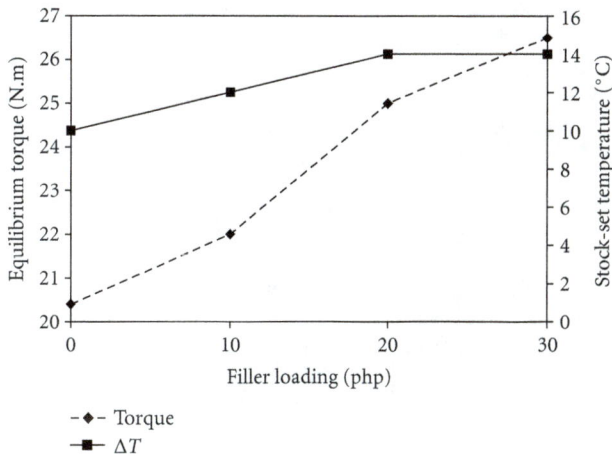

FIGURE 3: The influence of WF loading on the equilibrium torque and stock temperature of PV/NBR alloy.

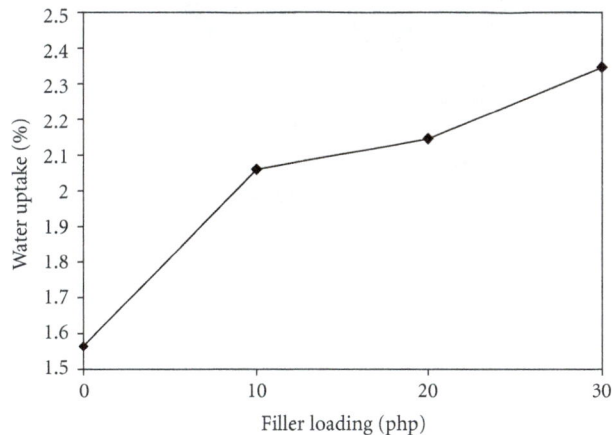

FIGURE 4: The influence of WF loading on % water uptake of PV/NBR alloy.

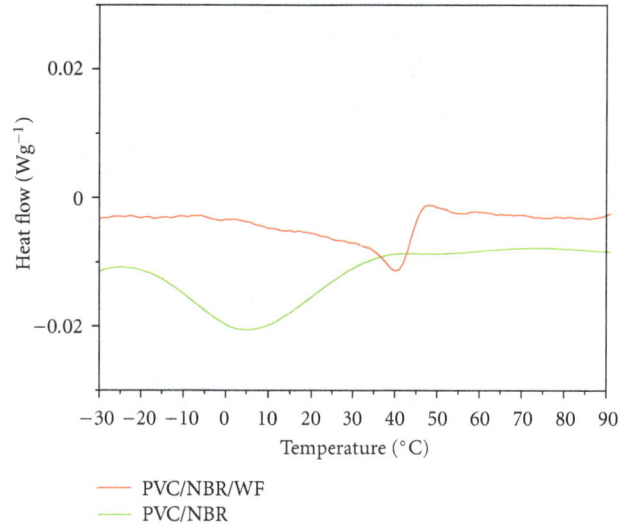

FIGURE 5: DSC traces of PVC/NBR alloy with and without WF.

melt strength. Such finding suggests that certain types of interactions between the filler and the matrix have occurred, such interactions are expected to slightly restrict the mobility of the alloy chains and increased the torque. Figure 4 shows the water absorption behavior of the PVC/NBR alloy as a function of filler loading. It can be seen that water absorption has increased with WF content. This is mainly due to hydroxyl content of the WF. The low uptake in case of the unfilled composite could be due to the hydrophobic nature of the polymer.

3.1. Thermal Analysis. Figure 5 illustrates DSC curves (first run) for PVC/NBR control and for the composite with WF at 20 php loading of filler loading. Looking at the DSC curves presented in Figure 5 at the −40–90°C temperature interval, one can see that the endotherm peak was detected. Such peak was shifted to the right by the incorporation of the WF loading. One may conclude that this peak is caused by

molecular relaxation of the PVC/NBR alloy. Note that the addition of WF has delayed the molecular relaxation of the alloy. This should be related to the interaction between the WF and the PVC/NBR alloy via polar-polar interaction as reported earlier [9, 10].

3.2. SEM. Interfacial interactions and the strength of adhesion determine micromechanical deformation processes and the failure mode of the composites [11, 12]. The SEM micrographs that were taken from the surfaces of broken specimens offer indirect information about the failure mode and bonding quality. Figure 6(a) presents the fracture surface of the plain blend. It can bee seen that the system is one phase with some particles that come from ingredients added to the PVC compound such as stabilizers and so forth. Figure 6(b) shows the fracture surface of sample with 10 php filler. It can be seen that the wood is covered with the polymer and the relatively small number of holes related to debonding or fiber pull-out indicates good adhesion. On the other hand, the opposite is observed in composites prepared with the higher amounts of filler doses, namely, 20 and 30 php loading, respectively, as shown in Figures 6(c) and 6(d). The number of debonded particles is quite large, the contours of particles remaining on the surface of the matrix are sharp, and adhesion seems to be poor, at least compared to Figure 6(b).

4. Conclusions

Based on this paper it can be concluded that pristine WF has improved the tensile modulus of the blend whereas the tensile strength remained more or less the same. It also can be concluded that the filler has good degree of interactions as indicated by the torque data obtained from the Brabender plasticorder. The DSC traces showed that the molecular relaxation of the blend was hindered with the presence of the WF. SEM micrographs showed that the failure mode was due

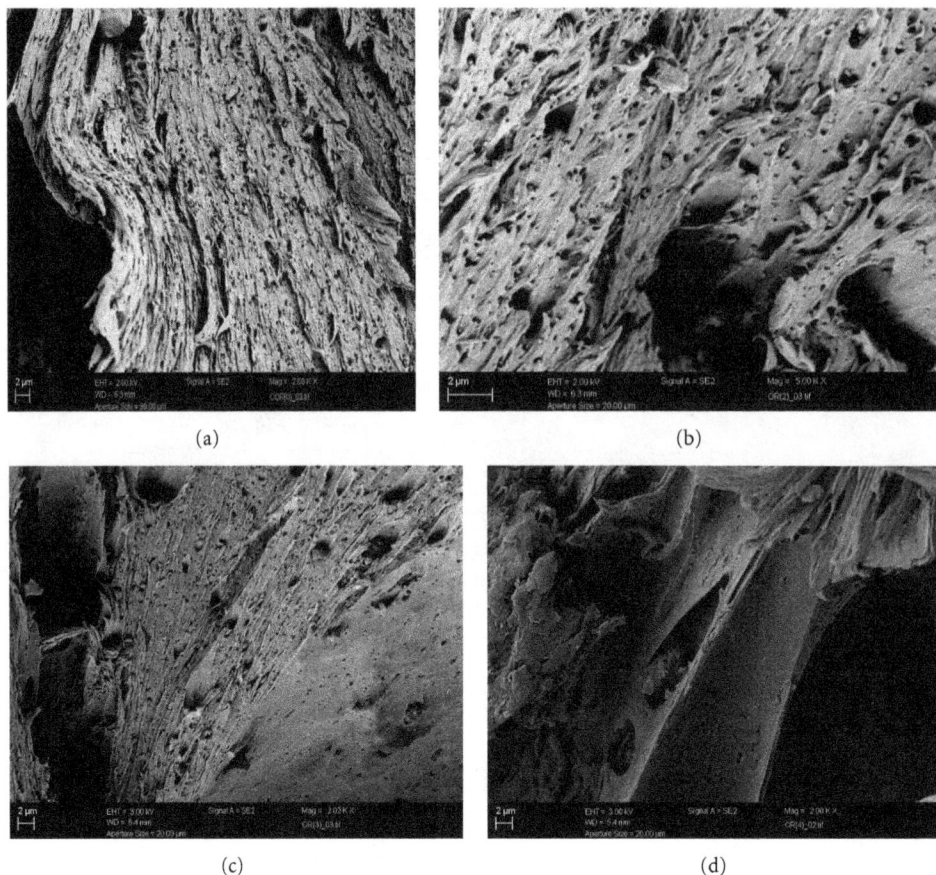

(a)

(b)

(c)

(d)

FIGURE 6: Scanning electron microscope images of (a) plain PVC/NBR alloy (b) PVC/NBR at 10 php filler loading, (c) 20 php filler, and (d) 30 php filler.

to the pull-out of the filler; furthermore, higher loading of the WF has led to agglomeration.

References

[1] L. A. Utracki, D. J. Walsh, and R. A. Weiss, "Polymer alloys, blends, and Ionomers: an overview," in *Multiphase Polymers: Blends and Ionomers*, L. A. Utracki and R. A. Weiss, Eds., vol. 395 of *ACS symposium Series*, pp. 1–35, American Chemical Society, Washington, DC, USA, 1989.

[2] K. T. Varughese, P. P. De, S. K. Sanyal, and S. K. De, "Miscible blends from plasticized poly(vinyl chloride) and epoxidized natural rubber," *Journal of Applied Polymer Science*, vol. 37, no. 9, pp. 2537–2548, 1989.

[3] Y. Song, Q. Zheng, and C. Liu, "Green biocomposites from wheat gluten and hydroxyethyl cellulose: processing and properties," *Industrial Crops and Products*, vol. 28, no. 1, pp. 56–62, 2008.

[4] G. Siracusa, A. D. la Rosa, V. Siracusa, and M. Trovato, "Eco-compatible use of olive husk as filler in thermoplastic composites," *Journal of Polymers and the Environment*, vol. 9, no. 4, pp. 157–161, 2001.

[5] A. Abu Bakar, A. Hassan, and A. F. M. Yusof, "Effect of oil palm empty fruit bunch and acrylic impact modifier on mechanical properties and prcessability of unplasticized poly

[6] H. D. Rozman, G. S. Tay, R. N. Kumar, A. Abusamah, H. Ismail, and Z. A. Mohd, "The effect of oil extraction of the oil palm empty fruit bunch on the mechanical properties of polypropylene-oil palm empty fruit bunch-glass fibre hybrid composites," *Polymer-Plastics Technology and Engineering*, vol. 40, no. 2, pp. 103–115, 2001.

[7] Z. A. M. Ishak, A. Aminullah, H. Ismail, and H. D. Rozman, "Effect of silane-based coupling agents and acrylic acid based compatibilizers on mechanical properties of oil palm empty fruit bunch filled high-density polyethylene composites," *Journal of Applied Polymer Science*, vol. 68, no. 13, pp. 2189–2203, 1998.

[8] L. Avérous and F. le Digabel, "Properties of biocomposites based on lignocellulosic fillers," *Carbohydrate Polymers*, vol. 66, no. 4, pp. 480–493, 2006.

[9] A. Mousa, G. Heinrich, U. Gohs, R. Hässler, and U. Wagenknecht, "Application of renewable agro-waste-based olive pomace on the mechanical and thermal performance of toughened PVC," *Polymer-Plastics Technology and Engineering*, vol. 48, no. 10, pp. 1030–1040, 2009.

[10] A. Mousa, G. Heinrich, and U. Wagenknecht, "Thermoplastic composites based on renewable natural resources: unplasticized PVC/olive husk," *International Journal of Polymeric Materials*, vol. 59, no. 11, pp. 843–853, 2010.

(vinyl chloride) composites," *Polymer-Plastics Technology and Engineering*, vol. 44, pp. 1125–1137, 2005.

[11] T. G. Vladkova, P. D. Dineff, and D. N. Gospodinova, "Wood flour: a new filler for the rubber processing industry. II. Cure characteristics and mechanical properties of NBR compounds filled with corona-treated wood flour," *Journal of Applied Polymer Science*, vol. 91, no. 2, pp. 883–889, 2004.

[12] L. Fama, A. Mônica, B. Q. Bittante, P. J. A. Sobral, S. Goyanes, and L. N. Gerschenson, "Garlic powder and wheat bran as fillers: their effect on the physicochemical properties of edible biocomposites," *Materials Science and Engineering C*, vol. 30, no. 6, pp. 853–859, 2010.

4

Phosphorus Removal from Wastewater Using Oven-Dried Alum Sludge

Wadood T. Mohammed and Sarmad A. Rashid

Chemical Engineering Department, College of Engineering, University of Baghdad, Baghdad, Iraq

Correspondence should be addressed to Sarmad A. Rashid, sermed1972@yahoo.com

Academic Editor: See-Jo Kim

The present study deals with the removal of phosphorus from wastewater by using oven-dried alum sludge (ODS) as adsorbent that was collected from Al-Qadisiya treatment plant (Iraq); it was heated in an oven at 105°C for 24 h and then cooled at room temperature. The sludge particles were then crushed to produce a particle size of 0.5–4.75 mm. Two modes of operation are used, batch mode and fixed bed mode, in batch experiment the effect of oven-dried alum sludge doses 10–50 g/L, pH of solution 5–8 with constant initial phosphorus concentration of 5 mg/L, and constant particle size of 0.5 mm were studied. The results showed that the percent removal of phosphorus increases with the increase of oven-dried alum sludge dose, but pH of solution has insignificant effect. Batch kinetics experiments showed that equilibrium time was about 6 days. Adsorption capacity was plotted against equilibrium concentration, and isotherm models (Freundlich, Langmuir, and Freundlich-Langmuir) were used to correlate these results. In the fixed bed isothermal adsorption column, the effect of initial phosphorus concentration (C_o) 5 and 10 mg/L, particle size 2.36 and 4.75 mm, influent flow rate (Q) 6 and 10 L/hr, and bed depth (H) 0.15–0.415 m were studied. The results showed that the oven-dried alum sludge was effective in adsorbing phosphorus, and percent removal of phosphorus reaches 85% with increasing of contact time and adsorbent surface area (i.e., mass of adsorbent 50 g/L with different pH).

1. Introduction

Wastewater or contaminated water is a big environmental problem all over the world, in industrial plants; contaminants may be a result of side reactions, rendering the water stream an effluent status. These impurities are at low-level concentration but still need to be further reduced to levels acceptable by various destinations in the plant. Surface waters contain certain level of phosphorus (P) in various compounds, which is an important constituent of living organisms. In natural conditions the phosphorus concentration in water is balanced; that is, accessible mass of this constituent is close to the requirements of the ecological system. When the input of phosphorus to water is higher than it can be assimilated by a population of living organisms, the problem of excess phosphorus content occurs. Regulatory control on phosphorus disposal is evident all over the world recently [1–3]. Strict regulatory requirements decreased the permissible level of phosphorus concentration in wastewater at the point of disposal (i.e.,

1 mg/L). This has made it very important to find appropriate technological solution for treatment of wastewater prior to disposal.

Phosphorus removal is considered as a major challenge in wastewater treatment, particularly for small-scale wastewater treatment systems. Processes available for P-treatment are generally classified into three general categories: chemical, physical, or biological-based treatment systems. Among physical-chemical methods, phosphorus removal is achieved using ion exchange [4, 5], dissolved air flotation [6, 7], and membrane filtration [8, 9]. Filtration has been used either alone or in conjunction with a coagulation process as a means to remove phosphorus from wastewater [10, 11]. High-rate sedimentation has also been attempted in some studies [12, 13]. Among the various physical-chemical methods, coagulation with chemical precipitation and adsorption are the most common techniques being used for removing phosphorus. Enhanced biological methods for removing phosphorus are also used with success [14, 15]. For small-scale applications (e.g., aquaculture) biological methods may not be appropriate for phosphorus removal because of the low carbon concentrations, which increases cost and time involved in biological methods [16]. Alternatively, physical-chemical methods can offer advantages for small industries because of lower initial costs involvement. These methods are also easier to use and do not require high level of expertise to maintain. Physical-chemical methods can also accommodate recycling sludge to reduce further costs involved in handling sludge. However, finding an effective and feasible material is a significant challenge in physical-chemical approach. This problem has not been addressed so far as a complete solution. The key problem is to find a suitable material, which is easily available and effective to remove phosphorus from small-scale wastewater applications.

Biosolid management is considered very important, as there are considerable amounts of biosolids generated due to anthropogenic reasons. Alum sludge, a biosolid generated in the coagulation process in a water treatment plant, is one such type.

Divalent and trivalent cation-based materials are known to be effective for phosphorus removal. Therefore, aluminum-based residuals (i.e., alum sludge) are a viable option for being an effective phosphorus removal material. Alum is typically effective in phosphorus removal in chemical precipitation process [17]. Therefore, use of alum sludge can be effective for phosphorus removal. Air-dried alum sludge has also been attempted in limited manner by some researchers with success [18]. However, the use of waste material (alum sludge) not only can provide low-cost appropriate technological alternative for small-scale applications but also reduce hazard and cost related to the disposal of large amount of alum sludge.

The aim of this work was to investigate the effectiveness of oven-dried alum sludge for adsorption of orthophosphate from deionized water and to compare the results with other conventional adsorbent (i.e., activated carbon).

1.1. Adsorption Isotherm. The most common forms of adsorption isotherms used in chemical-environmental engineering are the Langmuir and the Brunauer, Emmett and Teller (BET) besides the Freundlich empirical model [19–23]. The Langmuir model can be described as

$$\frac{X}{m} = \frac{(abC_e)}{(1 + aC_e)},\tag{1}$$

where X = mass of solute adsorbed to the solid (mg), m = mass of adsorbent used (g), C_e = concentration of solute in solution at equilibrium (mg/L), a = Langmuir constant; the amount of solute adsorbed per unit weight of an adsorbent in forming a complete monolayer (L/mg), b = Langmuir constant (mg/g).

Assumptions made in developing the Langmuir model are as follows [19].

(a) The maximum adsorption corresponds to a saturated monolayer of solute molecules on the adsorbent surface.

(b) The energy of adsorbent is constant.

(c) There is no transmigration of adsorbate in the plane of the surface.

(d) The adsorption is reversible.

The Freundlich isotherm is an empirical model and was developed for heterogeneous surfaces. The Freundlich adsorption model is of the form [19]

$$\frac{X}{m} = kC_e^{1/n},\tag{2}$$

where k = Freundlich equilibrium constant indicative of adsorptive capacity, n = Freundlich constant indicative of adsorption intensity.

Combination of Langmuir-Freundlich isotherm model, that is, the SIPS model for single component adsorption [24] is

$$q_e = \frac{\left(bqmC_e^{1/n}\right)}{\left(1 + bC_e^{1/n}\right)}.\tag{3}$$

The BET isotherm model is of the form

$$\frac{X}{m} = \frac{(AC_eX_m)}{\left[(C_s - C_e)\left\{1 + (A - 1)^{C_e/C_s}\right\}\right]},\tag{4}$$

where A = a constant describing energy interaction between the solute and the adsorbent surface, X_m = amount of solute adsorbed in forming a complete monolayer (M/M), C_e = concentration of solute in solution at equilibrium, C_s = saturation concentration of solute in solution.

The BET isotherm model has been developed on the following assumptions [19].

(a) The model describes multilayer adsorption at the adsorbent surface.

(b) The Langmuir model applies to each layer.

(c) A given layer needs not complete formation prior to initiation of subsequent layers.

2. Experimental

2.1. Alum Sludge. Alum sludge is a waste material generated during the coagulation/sedimentation process in a drinking water treatment plant. Inorganic materials in alum sludge are presented in Table 1.

Alum sludge used in this research was heated in an oven at 105°C for 24 hours. The dried sludge was then cooled to room temperature. The sludge particles were then crushed to produce a particle size of (0.5, 2.36, and 4.75) mm. The physical properties are listed in Table 2.

2.2. Granulated Activated Carbon (GAC). Granulated activated carbon (GAC) is one of the most widely used adsorbents for organic, metallic, and inorganic contaminants in water. It was supplied by Unicarbo, Italians to the Iraqi local markets. The physical properties are listed in Table 3 [25].

2.3. Adsorbate. Orthophosphate (potassium dihydrogen orthophosphate KH_2PO_4) was used to prepare a phosphorus solution. Orthophosphate was chosen as it is the key species of phosphorus in most wastewaters. The physical properties of KH_2PO_4 are listed in Table 4.

TABLE 1: Inorganic materials in alum sludge.

Constituent	Weight percent
Aluminum	3.38%
Iron	0.819%
Manganese	0.16%
Chromium	0.013%
Vanadium	0.002%
Zinc	0.0098%
Lead	0.0001%
Barium	0.0001%
Arsenic	0.0002%

TABLE 2: Physical properties of oven dried alum sludge.

Item name	Oven dried alum sludge
Bulk density (kg/m^3)	786.7
Particle porosity	0.7
Bed porosity	0.65
BET surface area (m^2/g)	191

TABLE 3: Physical properties of activated carbon.

Item name	Granulated activated carbon
Base	Coconut shell
Apparent density (kg/m^3)	480–490
Bulk density (kg/m^3)	770
BET surface area (m^2/g)	1000
Particle porosity	0.5
Bed porosity	0.4
Ash content (%)	5 (max)
Iodine number (mg/g)*	1100–1130
pH**	10.2–10.6

*The iodine number refers to the milligrams of a 0.02 normal iodine solution per gram of GAC during a standard test (ASTM D4607).
**The pH of a solution of distilled water with a specific dosage of GAC according to a standard test (ASTM D3838).

TABLE 4: Main properties of adsorbate.

Name of component	Potassium dihydrogen orthophosphate
Chemical symbol	KH_2PO_4
Name of company	The British Drug Houses LTD/England
Molecular weight	136.09 kg/kg·mole
Assay (acidimetric)	99 to 101 per cent
Chloride (Cl)	Not more than 0.01 per cent
Sodium (Na)	Not more than 0.2 per cent
Sulphate (SO_4)	Not more than 0.05 per cent
pH (1 percent solution)	4.5 to 4.7

2.4. Wastewater. Deionized water was spiked with KH_2PO_4 to prepare a phosphorus solution of (5 and 10) mg/L; the masses used are listed in Table 5. The phosphorus concentration was chosen as typical phosphorus concentration in many wastewaters.

TABLE 5: Phosphorus solution.

Phosphorus concentration (mg/L)	KH_2PO_4 to be added (g)
5	0.00715
10	0.0143

2.5. Adsorption Column. The fixed bed adsorber studies were carried out in Q.V.F. glass column of 2 in. (50.8 mm) I.D. and (30 and 50) cm in height.

The oven-dried alum sludge was confined in the column by fine stainless steel screen at the bottom and a glass cylindrical packing at the top of the bed to ensure a uniform distribution of influent through the alum sludge. The influent solution was introduced to the column through a perforated plate, fixed at the top of the column.

2.6. Experimental Arrangements. The schematic representation of experimental equipment is shown in Figure 1.

3. Experimental Work

3.1. Batch Experiments. Batch experiments were used to obtain the equilibrium isotherm curves and then the equilibrium data. In batch mode the following experiments were carried out:

(i) effect of oven-dried alum sludge weight on adsorption process,

(ii) effect of pH of solution on adsorption process,

(iii) equilibrium isotherm experiments.

All experiments were carried out at 25°C ± 1. The desired pH was adjusted using 0.1 M NaOH and HCl.

Five 1 L flasks were used for each pH solution (i.e., 5, 6, 7, and 8). For experiments conducted with an initial phosphorus concentration of 5 mg/L, alum sludge was used in concentration of (10, 20, 30, 40 and 50) g/L. Samples were collected from the flasks and after 7.5 days tested for total orthophosphate phosphorus concentration. Phosphorus that was lost from the solution was assumed to be adsorbed onto the adsorbents. Data obtained from batch tests conducted on deionized water fitted to Freundlich, Langmuir, and Freundlich-Langmuir adsorption isotherm equations.

3.2. Fixed Bed Column Experiments. Column experiments were carried out to measure the breakthrough curves for the systems. Experiments were carried out at various pH, initial phosphorus concentrations (C_o), particle size, flow rate (Q), and bed depth (H).

4. Results

4.1. Batch Experiments

4.1.1. Effect of Mass of Oven-Dried Alum Sludge on the Adsorption Process. The results of the dependence of phosphorus on the mass of oven-dried alum sludge of size 0.5 mm at 25°C

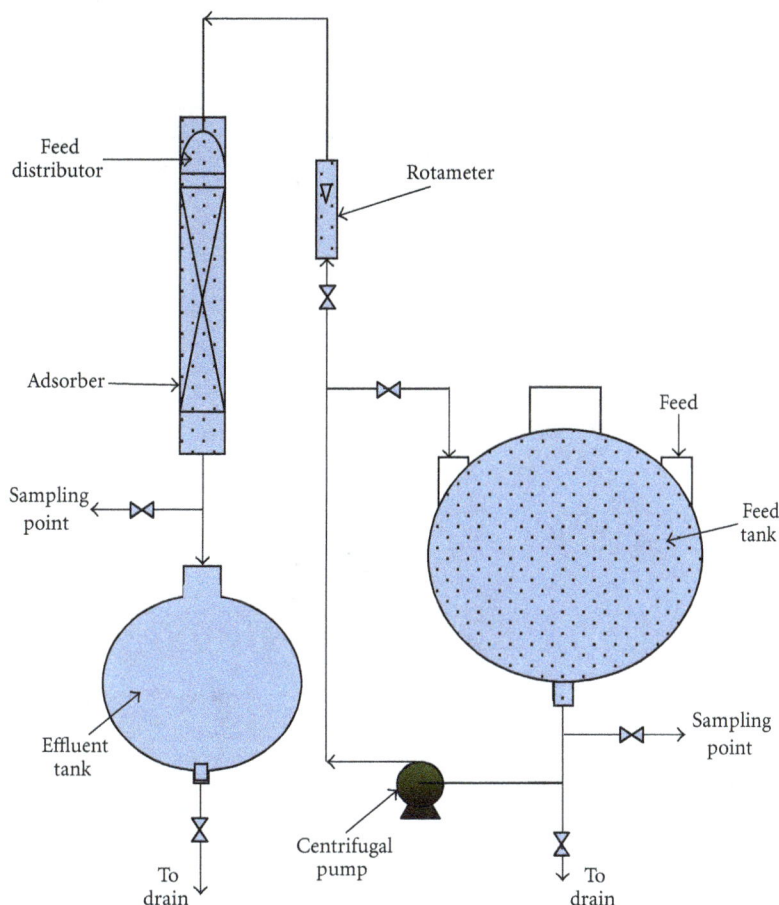

FIGURE 1: Schematic representation of experimental equipment.

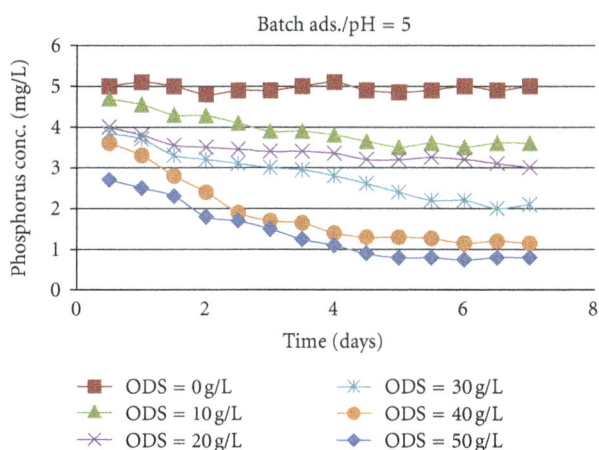

FIGURE 2: Change in phosphorus concentration with time of batch tests (C_o = 5 mg/L, pH = 5, Temp. = 25°C, particle size = 0.5 mm).

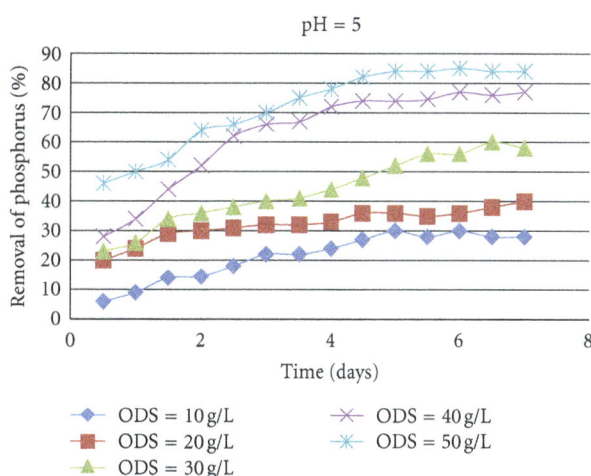

FIGURE 3: The effect of ODS on phosphorus removal (C_o = 5 mg/L, pH = 5, Temp. = 25°C, particle size = 0.5 mm).

are shown in Figures 2, 3, 4, 5, 6, 7, 8, and 9. These figures represent the plotting of the phosphorus concentration with time and the percentage removal of phosphorus against the mass of oven-dried alum sludge, respectively.

The percent removal of phosphorus increases with increasing weight of oven-dried alum sludge up to a certain value depending on adsorption sites. These figures can clearly show that the increase in the percent removal of phosphorus

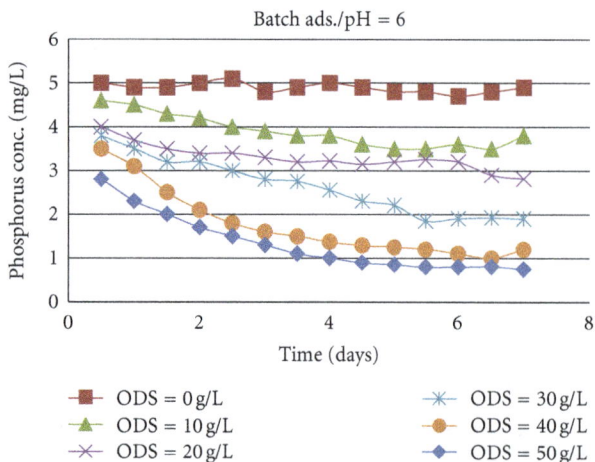

FIGURE 4: Change in phosphorus concentration with time of batch tests (C_o = 5 mg/L, pH = 6, Temp. = 25°C, particle size = 0.5 mm).

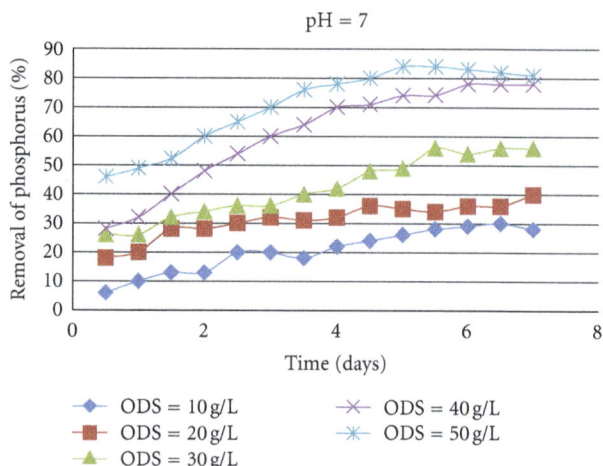

FIGURE 7: The effect of ODS on phosphorus removal (C_o = 5 mg/L, pH = 7, Temp. = 25°C, particle size = 0.5 mm).

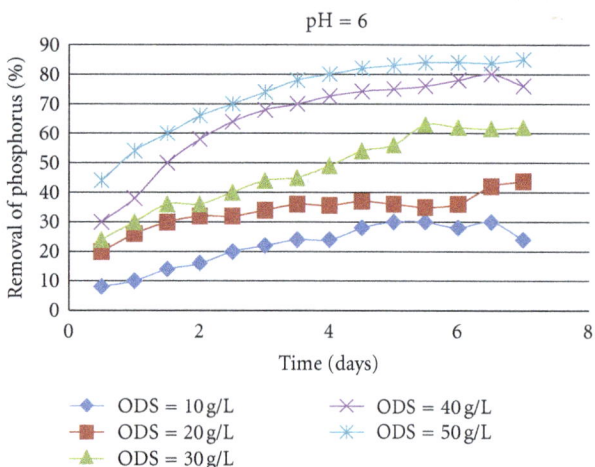

FIGURE 5: The effect of ODS on phosphorus removal (C_o = 5 mg/L, pH = 6, Temp. = 25°C, particle size = 0.5 mm).

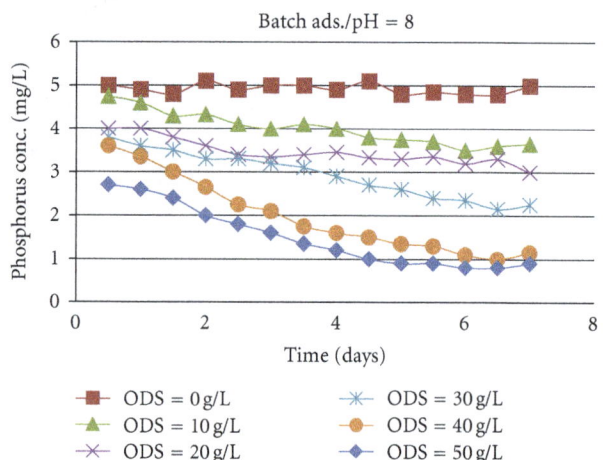

FIGURE 8: Change in phosphorus concentration with time of batch tests (C_o = 5 mg/L, pH = 8, Temp. = 25°C, particle size = 0.5 mm).

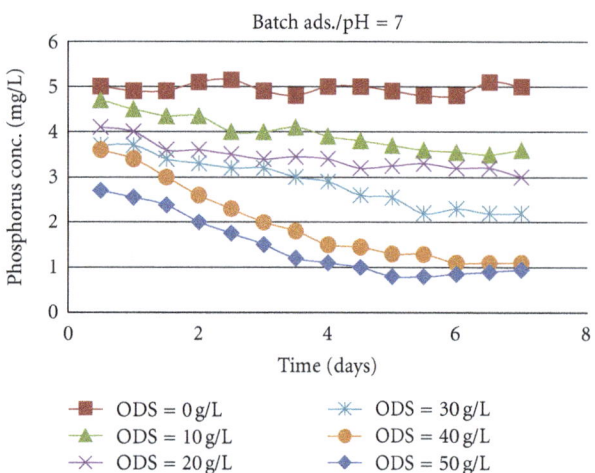

FIGURE 6: Change in phosphorus concentration with time of batch tests (C_o = 5 mg/L, pH = 7, Temp. = 25°C, particle size = 0.5 mm).

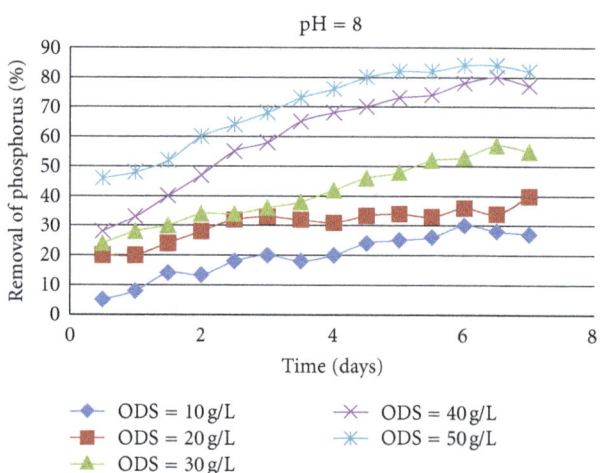

FIGURE 9: The effect of ODS on phosphorus removal (C_o = 5 mg/L, pH = 8, Temp. = 25°C, particle size = 0.5 mm).

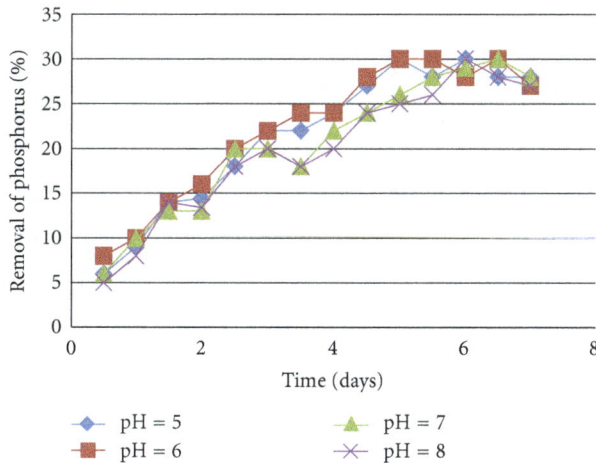

FIGURE 10: The effect of pH on adsorption of phosphorus by ODS (C_o = 5 mg/L, mass of sludge = 10 g/L, Temp. = 25°C).

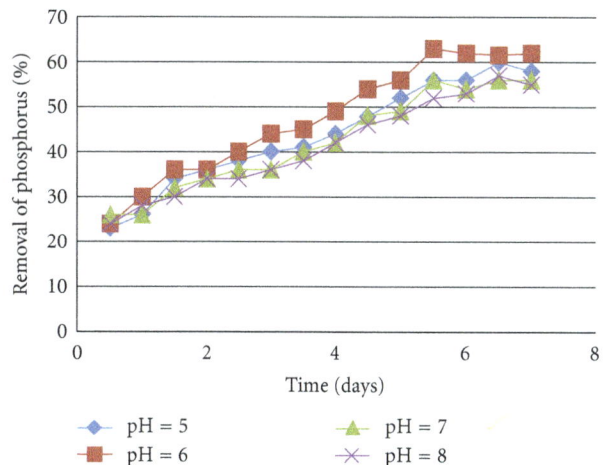

FIGURE 11: The effect of pH on adsorption of phosphorus by ODS (C_o = 5 mg/L, mass of sludge = 20 g/L, Temp. = 25°C).

is due to the greater availability of adsorption sites or surface area of adsorbent (oven-dried alum sludge).

4.1.2. Effect of pH on the Adsorption Process. The removal of phosphorus by using oven-dried alum sludge, as a function of pH, is presented in Figures 10, 11, 12, 13, and 14. Effect of pH, on the adsorption density is illustrated in Figure 15. pH had little effect on the adsorption density. However, solution pH of 6 appeared to produce maximum adsorption density in many of the experimental results.

The effluent pH was understandably dependent on the influent pH. An influent pH 5 produced effluent pH of (4.7–5.5). Similarly an influent pH 6 generated an effluent pH range (5.6–6.3), and an influent pH 7 generated an effluent pH range of (6.5–7.5). It was due to the adsorption and desorption of H$^+$ ions during the adsorption of phosphorus on alum sludge. An effluent pH below 4.5 is not suitable for disposal in surface water. The effluent pH can be increased prior to disposal in surface water. However, the cost of chemicals to reduce initial pH and to increase final pH and hazards of dealing with increased amount of sludge would pose negative interest for pH control.

In general, a pH value in the range of (6–9) is reasonable for wastewaters before disposal into surface water.

The removals of phosphorus on oven-dried alum sludge for all the pHs were more than (80)%.

4.2. Equilibrium Isotherm Experiments. The adsorption isotherm curves were obtained by plotting the weight of the solute adsorbed per unit weight of the adsorbent (q_e) against the equilibrium concentration of the solute (c_e). Figures 16, 17, 18, and 19 show the adsorption isotherm curves for phosphorus adsorption onto oven-dried alum sludge at 25°C. The obtained data of phosphorus was correlated with Langmuir, Freundlich, and Langmuir-Freundlich models. The parameters for each model were obtained from nonlinear statistical fit of the equation to the experimental

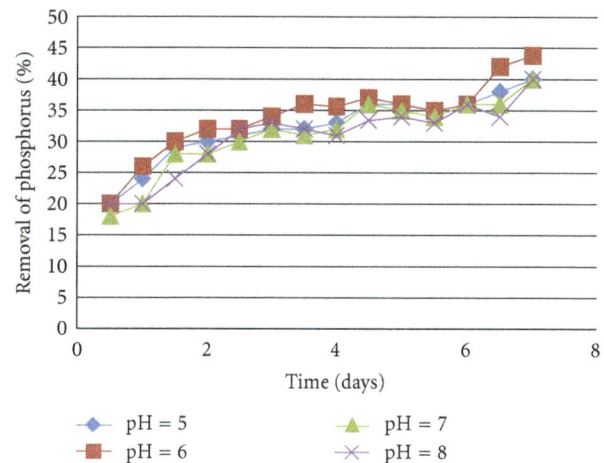

FIGURE 12: The effect of pH on adsorption of phosphorus by ODS (C_o = 5 mg/L, mass of sludge = 30 g/L, Temp. = 25°C).

data. All parameters with their correlation coefficients (R^2) are summarized in Table 6.

It is clear from the previous figures and table the following.

(i) The equilibrium isotherm is of favorable type, for being convex upward.

In order to assess the different isotherms and their ability to correlate with experimental results, the correlation coefficient was employed to ascertain the fit of each isotherm with experimental data. From Table 6, the correlation coefficient value was higher for Langmuir-Freundlich than other correlations. This indicates that the Langmuir-Freundlich isotherm is clearly the best fitting isotherm to the experimental data.

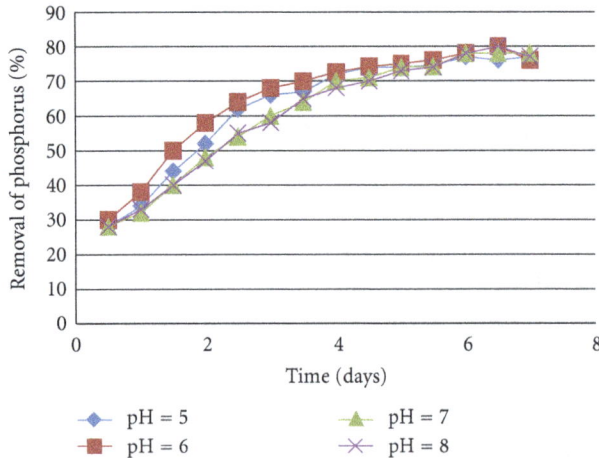

FIGURE 13: The effect of pH on adsorption of phosphorus by ODS (C_o = 5 mg/L, mass of sludge = 40 g/L, Temp. = 25°C).

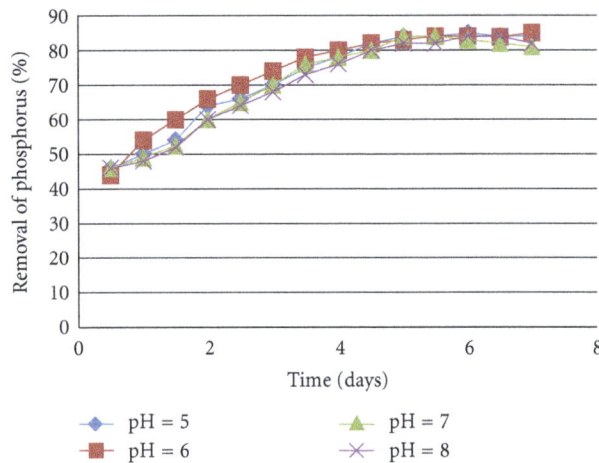

FIGURE 14: The effect of pH on adsorption of phosphorus by ODS (C_o = 5 mg/L, mass of sludge = 50 g/L, Temp. = 25°C).

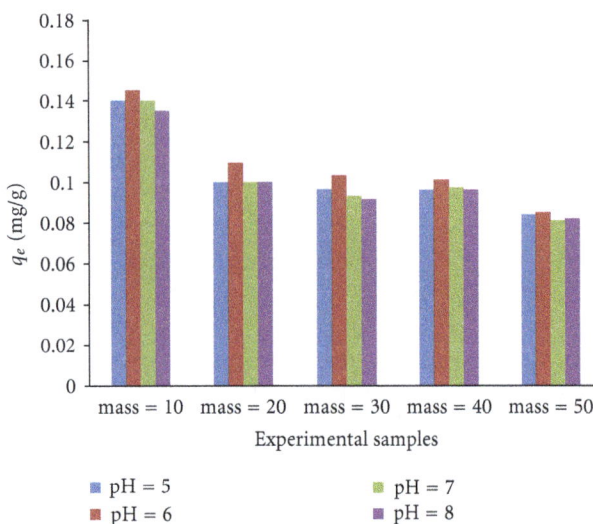

FIGURE 15: The effect of pH of solution on adsorption density.

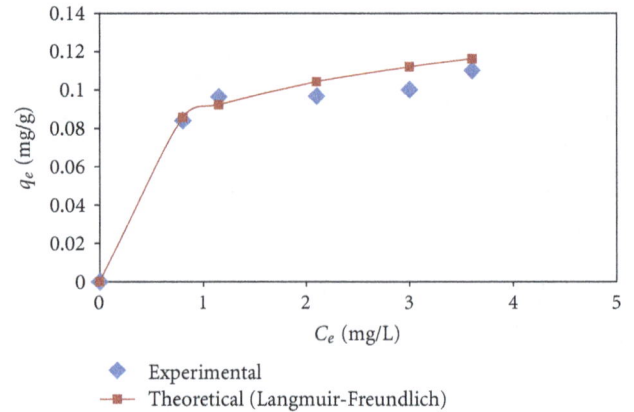

FIGURE 16: Adsorption isotherm for phosphorus onto ODS (C_o = 5 mg/L, pH = 5, Temp. = 25°C).

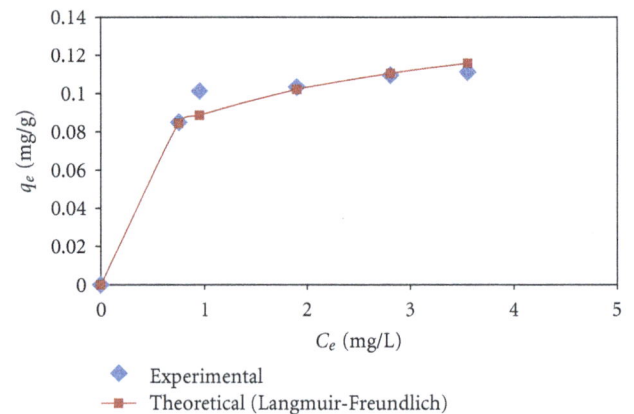

FIGURE 17: Adsorption isotherm for phosphorus onto ODS (C_o = 5 mg/L, pH = 6, Temp. = 25°C).

4.3. Fixed Bed Experiments

4.3.1. Effect of Initial Concentration. Experimental data for adsorption of phosphorus onto ODS at flow rate of 6 L/hr, bed depth of 0.25 m, and particle size of 2.36 mm are shown in Figure 20. It is clear from the breakthrough curves that an increase in the initial concentration makes the breakthrough curves much steeper, which would be anticipated with the basis of the increases in driving force for mass transfer with increases in concentration of solute in solution. The breakpoint was inversely related to the initial concentration; that is, the time required to reach saturation decreases with increasing the inlet solute concentration. This may also be explained by the fact that since the rate of diffusion is controlled by the concentration gradient, it takes a longer contact time to reach saturation for the case of low value of initial solute concentration.

4.3.2. Effect of Particle Size. The breakthrough curves as shown in Figure 21 were obtained for different particle size at constant initial concentration of phosphorus (10 mg/L), bed depth of oven-dried alum sludge (0.25 m), and constant flow rate (6 L/hr). The experimental results showed that fine

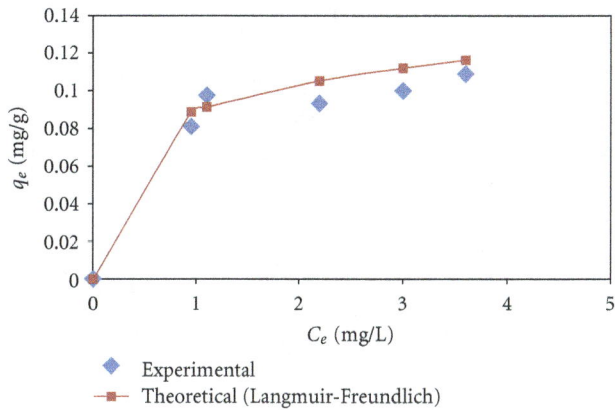

FIGURE 18: Adsorption isotherm for phosphorus onto ODS ($C_o = 5$ mg/L, pH = 7, Temp. = 25°C).

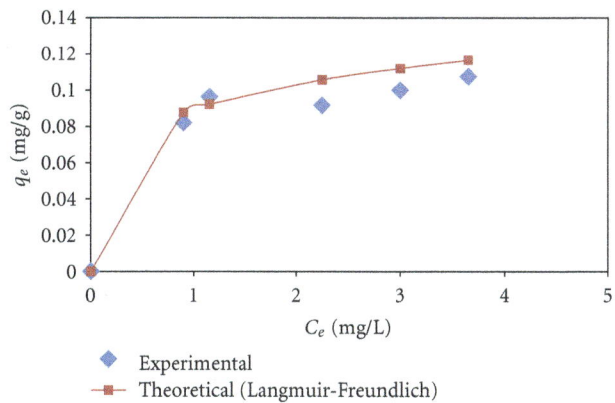

FIGURE 19: Adsorption isotherm for phosphorus onto ODS ($C_o = 5$ mg/L, pH = 8, Temp. = 25°C).

TABLE 6: Isotherm parameters for phosphorus adsorption onto ODS with the correlation coefficient.

Model	Parameters	Values
	a,	2.437878
Langmuir (1)	b,	0.128524
	Correlation coefficient	0.977
	K,	0.090537
Freundlich (2)	n,	2.92948
	Correlation coefficient	0.9946
	q_m,	3.675591
Combination of	b,	0.025254
Langmuir-Freundlich (3)	n,	4.793108
	Correlation coefficient	0.995

particle sizes showed a higher phosphorus removal than coarse particle sizes as illustrated in the figure. This was due to large surface area of fine particles.

4.3.3. Effect of Flow Rate and Bed Depth. In the design of a fixed bed adsorption column, the contact time is the most significant variable, and therefore the bed depth

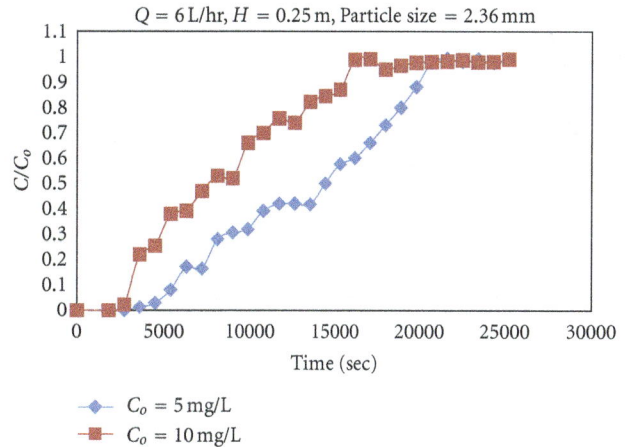

FIGURE 20: Experimental breakthrough curves for adsorption of phosphorus onto ODS at different initial concentration.

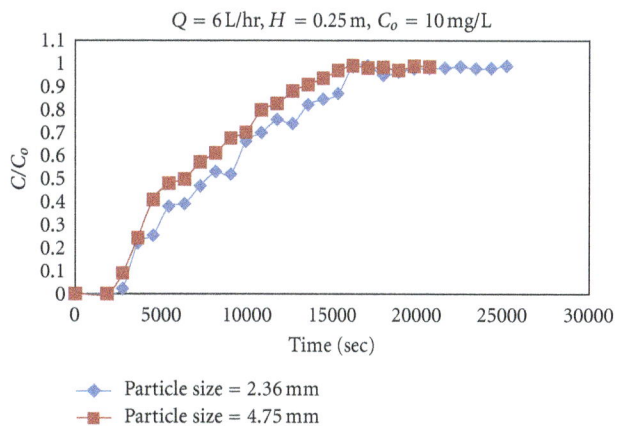

FIGURE 21: Experimental breakthrough curves for adsorption of phosphorus onto ODS at different particle size.

and the flow rate are the major design parameter. The effect of varying the volumetric flow rate was investigated. The experimental breakthrough curves are presented in Figure 22 in terms of C/C_o versus time at constant initial concentration of phosphorus (10 mg/L) and particle size (2.36 mm). Increasing the flow rate may be expected to make reduction of the surface film. Therefore, this will decrease the resistance to mass transfer and increase the mass transfer rate. Also, because the reduction in the surface film is due to the disturbance created when the film of the bed increased resulting from easy passage of the adsorbate molecules through the particles and entering easily to the pores, this decreased contact time between phosphorus and oven-dried alum sludge at high flow rate. The effect of bed depth was investigated for phosphorus adsorption onto oven-dried alum sludge; the experimental breakthrough curves are presented in Figure 23. The breakthrough curves were obtained for different bed depth of oven-dried alum sludge at constant flow rate and constant P concentration. It is clear that the increase in bed depth increases the breakthrough time and

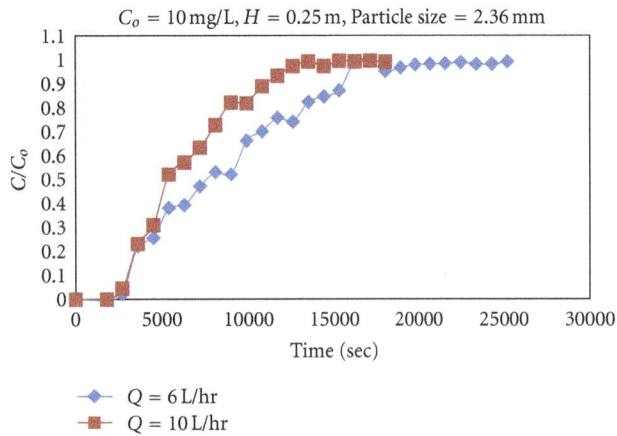

FIGURE 22: Experimental breakthrough curves for adsorption of phosphorus onto ODS at different flow rates.

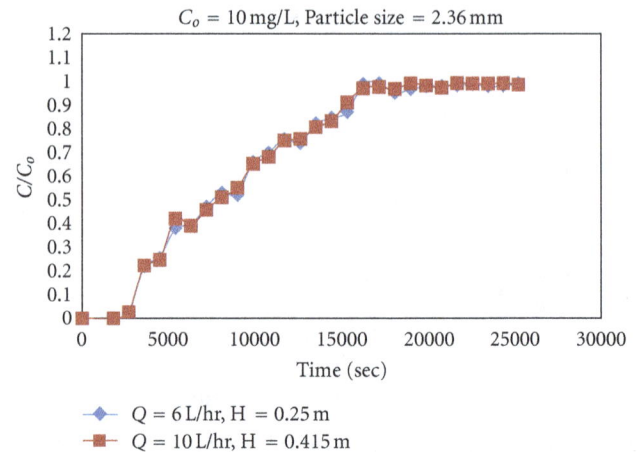

FIGURE 24: Experimental breakthrough curves for adsorption of phosphorus onto ODS at different residence time.

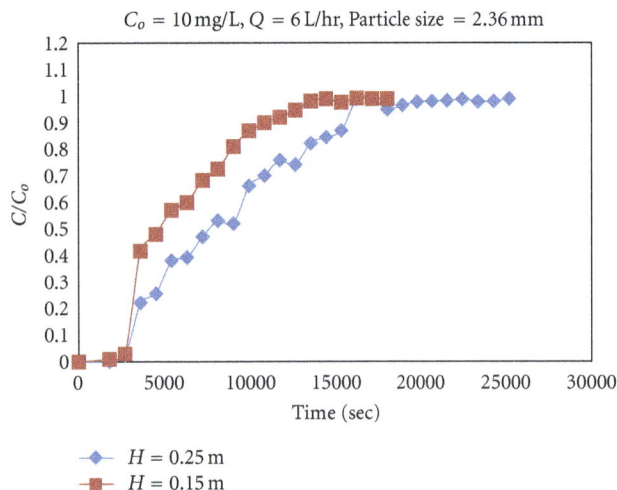

FIGURE 23: Experimental breakthrough curves for adsorption of phosphorus onto ODS at different bed depth.

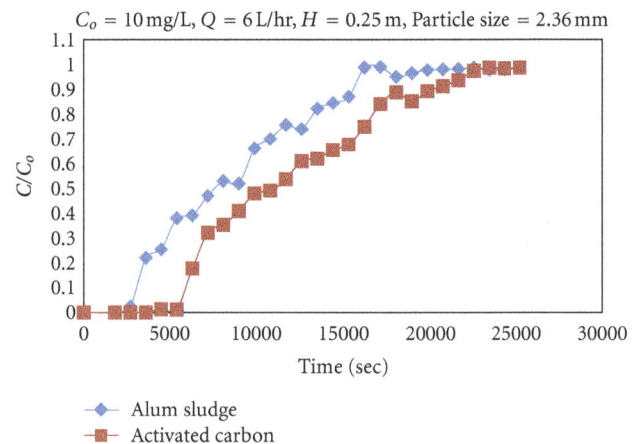

FIGURE 25: Comparison between ODS and granular-activated carbon for phosphorus adsorption.

the residence time of the solute in the column. A comparison is shown in Figure 24, where bed depths are different but residence time was kept constant by changing the flow rate. These results show that increases of fluid velocity had no significant effect. Thus, the residence time in the column is more important than fluid velocity in improving the removal efficiency. The results are obtained, here.

4.3.4. Oven-Dried Alum Sludge Compared to Granular Activated Carbon. The results obtained, here, are presented in Figure 25. Granular-activated carbon (GAC) was used in this study as a reference by which to compare adsorption with oven-dried alum sludge. The results showed that granular-activated carbon had the best phosphorus removal capability due to its higher surface area. These results were on the contrary with the results obtained by Maruf et al., [26] due to higher surface area of alum sludge that was collected from the

Lake Major Water Treatment Plant, Halifax Regional Municipality, Canada with respect to granular-activated carbon; also this was probably because of other adsorbates competing with phosphorus in secondary municipal wastewater for adsorption sites.

5. Conclusions

The present study has led to the following conclusions.

(1) Oven-dried alum sludge was effective in adsorbing phosphorus from deionized water.

(2) In batch experiment the percent removal of phosphorus increases (85%) with increasing in the oven-dried alum sludge dose (i.e., 50 g/L at deferent pH).

(3) Batch kinetics experiments showed that equilibrium time was about 6 days without mechanical mixing.

(4) The results showed that the model (Langmuir-Freundlich) gave good fitting for adsorption capacity.

(5) In fixed bed experiment, the percent removal of phosphorus increases with increasing contact time and adsorbent surface area.

Nomenclature

a : Langmuir constant (L/mg)
b: Langmuir constant (mg/g)
C: Concentration of solute in solution (mg/L)
k: Freundlich equilibrium constant
n: Freundlich constant
m: Mass of solute adsorbent (g)
H: Bed depth (m)
pH: Acidity
Q: Flow rate (L/hr)
q: Amount of metal ion adsorbed (mg/g)
R^2: Correlation coefficients
x: Mass of solute adsorbed (mg)
X_m: Amount of solute adsorbed in forming a complete monolayer (M/M).

Abbreviations

P: Phosphorus
ODS: Oven-dried alum
MTZ: Mass transfer zone
GAC: Granulated activated carbon
BET: The Brunauer, Emmett, and Teller
QVF: Quality vessels fabrication
ASTM: American society for testing and materials
USEPA: United states environmental protection agency
ID: Inner diameter.

Subscript

o : Initial
e : Equilibrium
s : Saturation.

References

[1] Environment Canada, "Proposed approach for wastewater effluent quality," Final Report, Environment Canada, 2000.

[2] USEPA, Environmental Assessment for the Final Effluent Limitations Guidelines, Pretreatment Standards for New and Existing Sources and New Source Performance Standards for the Centralized Waste Treatment, Prepared by Charles Tamulonis, 2000.

[3] Department of Justice, *Pollutant Substances Prevention Regulation*, Department of Justice, Ottawa, Canada, 2004.

[4] D. Zhao and A. K. Sengupta, "Ultimate removal of phosphate from wastewater using a new class of polymeric ion exchangers," *Water Research*, vol. 32, no. 5, pp. 1613–1625, 1998.

[5] L. Liberti, D. Petruzzelli, and L. De Florio, "REM NUT ion exchange plus struvite precipitation process," *Environmental Technology*, vol. 22, no. 11, pp. 1313–1324, 2001.

[6] R. G. Penetra, M. A. P. Reali, E. Foresti, and J. R. Campos, "Post-treatment of effluents from anaerobic reactor treating domestic sewage by dissolved-air flotation," *Water Science and Technology*, vol. 40, no. 8, pp. 137–143, 1999.

[7] P. Jokela, E. Ihalainen, J. Heinänen, and M. Viitasaari, "Dissolved air flotation treatment of concentrated fish farming wastewaters," *Water Science and Technology*, vol. 43, no. 8, pp. 115–121, 2001.

[8] K. C. Yu, J. S. Chang, I. P. Chen, D. J. Chang, C. Y. Chang, and S. H. Chen, "The removal of colloid and dissolved phosphorus by coagulation and membrane microfiltration," *Journal of Environmental Science and Health A*, vol. 35, no. 9, pp. 1603–1616, 2000.

[9] A. Dietze, R. Gnirß, and U. Wiesmann, "Phosphorus removal with membrane filtration for surface water treatment," *Water Science and Technology*, vol. 46, no. 4-5, pp. 257–264, 2002.

[10] W. Xie, M. Kondo, and Y. Naito, "Study on phosphorus removal using a new coagulation system," *Water Science and Technology*, vol. 30, no. 6, pp. 257–262, 1994.

[11] L. Jonsson, E. Plaza, and B. Hultman, "Experiences of nitrogen and phosphorus removal in deep-bed filters in the Stockholm area," *Water Science and Technology*, vol. 36, no. 1, pp. 183–190, 1997.

[12] F. Rogalla, G. Roudon, J. Sibony, and F. Blondeau, "Minimising nuisances by covering compact sewage treatment plants," *Water Science and Technology*, vol. 25, no. 4-5, pp. 363–374, 1992.

[13] S. Zeghal, N. Puznava, J. P. Subra, and P. Sauvegrain, "Process control for nutrients removal using lamella sedimentation and floating media filtration," *Water Science and Technology*, vol. 38, no. 3, pp. 227–235, 1998.

[14] D. G. Wareham, K. J. Hall, and D. S. Mavinic, "ORP screening protocol for biological phosphorus removal in sequencing batch reactors," *Canadian journal of civil engineering*, vol. 22, no. 2, pp. 260–269, 1995.

[15] N. R. Louzeiro, D. S. Mavinic, W. K. Oldham, A. Meisen, and I. S. Gardner, "Methanol-induced biological nutrient removal kinetics in a full-scale sequencing batch reactor," *Water Research*, vol. 36, no. 11, pp. 2721–2732, 2002.

[16] J. K. Park, J. Wang, and G. Novotny, "Wastewater characteristization for evaluation of biological phosphorus removal," Research Report 174, Department of Natural Resources, Wisconsin, Wis, USA, 1997.

[17] M. I. Aguilar, J. Sáez, M. Lloréns, A. Soler, and J. F. Ortuño, "Nutrient removal and sludge production in the coagulation-flocculation process," *Water Research*, vol. 36, no. 11, pp. 2910–2919, 2002.

[18] J. G. Kim, J. H. Kim, H. S. Moon, C. M. Chon, and J. S. Ahn, "Removal capacity of water plant alum sludge for phosphorus in aqueous solutions," *Chemical Speciation and Bioavailability*, vol. 14, no. 1–4, pp. 67–73, 2003.

[19] J. R. Weber and J. Walter, *Physicochemical Processes for Water Quality Control*, Wiley-Inter Science, New York, NY, USA, 1972.

[20] Metcalf and Eddy, *Waste Water Engineering Treatment, Disposal and Reuse*, McGrew Hill Book Company, New York, NY, USA, 1979.

[21] W. W. Eckenfelder Jr, *Application of Adsorption to Waste Water Treatment*, Enviro Press, Nashville, Tenn, USA, 1981.

[22] L. D. Benefield, J. L. Judkins, and B. L. Weanal, *Process Chemistry for Water and Wastewater Treatment*, Prentice Hall, New Jersey, NJ, USA, 1982.

[23] T. D. Reynolds and P. A. Richards, *Unit Operation and Processes in Environmental Engineering*, PWS publishing Co, Boston, Mass, USA, 1996.

[24] R. Sips, "Structure of a catalyst surface," *Journal of Chemical Physics*, vol. 16, pp. 490–495, 1984.

[25] S. E. Ebrahim, *Evaluation of mixture adsorption and glass bed for the removal of phenol and methylene blue from water [Ph.D. thesis]*, University of Baghdad, Baghdad, Iraq, 2008.

[26] M. Mortula, M. Gibbons, and G. A. Gagnon, "Phosphorus adsorption by naturally-occurring materials and industrial by-products," *Journal of Environmental Engineering and Science*, vol. 6, no. 2, pp. 157–164, 2007.

5

Hybrid Multiphase CFD Solver for Coupled Dispersed/Segregated Flows in Liquid-Liquid Extraction

Kent E. Wardle[1] and Henry G. Weller[2]

[1] Chemical Sciences and Engineering Division, Argonne National Laboratory, Argonne, IL 60439, USA
[2] OpenCFD Limited Bracknell, Berkshire RG12 1BW, UK

Correspondence should be addressed to Kent E. Wardle; kwardle@anl.gov

Academic Editor: Alírio Rodrigues

The flows in stage-wise liquid-liquid extraction devices include both phase segregated and dispersed flow regimes. As a additional layer of complexity, for extraction equipment such as the annular centrifugal contactor, free-surface flows also play a critical role in both the mixing and separation regions of the device and cannot be neglected. Traditionally, computional fluid dynamics (CFD) of multiphase systems is regime dependent—different methods are used for segregated and dispersed flows. A hybrid multiphase method based on the combination of an Eulerian multifluid solution framework (per-phase momentum equations) and sharp interface capturing using Volume of Fluid (VOF) on selected phase pairs has been developed using the open-source CFD toolkit OpenFOAM. Demonstration of the solver capability is presented through various examples relevant to liquid-liquid extraction device flows including three-phase, liquid-liquid-air simulations in which a sharp interface is maintained between each liquid and air, but dispersed phase modeling is used for the liquid-liquid interactions.

1. Introduction

While multiphase flows present unique challenges for computational fluid dynamics (CFD) simulation, a host of solution methods exist for simulation of well categorized flows. For "dispersed" flows in which one phase is continuous and the other is distributed in fine droplets, one can use Lagrangian particle tracking for small phase fractions (less than ~10%) in which each individual fluid particle is followed through the fluid in response to local flow conditions. For high phase fraction dispersed flows, a multi-fluid Eulerian-Eulerian solution method with interphase mass and momentum transfer can be applied. For stratified flows in which the fluid phases have a clearly defined phase interface, free-surface capturing methods such as Volume of Fluid (VOF) can be employed. Real flows, such as those encountered in liquid-liquid extraction devices, are not so easily categorized and can span multiple flow regimes (both spatially and temporally). In theory, interface capturing methods could be used for direct simulation of dispersed flows given that a mesh spacing of ~10x smaller than the smallest droplet can be used; however, accurate physical capturing of droplet-droplet interactions requires yet finer mesh resolution or droplet coalescence is severely overpredicted. In practice such meshing—and the small timesteps required (on the order of $1E-7$ s) for stable solution—is not feasible for realistic turbulent multiphase flows and will not be in the foreseeable future even on large computers unless CFD algorithm developments are made which allow significant timestep acceleration or time parallelization (spatial decomposition is the only option currently for CFD parallelization). This timestep limit is due to the fact that interface capturing methods require explicit solution and are limited by Courant number:

$$Cr = \frac{\Delta t}{\Delta x/\vec{u}} \approx 0.25. \qquad (1)$$

Thus, the timestep Δt is directly proportional to the mesh spacing Δx (u is the flow velocity)—that is, if the mesh spacing is cut in half, the timestep must essentially be decreased by the same margin. Consequently, for complex multiphase flows in which both dispersed flow and segregated flow regions are present one would like to couple these two methods into a single solver. In such a method, interface capturing would be used in regions where meshing is sufficient to resolve large droplets and bulk fluid-fluid interfaces or for phase pairs where interdispersion can be neglected; dispersed flow models would be used in regions where droplet characteristics move into the "subgrid" scale. As an example, for complex multiphase flows such as those found in liquid-liquid extraction devices where two immiscible liquids are mixed and air can also be present, one could employ sharp interface methods for certain phase pairs (e.g., liquid-air) and at the same time use dispersed modeling for others (e.g., liquid-liquid).

The idea of coupling these two methods for solution of such flows was explored by Cerne et al. [1]. They employed a simplified switching routine based on the gradient of the volume fraction across neighboring cells to flag cells as either VOF or two-fluid and solved the appropriate number of equations in each cell—resulting in complicated numerical issues due to solving models with different numbers of equations across the same domain [2]. To avoid such issues, this same research team has shifted toward multi-fluid-VOF coupling via the addition of interface capturing on top of an Eulerian multi-fluid solver. In this way, the multi-fluid formulation with momentum equations for each phase is applied across the entire domain and an interface sharpening algorithm (in their case a conservative level-set method which is similar in concept to the interface compression method described later) is applied for sharp interface regions [2, 3]. Strubelj and Tiselj [2] give a good overview of methods that have been employed for this coupling along with details regarding difficulties in coupling the phase momentum equations at the sharp interface (where the phase velocities should be equal). Again, the simple switching function of Cerne et al. [1] has been used by these authors who acknowledge its somewhat arbitrary nature and identify this as a key area of work to make the coupling more physically based.

A recent study by Yan and Che [4] has also attempted to couple multi-fluid and VOF methods with a "switching" mechanism based on the identification and transport of "large" (grid-resolved) and "small" (sub-grid) interface structures with models for exchange between the two. While this provides a means of phenomenological switching between the grid-resolved and subgrid scales and enables the use of accurate interface reconstruction methods (piece-wise linear interface construction (PLIC) was used), it requires the solution of two transport equations for each dispersable phase and is not easily generalizable to multiple phases beyond two.

The current study is part of a broader research effort to deliver computational tools for detailed simulation of liquid-liquid extraction (solvent extraction) unit operations with the aim of providing a pathway for prediction of key operational performance measures (e.g., stage efficiency, and other-phase carryover (backmixing)) for any given set of conditions. Such predictive capability will help inform process-level modeling tools and deliver insight into unit design and operation. To accomplish this, methods are required which can adequately predict liquid-liquid mixing and interfacial area generation as well as the formation and transport of

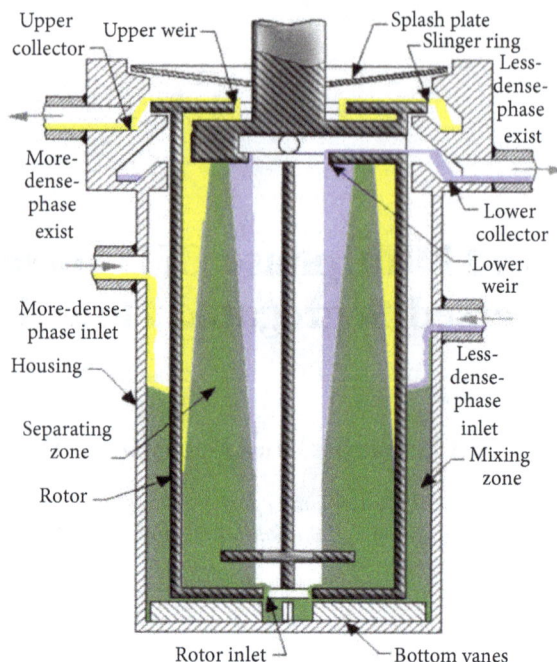

FIGURE 1: Sketch of an annular centrifugal contactor.

small droplets. Liquid-liquid contacting equipment used in solvent extraction processes has the dual purpose of mixing and separating two immiscible fluids. Consequently, such devices inherently encompass a wide variety of multiphase flow regimes from segregated to dispersed flow types. From a simulation perspective one might perform separate simulations for the mixing [5] and separation [6] functions of stage-wise extraction devices such as centrifugal contactors or mixer settlers, but in this case assumptions must be made regarding the coupling between the two regions. This is problematic, especially in the case of the annular centrifugal contactor (Figure 1). In this device, mixing occurs in a narrow annulus between the stationary outer housing and the inner rotating cylinder (which itself is the centrifuge). The connection between the two regions occurs at the bottom of the rotor where the dispersion of the two fluids enters the separating region in the hollow rotor. These two regions are not necessarily independent and the coupling is poorly understood. These goals and simulation challenges have led us towards the development of a hybrid method building upon the work of the others mentioned above and extended for application to the multiphase flow and operation of these devices.

Of the equipment types generally used for solvent extraction processing of used nuclear fuel, centrifugal contactors have the largest relative knowledge gap and at the same time the greatest opportunity for significant benefits to a future fuel cycle facility. Thus, the present focus has been on these devices for which this multi-fluid-VOF coupling methodology is of particular importance to capture both the liquid-liquid dispersion flow as well as the complex, dynamic fluid-rotor interaction. A thorough review of CFD modeling efforts for annular centrifugal contactors (also called annular centrifugal extractors) has recently been published by Vedan-tam et al. [7]. Three-phase liquid-liquid-air simulation of the flow in an annular centrifugal contactor using a VOF-based method with single momentum equation for the mixture has been performed previously by Wardle [8]. While simulations of flow in centrifugal contactor-related geometries from other authors are limited, there is one available study in which a multi-fluid solution method is used to simulate the separation of two liquids in a simplified rotor geometry [9]. Recent work by Li et al. [10] reports three-phase simulations in a coupled mixer/separator centrifugal contactor model using a pure multi-fluid approach. Curiously, the authors found very little mixing between the two liquids. This appears to have been due to the assumption of a large dispersed phase diameter—however, it may also have been related to smearing of the fluid-rotor contact resulting from diffuse liquid-air interfaces. Coupling of the multi-fluid method with interface capturing as developed in this current effort allows simulation of annular centrifugal contactor mixing zone and rotor flows in which a liquid-air free surface is captured.

2. Computational Methodology and Implementation

The implementation of the methodology described here has been done using the OpenFOAM CFD package (version 2.1) which provides a collection of libraries and utilities for constructing custom CFD solvers for a wide variety of applications. A working version of the solver named multi-phaseEulerFoam has been included as part of the release of version 2.1 of the toolkit and is available for download at http://www.openfoam.org/.

3. Multifluid Momentum Equations and Implementation of Interphase Drag Models

The multi-fluid model equations for incompressible, isother-mal flow are given by sets of mass and momentum equations for each phase k:

$$\frac{\partial \alpha_k}{\partial t} + \vec{u}_k \cdot \nabla \alpha_k = 0, \tag{2}$$

$$\frac{\partial \left(\rho_k \alpha_k \vec{u}_k \right)}{\partial t} + \left(\rho_k \alpha_k \vec{u}_k \cdot \nabla \right) \vec{u}_k$$
$$= -\alpha_k \nabla p + \nabla \cdot \left(\mu_k \alpha_k \nabla \vec{u}_k \right) + \rho_k \alpha_k \vec{g} + \vec{F}_{D,k} + \vec{F}_{s,k}, \tag{3}$$

where the density, phase fraction, and velocity for phase k are given by ρ_k, α_k, and $\vec{u}_k$, respectively, and $\vec{g}$ is gravity. The two interfacial forces are the inter-phase momentum transfer or drag force $\vec{F}_{D,k}$ and the surface tension force $\vec{F}_{s,k}$. For the examples shown here, only the drag force is accounted for—though surface tension capability (based on the continuum surface force model of Brackbill [12]) and surface contact

angle effects have also been subsequently added to the solver. The drag term $\vec{F}_{D,k}$ is given by

$$\vec{F}_{D,k} = \frac{3}{4} \rho_c \alpha_c \alpha_d C_D \frac{|\vec{u}_d - \vec{u}_c| \left(\vec{u}_d - \vec{u}_c \right)}{d_d} = \alpha_c \alpha_d K \left(\vec{u}_d - \vec{u}_c \right), \tag{4}$$

where the subscripts c and d denote the continuous and dispersed phase values and where K is

$$K = \frac{3}{4} \rho_c C_D \frac{|\vec{u}_d - \vec{u}_c|}{d_d}. \tag{5}$$

As implemented in the solver, the drag calculation is generic such that the model must simply return the value of K. A variety of correlations could be used for calculation of the drag coefficient C_d (in (5)) and several common models are available in OpenFOAM. As an example, the commonly used model of Schiller and Naumann [13] was used for the test cases here. In this model, the drag coefficient is a function of the Reynolds number Re according to

$$C_D = \begin{cases} \dfrac{24 \left(1 + 0.15 \text{Re}^{0.683} \right)}{\text{Re}}, & \text{Re} \le 1000, \\ 0.44, & \text{Re} > 1000, \end{cases} \tag{6}$$

where

$$\text{Re} = \frac{|\vec{u}_d - \vec{u}_c| \, d_d}{\nu_c} \tag{7}$$

in which ν_c is the continuous phase kinematic viscosity. In this solver, calculation of the drag coefficient can be done by specifying a dispersed phase or by independent calculation with each phase as the "dispersed phase" and the overall drag coefficient applied to the momentum equations taken as the volume fraction weighted average of the two values. This is a so-called blended scheme which is a useful approximation for flows with regions in which either phase is the primary phase.

A constant droplet diameter size (independently defined for each phase) is used in the work reported here, but models for variable droplet size are compatible with this flexible framework. Indeed models based on a reduced population balance method have been developed and will be reported separately.

3.1. Interface Capturing. The interface sharpening method-ology employed here is the interface compression method of Weller [14]. Gopala and van Wachem [15] give a useful comparison of this type of method with several other inter-face sharpening and reconstruction algorithms employed in a variety of CFD codes. The interface compression scheme of Weller [14] adds an additional "artificial" compression term to the LHS of the volume fraction transport equation (2) as

$$\frac{\partial \alpha_k}{\partial t} + \vec{u}_k \cdot \nabla \alpha_k + \nabla \cdot \left(\vec{u}_c \alpha_k \left(1 - \alpha_k \right) \right) = 0, \tag{8}$$

where the velocity $\vec{u}_c$ is applied normally to the interface to compress the volume fraction field and maintain a sharp interface. The $\alpha_k (1 - \alpha_k)$ term ensures the term is only active in the interface region. In addition, a bounded differencing scheme is employed for discretization of (8). The value for the artificial interface "compression velocity" $\vec{u}_c$ is given by

$$\vec{u}_c = \min \left(C_\alpha \, |\vec{u}|, \max (|\vec{u}|) \right) \frac{\nabla \alpha}{|\nabla \alpha|}. \tag{9}$$

The $\nabla \alpha / |\nabla \alpha|$ term gives the interface unit normal vector for the direction of the applied compression velocity. The magnitude of the velocity $|\vec{u}|$ is used since dispersion of the interface (which is being counteracted by the compression velocity) can only occur as fast as the magnitude of the local velocity in the worst case. The coefficient C_α is the primary means of controlling the interfacial compression. While C_α can mathematically be any value ≥ 0, if we restrict $C_\alpha \leq 1$, (9) reduces to

$$\vec{u}_c = C_\alpha \, |\vec{u}| \, \frac{\nabla \alpha}{|\nabla \alpha|} \tag{10}$$

and C_α is then simply a binary coefficient which switches interface sharpening on (1) or off (0). With C_α set to 0 for a given phase pair, there is no imposed interface compression resulting in phase dispersion according to the multi-fluid model. Conversely when it is set to 1, sharp interface capturing is applied and VOF-style phase fraction capturing occurs (forcing interface resolution on the mesh). The implementation of the solver is configured such that the interface compression coefficient C_α is defined and applied independently for all phase pairs. Thus, a sharp interface can be maintained at all interfaces between specific phases (e.g., air-water and air-oil) and dispersed phase modeling with no interface compression can be used for other phase pairs (e.g., water-oil).

In general, the interface compression method for interface capturing is not as physically accurate as interface reconstruction methods such as Piecewise Linear Interface Construction (PLIC). However, it is much easier to implement and performs faster, and most importantly, unlike PLIC it is mass conservative [15]. One problem that this method can suffer from is parasitic wavy currents at the interface which are particularly problematic for small-scale, surface tension driven flows (e.g., capillary rise), but become less important for advective flows such as is the current target of this work. The false interfacial currents can also be minimized by maintaining a low Cr number through subtimestepping and restricting $C_\alpha = 1$ as was done here. For a discussion of other limitations common to VOF methods see [16].

3.1.1. Dynamic C_α Switching.

With the computational framework as outlined above, it is possible to imagine a unified method which allows simulation of complex flows with any combination of regimes ranging from fully dispersed to fully segregated. However, as noted by the various researchers who have attempted coupling between multi-fluid and VOF methods [1, 2, 4], the key area of uncertainty is the method by which switching occurs between the dispersed and segregated models. As outlined above, the current implementation could be made to allow for dynamic switching of the

interface sharpening only in regions where the flow is segregated through implementation of a spatially nonuniform C_α field(s). Others have suggested that one could choose to switch according to some predetermined flow regime map [2]. Such an approach, however, would likely be useful only for simple geometry flows (e.g., pipe flow) and a more general physics-based approach is needed for application to a broader class of flows such as those under investigation here. From a more general physical perspective, one would like to use the sharp interface method in regions of the flow. From a more general physical perspective, one would like to use the sharp interface method in regions of the flow where the actual droplet size and corresponding local mesh resolution is sufficient for accurate capture of droplet curvature. Conversely, where the droplet size falls below the mesh-resolvable scale, inter-phase dispersion modeling would be employed (sharpening deactivated). Such a scheme would require prediction of local droplet size based on a population balance or similar method and comparison with local mesh spacing using some cutoff criteria.

For the present implementation limited testing was done for a two-phase version of the solver using a switching function based on the work by Cerne et al. [1] and later used by others [3]. This technique applies a switch according to the magnitude of the gradient of the volume fraction itself—with the assumption being that when the gradient of the volume fraction becomes smaller than some cutoff value it is an indication of actual phase dispersion and interface sharpening is turned off. In this case, the normalized magnitude of the gradient of the volume fraction (γ) is used as defined by

$$\gamma = \frac{|\nabla \alpha|}{\max (|\nabla \alpha|)}. \tag{11}$$

Thus, when $\gamma \geq \gamma^*$ the value of C_α is set to 1. A cutoff value of 0.4 is recommended by Cerne et al. [1]; however, the formulation used here γ has been normalized to 1 so the corresponding value may be somewhat different. One drawback of this mechanism for switching is that it is somewhat of a "self-fulfilling prophecy." Where the interface is already sharp (steep gradient in the volume fraction α), you apply interface compression to keep it sharp, and where it is dispersed, you let it stay dispersed. Even so, it gives a reasonable preliminary model for the coupling and has been implemented and tested for a two-phase version of the solver only. An example case is described in Section 4.1.

For the majority of examples reported here, which are three-phase liquid-liquid-air cases, it was assumed that dispersion of air into the liquids can be neglected and thus the C_α parameter was fixed for each phase pair. A value of 1 (interface capturing on) was used for any air-liquid interfaces and interface capturing was turned off ($C_\alpha = 0$) for the liquid-liquid interfaces. In this case, no entrained air will be captured and the air-liquid interface will be everywhere sharp while the liquid phases will be allowed to interdisperse and no sharp interfaces will be imposed. This assumes that the entrainment of air has a negligible effect on the flow and mixing of the two liquids.

3.2. Solution Procedure for Multifluid-VOF Coupling. The general solution procedure for the hybrid solver using the equations above is as follows:

(1) update timestep according to Courant number limit (ratio of timestep to interface transit time in cell);

(2) solve coupled set of volume fraction equations with interface sharpening for selected phase pairs ((8) with multiple subtimesteps);

(3) compute drag coefficients;

(4) construct equation set for phase velocities and solve for preliminary values;

(5) solve pressure-velocity coupling according to Pressure Implicit with Splitting of Operators (PISO) algorithm:

    (a) compute mass fluxes at cell faces;

    (b) define and solve pressure equation (repeat multiple times for non-orthogonal mesh corrector steps);

    (c) correct fluxes;

    (d) correct velocities and apply BCs;

    (e) repeat for number of PISO corrector steps;

(6) compute turbulence and correct velocities;

(7) repeat from 1 for next timestep.

3.2.1. Numerical Considerations for Stability of Momentum Coupling and Phase Conservation. In the limit of a sharp interface, the velocities on either side of the interface must be equal to meet the so-called no-slip interface condition. This is an inherent feature of traditional VOF simulations as all phases share a single momentum equation and thus the phase velocities are the same everywhere. Imposition of a sharp interface through the addition of interface compression "on top" of a multi-fluid formulation in which each phase has its own momentum equation requires that an additional artificial drag is imposed to equalize the velocities at the interface. In the work of Strubelj and Tiselj [2] in which multifluid-VOF coupling was performed, an arbitrary function proportional to the inverse of the time step divided by 100 (resulting in a large value) was imposed to force large interphase drag coefficients at the interface. In this case, rather than devise some arbitrary formulation, small "residual drag" and "residual phase fraction" constants were added for each phase pair (typically both equal to $1E-3$) to stabilize the phase momentum coupling. These added residual values were only used in calculation of the drag for momentum coupling stability and did not affect actual phase fractions or overall phase conservation.

In order to ensure phase conservation for the coupled phase fractions with added interface sharpening, it was necessary to incorporate limiters on the phase fraction as well as on the sum of the phase fractions prior to the explicit solution of the phase fraction equation system. These additional limiters have been incorporated in a new multiphase implementation of the Multidimensional Universal Limiter

with Explicit Solution (MULES) solver framework within OpenFOAM. This multiphase MULES implementation used in multiphaseEulerFoam is also leveraged for enhancing the phase conservation performance of the n-phase VOF-only solver multiphaseInterFoam. As in the case of the standard VOF solver, solution of the volume fraction transport equations was done using subtimestepping over several subintervals of the overall time step to maintain solution stability according to the Courant number limit (1) while maximizing overall timestep to minimize time to solution for the transient solver. It was found that an overall Cr number limit of 1.5 (based on velocities near sharp interface) with 5 subtimesteps could deliver stable results.

4. Results: Example Cases

The following example cases demonstrate the capability of the simulation methods to capture, on a per-phase-pair basis, both dispersed and segregated flows. Only the first case, on the breaking of a dam considers treatment of the sharpening coefficient C_α as a volumetric field with local, dynamic switching based on the gradient of the volume fraction as mentioned above. In all other cases, the value of C_α is generally set to 1 (imposed interface sharpening) for liquid-air interfaces and to 0 for liquid-liquid phase interactions. The properties used are representative of water, oil, and air at room temperature conditions. As done in the work of Padial-Collins et al. [9], a constant droplet size of 150 microns was assumed for the liquids. A value of 1 mm was used for air. Interphase drag was treated via the blended method mentioned earlier. Visualizations were done using ParaView version 3.12.

4.1. Liquid-Liquid "Column" Collapse. This example is a modification to the classic collapsing liquid column 2D test case in this case for a liquid-liquid system with two initial regions of dispersed phase volume fraction as shown in Figure 2. The domain is 58.4 cm square, with the short barrier on the bottom surface having a width of 2.4 cm and a height of 4.8 cm. Figure 3 shows a comparison of successive time snapshots of simulations having set the interface sharpening coefficient to 0 (VOF behavior) and 1 (multi-fluid behavior). The behavior is as expected for the two cases—that is, with $C_\alpha = 1$, immediately upon startup droplets with a characteristic size similar to the mesh size are formed, and despite this, overall segregation of the phases is relatively slow. When interface sharpening is not imposed ($C_\alpha = 0$) and the multi-fluid behavior is governed by the interphase drag correlation, separation of the phases appears to be more physical and occurs on a faster time scale.

A variation of the above case was performed starting from the same initial state but in which the value of C_α was allowed to vary locally (as 0 or 1 only) according to (11). Successive time snapshots of phase fraction and the sharpening coefficient field are shown in Figure 4. Note that the interphase drag has been increased in this case (through a larger assumed droplet size) resulting in slightly slower separation dynamics as compared to the earlier case shown in Figure 3.

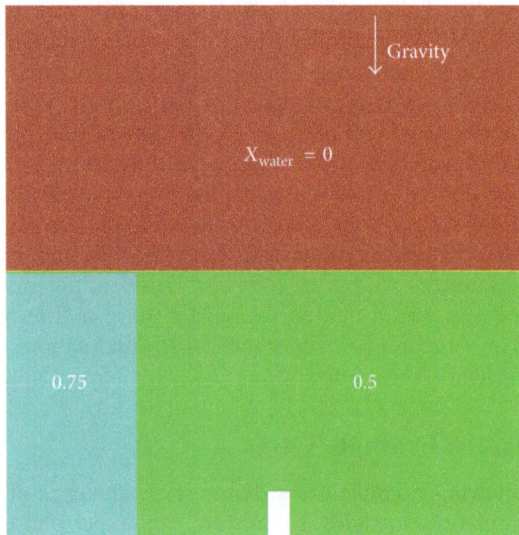

FIGURE 2: Initial condition of collapsing liquid-liquid dispersion test case.

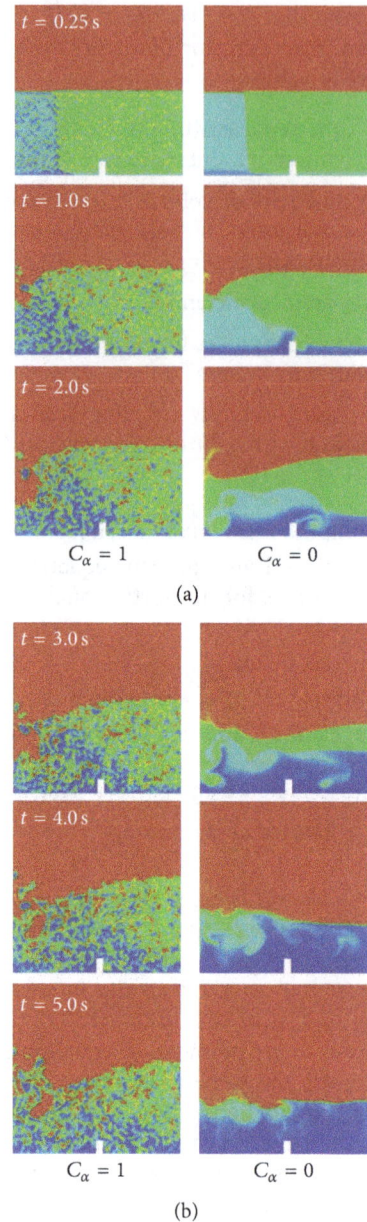

It was observed that around $t = 6$ s the first region of active interface sharpening appears, but left-right interface motion is sufficient that the sharpened region is not maintained. Between 10 and 12 s, a stable, sharpened interface appears and grows until it covers the length of the phase interface around 16-17 s.

This example demonstrates the functionality of dynamic interface sharpening switching based on the volume fraction gradient for a simple test case. As noted earlier, a switching function of this type is somewhat arbitrary. Ideally, one would like a physical basis for governing switching according to a comparison of the local predicted droplet size and the local mesh spacing—where mesh resolution is sufficient to resolve droplets adequately, sharpening is activated, and where droplet size falls into the subgrid scale, sharpening is deactivated. One could imagine a very flexible model which could simulate multiple flow regimes in this way in a multi-scale manner with the multifluid method capturing sub-grid phase particle transport analogous in principle to the idea of the sub-grid scale modeling done in Large Eddy Simulation (LES).

4.2. Horizontal Settler. The next test case simulates the separation of a 50 : 50 liquid-liquid dispersion entering into a 2D, horizontal settler as shown in Figure 5 where the initial state is stratified layers of oil (red) and water (blue). Gravity acts in the vertical direction. Flows of each phase exit from the corresponding surfaces in the upper right of the domain. The length of the main body of the domain is 10 cm and the overall height is 2.25 cm. The development of a "dispersion band" between the regions of separated phases was observed. The dispersion band was not static, but was found to be disturbed by longitudinal waves generated by a periodic vortex at the back of the so-called dispersion disk (wall just upstream of the inlet). The dispersion disk is placed just upstream of the inlet to direct the entering dispersion toward the central

FIGURE 3: Comparison of simulation using the solver with $C_\alpha = 1$ (left columns) and $C_\alpha = 0$ (right columns) showing the behavior of VOF versus Euler-Euler. Red is oil and blue is water.

vertical position of the separating region and try to minimize downstream disturbances. Simulations were done with no dispersion disk in order to verify the effectiveness of this feature. It was indeed observed that the overall width of the dispersion band is greater without the dispersion disk as inlet disturbances propagate farther downstream and disrupt the separation of the two phases leading to more entrainment in the exit streams.

4.3. Annular Mixer. As noted above, a principle application requiring the capability of this hybrid solution method is the liquid-liquid-air flow in an annular centrifugal contactor. This device (Figure 1) consists of an annular region with

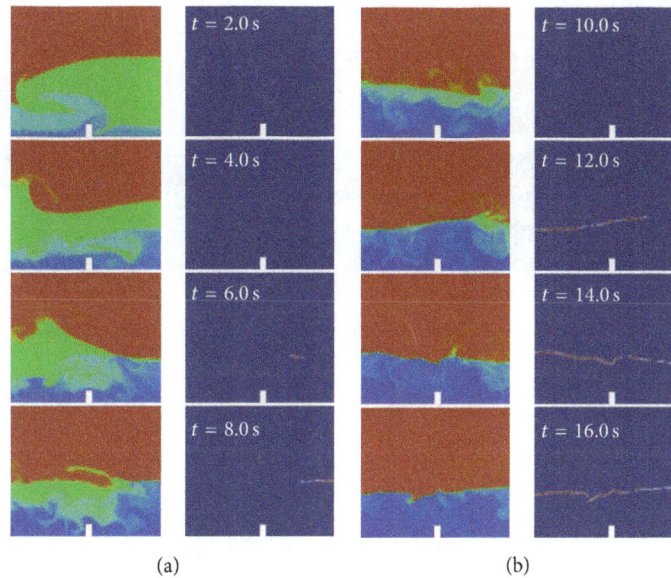

FIGURE 4: Time sequence of the volume fraction field (left of each column) and the C_α field (right of each column) showing the region of active interface sharpening showing the evolution of the separation process and appearance of a sharp interface.

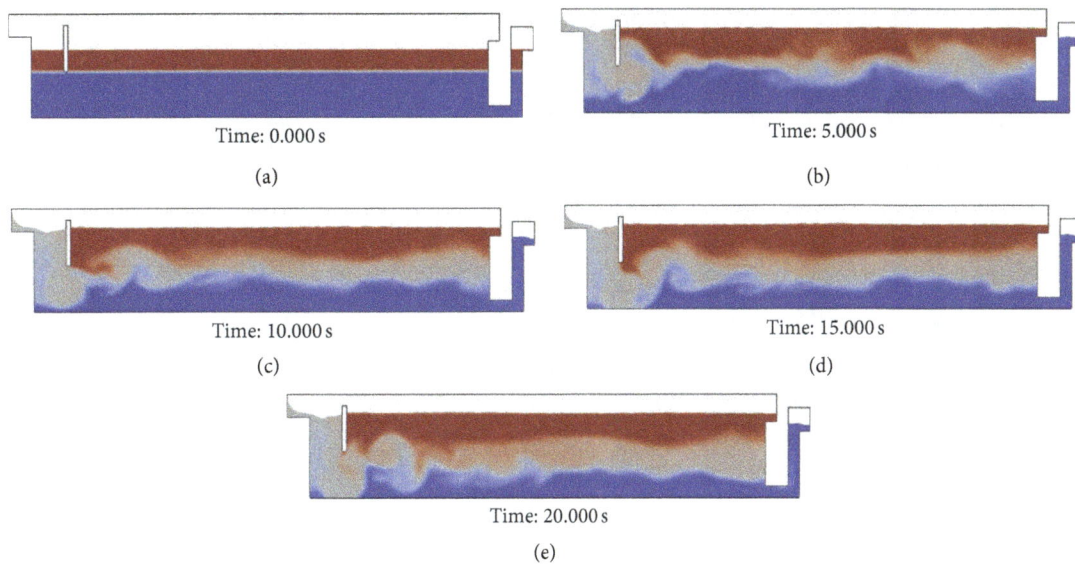

FIGURE 5: Time evolution of phase fraction field for a separating liquid-liquid dispersion in a horizontal settler.

a rotating inner cylinder and stationary outer cylinder in which the two immiscible liquids are mixed in the presence of a free surface. This complicates the physics significantly, requiring sharp interface capturing to accurately predict the intermittent liquid-rotor contact [17]. At the same time, dispersed phase modeling is needed to predict the mixing and flow of the liquid-liquid dispersion.

To demonstrate the capability of the solver to capture such flow dynamics, simulations were conducted in an idealized annular mixer both for a 2D, axisymmetric case and a fully 3D case. In both cases, the inner radius is 2.54 cm and the outer 3.17 cm (annular gap of 0.63 cm) and the height of the annulus is 7 cm. The top surface is open to air at constant

atmospheric pressure and the bottom surface is treated as a wall. Unless otherwise stated, the rotation rate of the inner cylinder is 3600 RPM (377 rad/s) resulting in a surface velocity of 9.56 m/s. Turbulence was treated using Large Eddy Simulation (LES) with the Smagorinsky sub-grid model. A uniform quadrilateral mesh was used for the 2D model with spacing of 0.2 mm (32 cells across the annular gap). In order to explore mesh dependency of the new solver, additional simulations were done with mesh spacings of 0.4 mm (coarse) and 0.1 mm (fine). The relative mesh spacings for the three sizes can be seen in Figure 6. For the 3D model, the base case simulation was done with a mesh spacing ~0.4 mm (15 hexahedral cells across the annular gap, 675 K cells total)

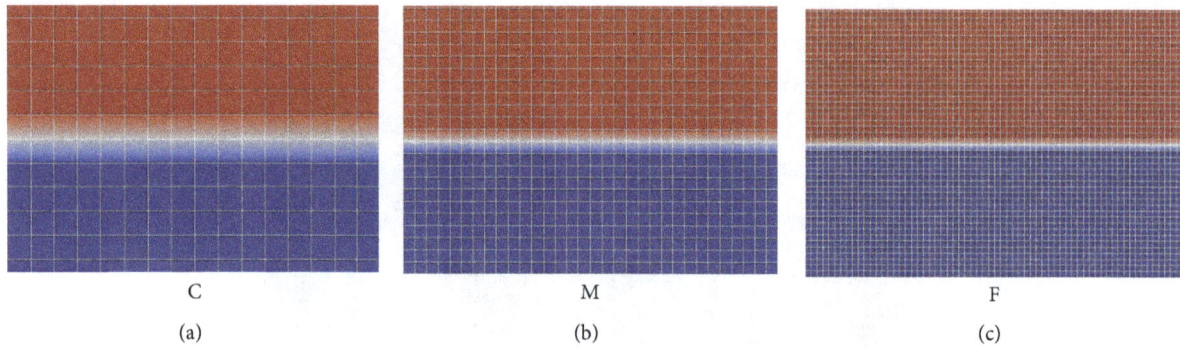

FIGURE 6: Relative mesh spacings for the 0.4 mm (C), 0.2 mm (M), and 0.1 mm (F) meshes of the 2D annular geometry. Only a short vertical section showing the initial liquid-liquid interface is shown.

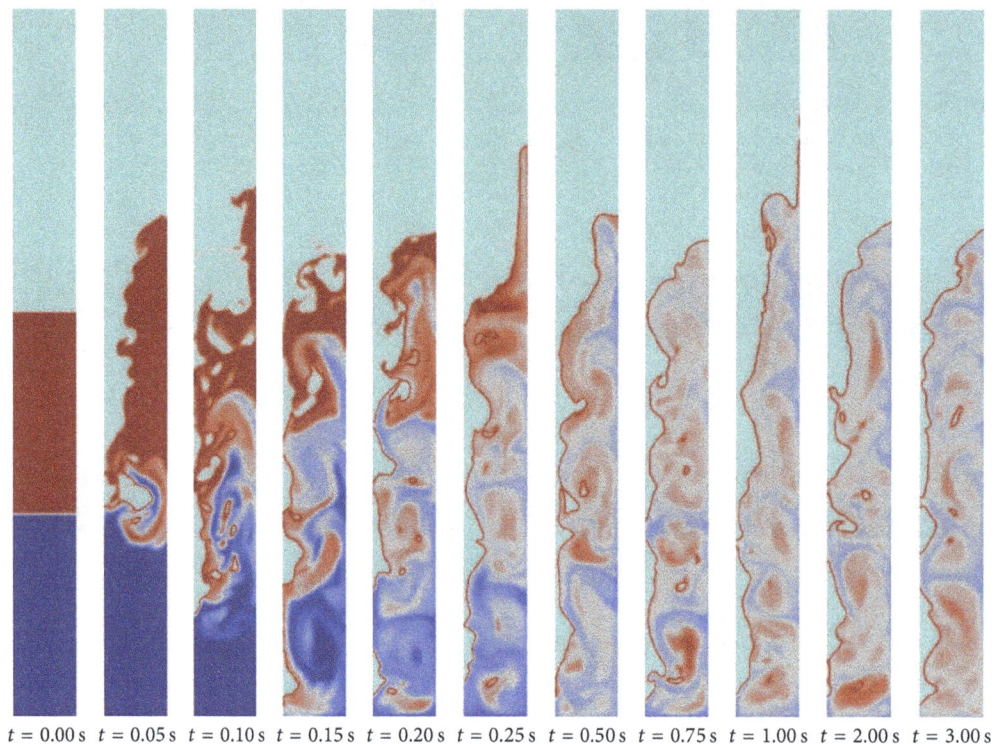

$t = 0.00$ s $t = 0.05$ s $t = 0.10$ s $t = 0.15$ s $t = 0.20$ s $t = 0.25$ s $t = 0.50$ s $t = 0.75$ s $t = 1.00$ s $t = 2.00$ s $t = 3.00$ s

FIGURE 7: Sequence of snapshots of phase fraction for water (blue), oil (red), and air (cyan) in the 2D axisymmetric annular mixer geometry at 3600 RPM.

though for comparison an additional run was also done for a finer mesh (~0.25 mm, 2.4 M cells).

4.3.1. 2D, Axisymmetric Model. Figure 7 gives a time series of snapshots showing volume fractions for water (blue), oil (red), and air (cyan) from startup through $t = 3.0$ s for simulation on the medium mesh refinement (Figure 6(M)). It is clear that even for this very turbulent flow, the hybrid solver is able to maintain a sharp interface for the liquid-air free surface while at the same time allowing phase inter-dispersion for the two liquids. Unlike the simplifications often used by other CFD studies of this type of flow (e.g., [18]), the presence of air and the existence of the free surface has a significant impact on the characteristics of the flow

and breaks down any Taylor-Couette vortices that would be present in the liquid-liquid only case. It was observed that there was one relatively stable vortex at the bottom which was characterized by a light-phase rich region at the center and rotation in the clockwise direction (flow inward along the bottom surface). The companion vortex (counterclockwise rotation) just above this lower one was found to be periodically formed and then break away and travel upward as the liquids are spun off the rotor and move toward a maximum height on the outer wall.

Sharp interface capturing methods such as the VOF method used here are inherently mesh dependent; the finer the mesh, the finer the interface features that can be captured. In order to explore the mesh dependency that the hybrid

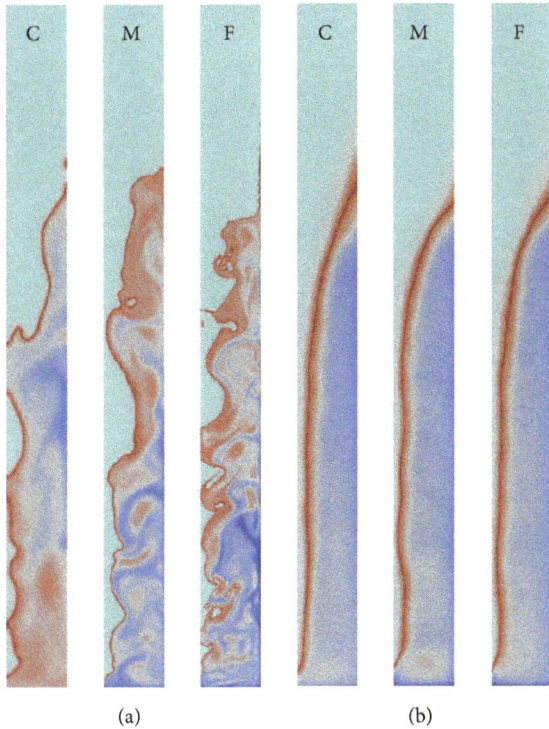

FIGURE 8: Snapshots of the phase fractions (a) at $t = 0.355$ s after startup and (b) for a time average over the period from $t = 2$ s to $t = 5$ s for the three meshes.

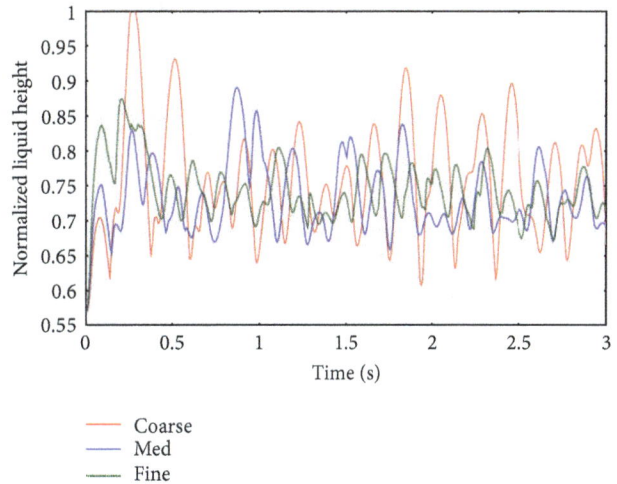

FIGURE 9: Plot of liquid height on outer wall (stationary) for the 2D annular mixer simulations at 3600 RPM on the three mesh densities (Figure 6).

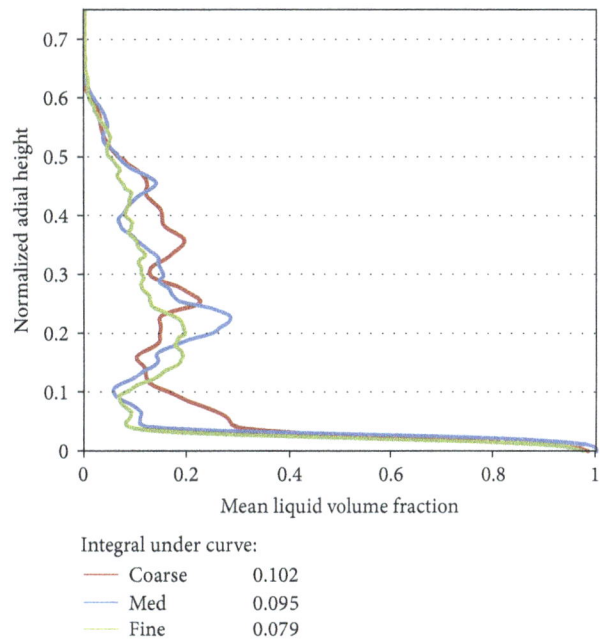

FIGURE 10: Plot of time-averaged liquid fraction on the rotor side (as a function of normalized height). Integrated values for the total liquid "coverage" are shown in the legend.

solver inherits from the VOF method, simulations for the 2D axisymmetric annular mixer model were performed on two additional meshes—one twice as coarse and one twice as fine as the base case. Snapshots of the phase fractions at $t = 0.355$ s after startup (a) and for a time average over the period from $t = 2$ s to $t = 5$ s (b) are shown in Figure 8. Even for the relatively short time after start-up shown in Figure 8(a) the flows observed for the simulations on the three meshes have clear differences. Additionally, the differences are apparent not only in the shape of the liquid-air free surface, which is to be expected, but also in the multi-fluid dispersion behavior between the two liquids. Comparison of the time-averaged behavior (Figure 8(b)), however, shows better general consistency in the overall liquid height. The Taylor-Couette vortex near the bottom mentioned previously was found to be most prominent in the medium mesh case.

Though it is not readily apparent from the snapshots in Figure 7, as has been observed previously for flow in this configuration at such conditions [17], the overall height of the liquid on the outer wall exhibited an oscillatory behavior. Figure 9 shows a plot of the liquid height on the outer wall versus time for the 2D axisymmetric simulation for the three different meshes. As noted previously, there is significant variation in the temporal evolution of the flow on the three mesh densities while the general behavior is similar. As will be shown later, the oscillations in liquid height for the 3D simulations exhibited a much more clear periodicity; for the 2D simulations, the oscillation magnitude in liquid height

was largest for the most coarse mesh. The minimum height of the oscillation of the liquid corresponds with a maximum contact area between the liquid and the rotor after which the liquid is accelerated and spun out and up the housing wall leading to a maximum liquid height corresponding to a minimum in fluid-rotor contact. The amount of overall contact between the liquid and the rotor has an impact on the level of mixing that occurs between the two liquid phases. Figure 10 shows a plot of the time-averaged liquid contact on the rotor side and integrated values corresponding to the fractional liquid "coverage" in each case. While not

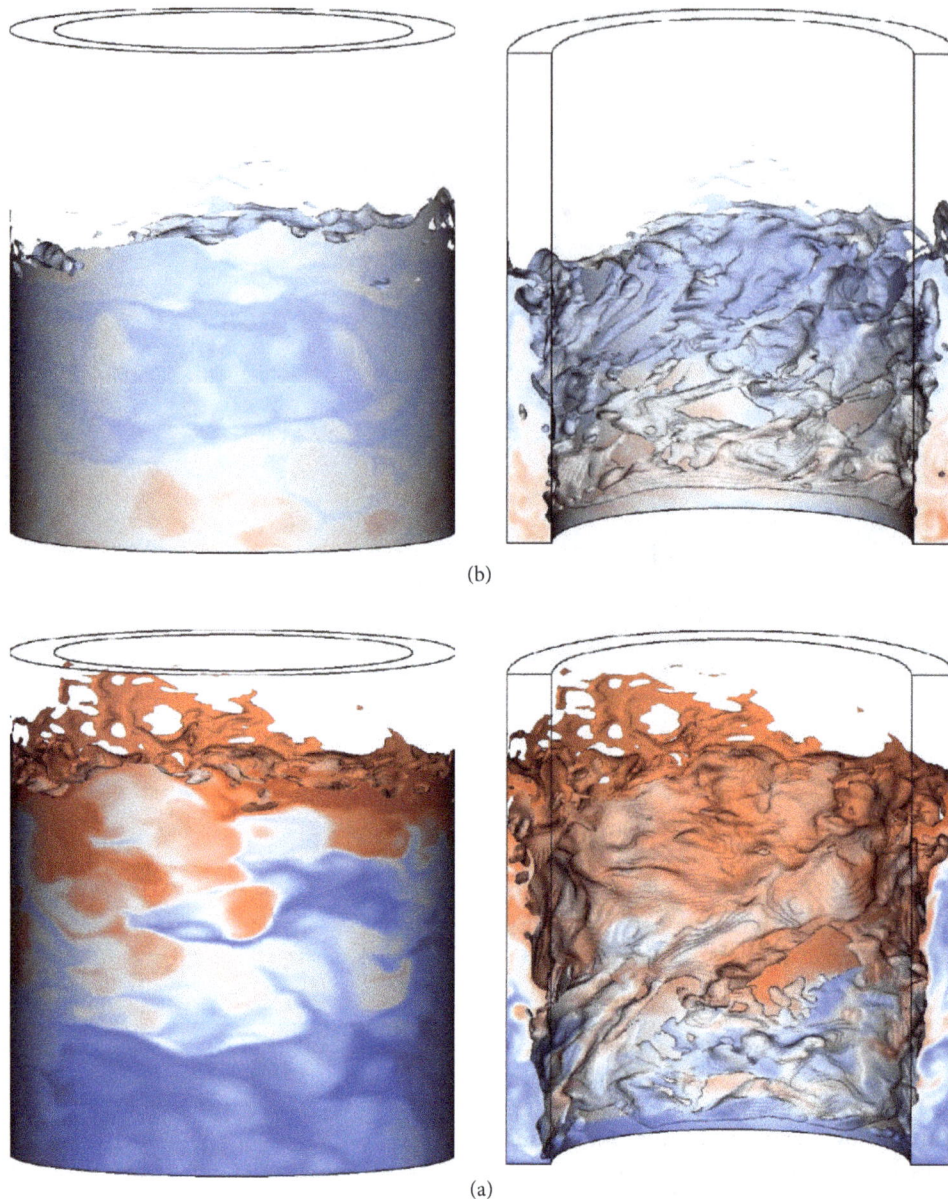

(b)

(a)

FIGURE 11: Snapshots of the liquid phase fractions in the 3D annular mixer model at (a) an early time (~0.25 s) and ~3 s after startup (b) 3600 RPM with the left images showing a side view and the right a cross-section.

unexpected, there are complex mesh dependencies for the hybrid solver which require additional investigation.

In order to provide a point of reference for the additional computational cost of the multi-fluid hybrid scheme versus an all-VOF simulation (single, shared momentum equation and sharp interfaces everywhere), a simulation was done using OpenFOAM's multiphase-capable VOF solver multiphaseInterFoam for the fine mesh case (44,800 hexahedral cells) of the 2D annular mixer problem described here. The simulation was done using the exact same solver settings (discretization schemes, number of subtimesteps on volume fraction solutions, etc.) and the same number of parallel processors (12 cores were used in this case). The simulations were compared out to 1 second of flow and it was found

that the hybrid multi-fluid/VOF solver is only 39% more costly *per timestep* than the comparable case with the all-VOF solver (8.69 CPU seconds/step versus 6.23 CPU seconds/step for all-VOF solver). As the bulk of the computational time is spent in the solution of the volume fractions and more so in the pressure-velocity coupling (Items #2 and #5, resp., in the solution procedure given in Section 3.2), only a modest computational cost is incurred due to the additional phase momentum equations and momentum coupling in the multi-fluid formulation that is not required in the VOF-only solver.

4.3.2. 3D Model. Figure 11 shows snapshots of the liquid phase fractions in the 3D annular mixer model soon after start-up (~0.25 s) and ~3 s after startup at 3600 RPM. In

TABLE 1: Liquid height oscillation frequency (cycles/s) at 3600 RPM.

Blended drag		AQ disp. drag		Experiment [11]	
Base mesh	Fine mesh	$d_{aq} = 150\,\mu m$	$d_{aq} = 50\,\mu m$	4 V, 500 mL/min	CV, 1000 L/min
4.56 ± 0.16	4.65 ± 0.20	4.62 ± 0.13	3.99 ± 0.17	4.91 ± 0.21	5.29 ± 0.29

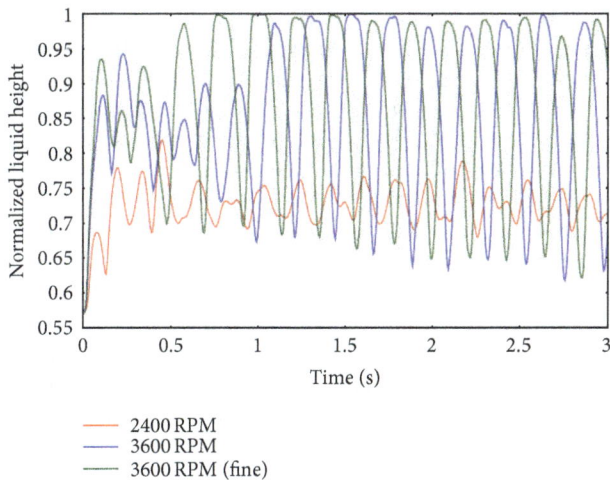

FIGURE 12: Plot of liquid height for the base case mesh at 2400 RPM and 3600 RPM and the fine mesh at 3600 RPM.

contrast with the 2D axisymmetric approximation, there is significant azimuthal variation in the flow and liquid-rotor contact. The liquid-liquid dispersion also exhibited a steady height oscillation due to being periodically thrown off the rotor and up the outer wall. A plot of the liquid height on the outer wall (azimuthal average) for two rotor speeds (2400 and 3600 RPM) are shown in Figure 12 along with the 3600 RPM value for the fine mesh for comparison. At the lower rotor speed, the liquid exhibits only minimal height variation with some periodicity while at the high rotor speed a steady oscillation develops ~1 s after startup. Table 1 gives a comparison of the liquid height oscillation frequency (calculated from trough-trough times for the ~6 cycles between 1.5–3 s) for the 3D model with different variations: the base case blended drag model with the droplet diameters for both liquid phases at 150 microns, aqueous dispersed with 150 micron droplet size, and aqueous dispersed with a 50 micron droplet size. The results for the fine mesh (blended case only) are also shown for comparison. There is not a very strong dependence on oscillation frequency for blended versus aqueous dispersed at the same droplet diameter. For the smaller droplet diameter, the oscillation frequency was found to decrease slightly. In terms of mesh dependency, despite oscillation "phase" offset due to differences at early times, the fully developed oscillation frequency is very comparable for the two mesh densities.

A companion experimental effort has also been initiated to provide means of validating the advanced simulation capability presented by this new solver for flows in actual annular centrifugal contactor equipment. While the simulations presented here are for a simple annular mixer, it was

found that certain characteristics of the flow showed good agreement with preliminary experimental observations and a brief comparison will be made here. Experiments were conducted using a CINC-V2 centrifugal contactor modified with a quartz outer cylinder as reported by Wardle et al. [17]. The tests included here were done using a liquid-liquid system consisting of 1 M nitric acid with 1 M aluminum nitrate as the aqueous phase (ρ = 1.17) and 40% (by volume) tributyl phosphate in dodecane (ρ = 0.85). The aqueous phase was dyed with methylene blue to aid in visual phase discrimination. While tests were done for a variety of housing vane types and inlet flow rates, conditions were selected for comparison here which gave a comparable annular liquid height in the mixing zone. For additional details regarding the experiments, please see Wardle 2012 [11].

It was observed that while the 3D annular mixer geometry has a fixed volume with no inlet/outlet flow, the general features of the flow compared favorably with observations in experiment. Figure 13 shows a comparison snapshot of the liquid-liquid flow as viewed from the housing side from the simulation (a) and a high-speed image from experiment (b). Both images are near a minimum liquid height and show similar banding and tendrils of the dispersed, heavy phase (blue). In both simulation and experiment, it was observed that as the dispersion approached a minimum liquid height, aqueous phase striations such as these appeared. Additionally, contact with the rotor for the collapsing liquid resulted in a sharp "spurting" of air across a line near the top of the main body of liquid followed by the rise of the liquid up the outer wall. In addition to such qualitative comparisons, the oscillation frequency of the liquid surface was also found to be quite comparable quantitatively under conditions of similar liquid height. As reported previously in Table 1, the mean oscillation frequency from simulation was found to be 4.6 Hz. From high-speed video analysis, the frequency observed in experiments was slightly higher at 4.9–5.3 Hz. Values from two different cases (4 straight mixing vanes (4V) and 8 curved mixing vanes (CV)) with similar liquid height are given in Table 1. Note that this is also quite comparable to what was found previously for similar conditions under single-phase flow [19] which ranged from 4.5 to 5.4 Hz at 3600 RPM depending on the total feed flow rate—with high frequency being observed at lower flow rates (lower overall liquid height). This seems to demonstrate that the phenomenon of liquid height oscillation (and the corresponding periodic liquid–rotor contact) is more a function of the annular geometry and rotor speed than it is of the fluid properties, flow configuration (whether net axial flow is present or not), or geometry beneath the rotor—provided that the conditions and configuration result in comparable liquid volume (height) in the annular region.

(a)

(b)

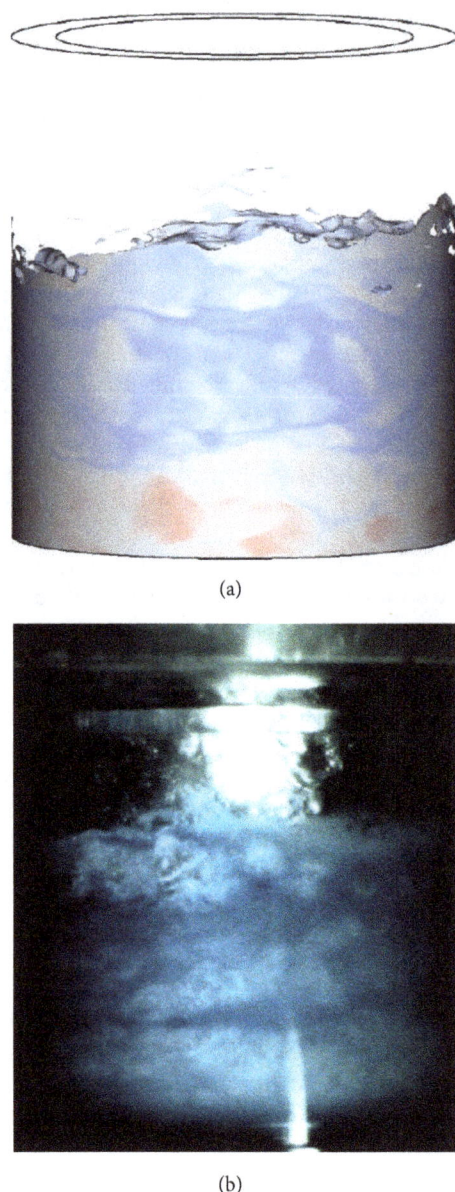

FIGURE 13: Comparison of snapshots from CFD simulation (a) of 3D annular mixer (fine mesh) and experiment [11] (b) in an actual CINC-V2 centrifugal contactor at a comparable overall liquid height showing similar aqueous phase (blue) striations.

5. Conclusions

A hybrid multiphase CFD solver has been developed which combines the Euler-Euler multi-fluid methodology with VOF-type sharp interface capturing on selected phase-pair interfaces. A variety of examples of cases have been presented here which are relevant to liquid-liquid extraction and demonstrate the functionality and flexibility of the new solver. The multiphase flow simulation capability described here has application to a variety of complex flows which span multiple regimes from fully segregated to fully dispersed. While the target application is flow in liquid-liquid extraction devices, this methodology could be useful to the simulation

of other multiphase flows which are currently restricted to a single flow regime. For example, coupling these methods with heat transfer and phase change could provide a tool for simulations of gas-liquid channel flows such as those seen in nuclear reactors spanning bubbly, slug, churn-turbulent, and annular flows.

The primary goal of the overall research effort of which this work is a part is the prediction of mass-transfer efficiency in stage-wise liquid-liquid extraction devices. This requires prediction of liquid-liquid interfacial area and consequently capturing of the dispersed phase droplet size distribution in some manner. The solver presented here provides a flexible foundation for building in the necessary models from any of the available methods. In the solver as part of the OpenFOAM release, the droplet diameter of each phase is a field variable and has been implemented as a callable library such that additional droplet diameter models can be easily implemented and selected at runtime. Extension of the solver to include variable droplet size using the reduced population balance method of Attarakih et al. [20] as implemented in [21] has recently been performed and exploration of droplet breakup and coalescence models is underway with very promising results.

Disclosure

The submitted paper has been created by UChicago Argonne, LLC, Operator of Argonne National Laboratory ("Argonne"). Argonne, a U.S. Department of Energy Office of Science Laboratory, is operated under Contract no. DE-AC02-06CH11357. The US Government retains for itself, and others acting on its behalf, a paid-up nonexclusive, irrevocable worldwide license in the said paper to reproduce, prepare derivative works, distribute copies to the public, and perform publicly and display publicly, by or on behalf of the government.

Acknowledgments

The authors gratefully acknowledge the use of the Fusion Linux cluster at Argonne National Laboratory and the Fission Linux cluster at Idaho National Laboratory for computational resources. K. E. Wardle was supported by the US DOE Office of Nuclear Energy's Fuel Cycle R&D Program in the area of Separations. K. E. Wardle would also like to thank Jing Gao of the University of Illinois at Chicago for experimental contributions as a summer research aide.

References

[1] G. Cerne, S. Petelin, and I. Tiselj, "Coupling of the interface tracking and the two-fluid models for the simulation of incompressible two-phase flow," *Journal of Computational Physics*, vol. 171, no. 2, pp. 776–804, 2001.

[2] L. Strubelj and I. Tiselj, "Two-fluid model with interface sharpening," *International Journal for Numerical Methods in Engineering*, vol. 85, pp. 575–590, 2011.

[3] L. Štrubelj, I. Tiselj, and B. Mavko, "Simulations of free surface flows with implementation of surface tension and interface

sharpening in the two-fluid model," *International Journal of Heat and Fluid Flow*, vol. 30, no. 4, pp. 741–750, 2009.

[4] K. Yan and D. Che, "A coupled model for simulation of the gas-liquid two-phase flow with complex flow patterns," *International Journal of Multiphase Flow*, vol. 36, no. 4, pp. 333–348, 2010.

[5] K. E. Wardle, T. R. Allen, M. H. Anderson, and R. E. Swaney, "Analysis of the effect of mixing vane geometry on the flow in an annular centrifugal contactor," *AIChE Journal*, vol. 55, no. 9, pp. 2244–2259, 2009.

[6] K. E. Wardle, T. R. Allen, and R. Swaney, "CFD simulation of the separation zone of an annular centrifugal contactor," *Separation Science and Technology*, vol. 44, no. 3, pp. 517–542, 2009.

[7] S. Vedantam, K. E. Wardle, T. V. Tamhane, V. V. Ranade, and J. B. Joshi, "CFD simulation of annular centrifugal extractors," *International Journal of Chemical Engineering*, vol. 2012, Article ID 759397, 31 pages, 2012.

[8] K. E. Wardle, "Open-source CFD simulations of liquid-liquid flow in the annular centrifugal contactor," *Separation Science and Technology*, vol. 46, no. 15, pp. 2409–2417, 2011.

[9] N. T. Padial-Collins, D. Z. Zhang, Q. Zou, X. Ma, and W. B. VanderHeyden, "Centrifugal contactors: separation of an aqueous and an organic stream in the rotor zone (LA-UR-05-7800)," *Separation Science and Technology*, vol. 41, no. 6, pp. 1001–1023, 2006.

[10] S. Li, W. Duan, J. Chen, and J. Wang, "CFD simulation of gas-liquid-liquid three-phase flow in an annular centrifugal contactor," *Industrial & Engineering Chemistry Research*, vol. 51, pp. 11245–11253, 2012.

[11] K. E. Wardle, "FY12 summary report on liquid-liquid contactor experiments for CFD model validation," Tech. Rep., Argonne National Laboratory, 2012.

[12] J. Brackbill, D. Kothe, and C. Zemach, "A continuum method for modeling surface tension," *Journal of Computational Physics*, vol. 100, no. 2, pp. 335–354, 1992.

[13] L. Schiller and Z. Naumann, "A drag coefficient corellation," *Zeitschrift des Vereins Deutscher Ingenieure*, vol. 77, p. 318, 1935.

[14] H. G. Weller, "A new approach to VOF-based interface capturing methods for incompressible and compressible flow," Tech. Rep., OpenCFD, 2008.

[15] V. R. Gopala and B. G. M. van Wachem, "Volume of fluid methods for immiscible-fluid and free-surface flows," *Chemical Engineering Journal*, vol. 141, no. 1–3, pp. 204–221, 2008.

[16] G. Erne, S. Petelin, and I. Tiselj, "Numerical errors of the volume-of-fluid interface tracking algorithm," *International Journal for Numerical Methods in Fluids*, vol. 38, no. 4, pp. 329–350, 2002.

[17] K. E. Wardle, T. R. Allen, M. H. Anderson, and R. E. Swaney, "Free surface flow in the mixing zone of an annular centrifugal contactor," *AIChE Journal*, vol. 54, no. 1, pp. 74–85, 2008.

[18] M. J. Sathe, S. S. Deshmukh, J. B. Joshi, and S. B. Koganti, "Computational fluid dynamics simulation and experimental investigation: study of two-phase liquid-liquid flow in a vertical Taylor-couette contactor," *Industrial and Engineering Chemistry Research*, vol. 49, no. 1, pp. 14–28, 2010.

[19] K. E. Wardle, T. R. Allen, M. H. Anderson, and R. E. Swaney, "Experimental study of the hydraulic operation of an annular centrifugal contactor with various mixing vane geometries," *AIChE Journal*, vol. 56, no. 8, pp. 1960–1974, 2010.

[20] M. Attarakih, M. Jaradat, C. Drumm et al., "Solution of the population balance equation using the one primary and one secondary particle method," in *Proceedings of the 19th European Symposium on Computer Aided Process Engineering (ESCAPE '09)*, 2009.

[21] C. Drumm, M. Attarakih, M. W. Hlawitschka, and H. J. Bart, "One-group reduced population balance model for CFD simulation of a pilot-plant extraction column," *Industrial and Engineering Chemistry Research*, vol. 49, no. 7, pp. 3442–3452, 2010.

6

Measurement of Membrane Characteristics Using the Phenomenological Equation and the Overall Mass Transport Equation in Ion-Exchange Membrane Electrodialysis of Saline Water

Yoshinobu Tanaka

IEM Research, 1-46-3 Kamiya, Ushiku-shi, Ibaraki 300-1216, Japan

Correspondence should be addressed to Yoshinobu Tanaka, fwis1202@mb.infoweb.ne.jp

Academic Editor: Seung Hyeon Moon

The overall membrane pair characteristics included in the overall mass transport equation are understandable using the phenomenological equations expressed in the irreversible thermodynamics. In this investigation, the overall membrane pair characteristics (overall transport number λ, overall solute permeability μ, overall electro-osmotic permeability ϕ and overall hydraulic permeability ρ) were measured by seawater electrodialysis changing current density, temperature and salt concentration, and it was found that μ occasionally takes minus value. For understanding the above phenomenon, new concept of the overall concentration reflection coefficient σ^* is introduced from the phenomenological equation. This is the aim of this investigation. σ^* is defined for describing the permselectivity between solutes and water molecules in the electrodialysis system just after an electric current interruption. σ^* is expressed by the function of μ and ρ. σ^* is generally larger than 1 and μ is positive, but occasionally σ^* becomes less than 1 and μ becomes negative. Negative μ means that ions are transferred with water molecules (solvent) from desalting cells toward concentrating cells just after an electric current interruption, indicating up-hill transport or coupled transport between water molecules and solutes.

1. Introduction

Mass transport across the membrane must be discussed fundamentally on the basis of the thermodynamics because the thermodynamics describes the rule of energy changes inevitably generating in the mass transport. However, the classical thermodynamics discusses only reversible phenomena and it does not treat transport rate. The irreversible thermodynamics came to succeed in discussing the transport rate by introducing the concept of "time" in its system [1–4]. Basic theory of the irreversible thermodynamics is established on the assumption of "microscopic irreversibility" [5–7]. This assumption holds more strictly in the circumstance being more close to equilibrium states. The actual electrodialysis process is not formed in the equilibrium states, so that the irreversible thermodynamics is assumed to exhibit only approximated meaning in the

electrodialysis system. However, the irreversible thermodynamics is considered to be applicable in the circumstances being apart to some extent from equilibrium states [8, 9]. The irreversible thermodynamics is the fundamental rule of mass transport and it is expressed by the functions including phenomenological coefficients. On the other hand, the performance of an electrodialyzer is expressed by the functions including parameters such as electrodialysis conditions and process specifications. These parameters cannot be discussed directly based on the irreversible thermodynamics. In the previous investigation, the overall mass transport equation was related to the irreversible thermodynamics and developed for analyzing the performance of an electrodialyzer [10]. It was successfully employed in the computer simulation of an electrodialysis process [11–13]. The overall mass transport equation includes the parameters such as the overall transport number λ, the overall solute permeability

Measurement of Membrane Characteristics Using the Phenomenological Equation and the Overall Mass Transport
Equation in Ion-Exchange Membrane Electrodialysis of Saline Water

51

μ, the overall electro-osmotic permeability ϕ, and the overall hydraulic permeability ρ. Among these parameters, μ is very small values and further in some cases, μ was found to take negative values. In such cases, μ has been conventionally neglected by setting $\mu = 0$. The aim of this investigation is to establish reasonable explanation of the negative μ. For understanding this phenomenon, the concept of the overall concentration reflection coefficient σ^* [10] must be supplied.

2. Theoretical

2.1. Phenomenological Equation and Overall Mass Transport Equation. When two kinds of ions are transported with solvent in a solution, their fluxes influence each other, because the fluxes and driving forces are not independent and coupled together. The interactions are presented in the irreversible thermodynamics. Kedem and Katchalsky introduced the following phenomenological equation expressing the electric current I, volume flow of a solution J and mass flux of component i; J_i in a membrane system [14],

$$I = L_E \Delta\psi + L_{EP}\Delta P + \sum_i L_{Ei}\Delta\mu_i,$$

$$J = L_{PE}\Delta\psi + L_P\Delta P + \sum_i L_{Pi}\Delta\mu_i, \qquad (1)$$

$$J_i = L_{iE}\Delta\psi + L_{iP}\Delta P + \sum_k L_{ik}\Delta\mu_k$$

$\Delta\psi$ is a potential difference, ΔP is a pressure difference, and $\Delta\mu_i$ is a chemical potential difference across the membrane. House [15] discussed an electrokinetic phenomenon based on the approaches of Kedem and Katchalsky [14]. Referring to this suggestion, Schultz discussed the principle of salt and solvent (water) transport in a two-cell electrodialysis system (cell I and cell II) incorporated with a cation-exchange membrane and introduced the following equations expressing the flux of solutes $J_{S,K}$ and a solution $J_{V,K}$ across the membrane [16],

$$J_{S,K} = t_K\frac{i}{F} + \{\omega_K - L_{P,K}\sigma_K(1 - \sigma_K)C_S^*\}RT\Delta C_S,$$
$$J_{V,K} = \beta_K i - \sigma_K L_{P,K}RT\Delta C_S \qquad (2)$$

in which t is the transport number, ω is the solute permeability, β is the electro-osmotic permeability, L_P is the hydraulic permeability, σ_K is the reflection coefficient of the cation-exchange membrane, and these phenomenological coefficients are membrane characteristics. R is the gas constant, F is the Faraday constant, and T is the absolute temperature. $\Delta C_S = (C_S^{II} - C_S^{I})$ is salt concentration difference between cell II and cell I. C_S^* is the logarithmic mean concentration across the membrane defined by

$$C_S^* = \frac{\Delta C_S}{\ln\left(C_S^{II}/C_S^{I}\right)}. \qquad (3)$$

The principle of separating salt from water is introduced based on the theory mentioned above using a three-cell

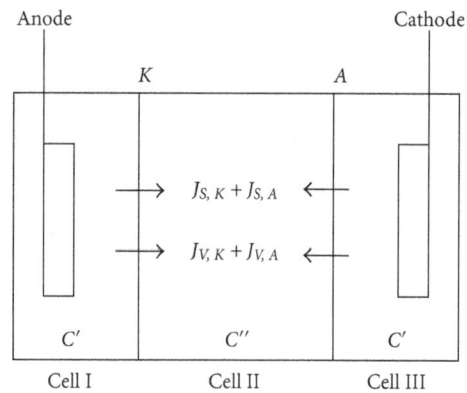

K: Cation-exchange membrane
A: Anion-exchange membrane

FIGURE 1: Three-cell electrodialysis system.

electrodialysis system (Figure 1). The system consists of a central cell (cell II) and electrode cells (cell I and cell III) placed on both outsides of cell II. A cation-exchange membrane (K) is placed between cell I and cell II, and an anion-exchange membrane (A) is placed between cell II and cell III. Supplying a salt solution into cell I and cell III, constant current density i is applied and a salt solution being collected in cell II is taken out from the system until the salt concentration in cell II reaches steady constant. Salt accumulation $J_{S,K} + J_{S,A}$ and solution accumulation $J_{V,K} + J_{V,A}$ in cell II in the steady state are given by the following equation introduced from (2) [10]:

$$J_{S,K} + J_{S,A} = (t_K + t_A - 1)\frac{i}{F} - RT$$
$$\times [(\omega_K + \omega_A) - \{L_{P,K}\sigma_K(1 - \sigma_K) \qquad (4)$$
$$+ L_{P,A}\sigma_A(1 - \sigma_A)\}C_S^*]\Delta C,$$
$$J_{V,K} + J_{V,A} = (\beta_K + \beta_A)i + RT(\sigma_K L_{P,K} + \sigma_A L_{P,A})\Delta C.$$

Here, we put $\Delta C = C'' - C' = \Delta C_S$. t_K and t_A are transport number of counter ions of a cation- and an anion-exchange membrane respectively. Subscripts K and A denote a cation- and an anion-exchange membrane. Superscripts $'$ and $''$ denote desalting sides (cells I and III) and a concentrating side (cell II), respectively. C_S^* is logarithmic mean concentration defined by (3).

The overall mass transport equation was developed from electrodialysis experiments as described in Section 3 experiment. Namely, a strong electrolyte solution is supplied to an electrodialyzer illustrated in Figure 2 keeping the linear solution velocities in desalting cells to be constant. Passing a constant electric current i (A/cm^2) through electrodes, concentrate is extracted from concentrating cells. After the electrolyte concentration reaches constant (C''), salt concentration in desalting cells C'(equiv./cm^3), ion flux J_S (equiv./cm^{2}s) and volume flux J_V (cm^3/cm^{2}s) transported across a membrane pair are measured. Plotting J_S/i and J_V/i against $(C'' - C')/i = \Delta C/i$ creates linear lines as exemplified

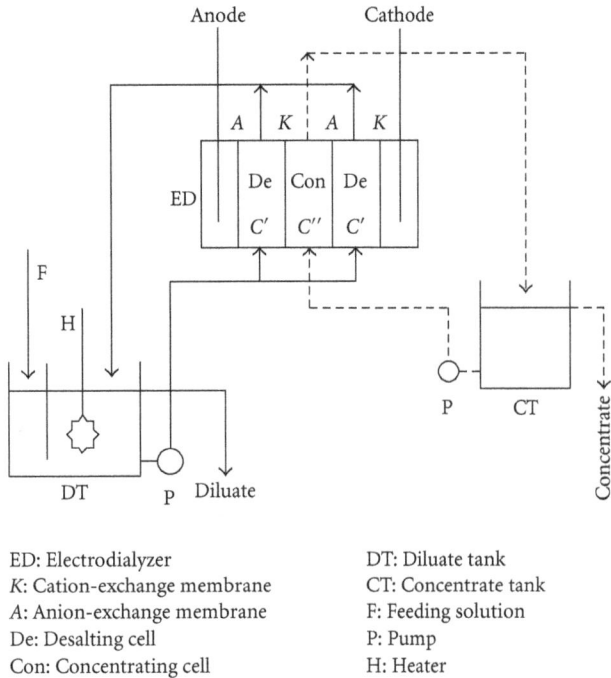

ED: Electrodialyzer   DT: Diluate tank
K: Cation-exchange membrane   CT: Concentrate tank
A: Anion-exchange membrane   F: Feeding solution
De: Desalting cell   P: Pump
Con: Concentrating cell   H: Heater

FIGURE 2: Electrodialysis process.

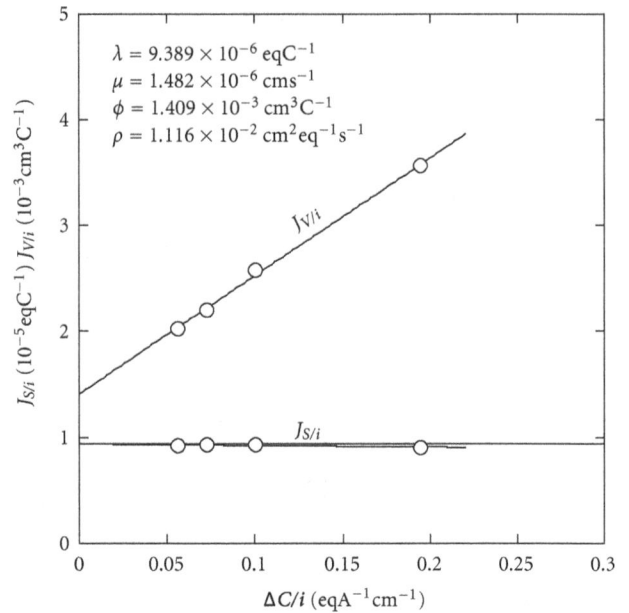

FIGURE 3: J_S/i versus $\Delta C/i$ plot and J_V/i versus $\Delta C/i$ plot. Selemion CMR/ASR 25°C.

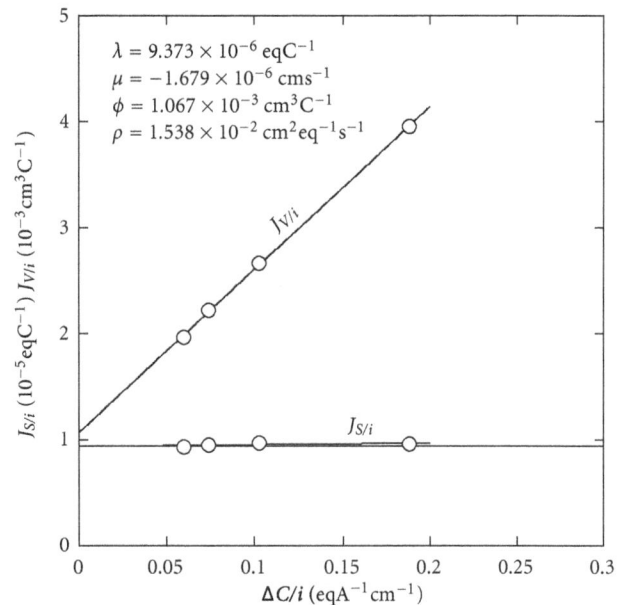

FIGURE 4: J_S/i versus $\Delta C/i$ plot and J_V/i versus $\Delta C/i$ plot. Neosepta CIM/ACS3 35°C.

in Figures 3 and 4. These lines are expressed by the following overall mass transport equation [10]:

$$J_S = C''J_V = \lambda i - \mu(C'' - C') = \lambda i - \mu\Delta C, \qquad (5)$$

$$J_V = \phi i + \rho(C'' - C') = \phi i + \rho\Delta C, \qquad (6)$$

in which λ (eqC^{-1}) is the overall transport number, μ (cm s^{-1}) is the overall solute permeability, ϕ (cm^{3}C^{-1}) is the overall electro-osmotic permeability, and ρ (cm^{4}eq^{-1}s^{-1}) is the overall hydraulic permeability. These parameters are termed the overall membrane pair characteristics and they are measured easily from the intercepts and gradients of the linear lines. The term "overall" means that the coefficients express the contributions of a cation- and an anion-exchange membrane. It means also that the coefficients express the contributions of many types of ions dissolving in a strong electrolyte solution. Terms λi and $\mu\Delta C$ in (5) stand for migration and diffusion of ions. Terms ϕi and $\rho\Delta C$ in (6) correspond to electro-osmosis and hydraulic osmosis of a solution. Osmosis refers to the movement of solvent in the original definition. However, the osmosis referred in (6) consists of the osmotic flow of solvent and of the volume flow associated with solutes.

The overall mass transport equation does not hold in the equilibrium state because it was developed from saline water electrodialysis experiments. It describes the mass transport in the nonequilibrium state. It is conceptually simple and available for discussing electrodialysis phenomena without contradictions.

In the electrodialysis experiments, the solutions supplied to desalting cells are mixed vigorously by spacers integrated. Thus, the boundary layer effects on the membrane surfaces are assumed to be negligible. The influence of concentration polarization on the overall membrane pair characteristics were discussed in the previous investigation [10] and concluded that the membrane pair characteristics are not influenced by the concentration polarization.

2.2. Overall Membrane Pair Characteristics and Phenomenological Coefficients. The overall mass transport equation ((5)

and (6)) is substantially identical to the phenomenological equation expressed in irreversible thermodynamics (4):

$$J_S = J_{S,K} + J_{S,A},$$
$$J_V = J_{V,K} + J_{V,A}. \tag{7}$$

From (7), the overall membrane pair characteristics λ, μ, ϕ and ρ are presented by the functions of the membrane characteristics defined in the phenomenological equation as follows:

$$\lambda = \frac{t_K + t_A - 1}{F} = \left(\frac{J_S}{i}\right)_{\Delta C=0} \tag{8}$$

$$\mu = RT[(\omega_K + \omega_A)$$
$$- \{L_{P,K}\sigma_K(1 - \sigma_K) + L_{P,A}\sigma_A(1 - \sigma_A)\}C_S^*]$$
$$= RT[(\omega_K + \omega_A) - \sigma(1 - \sigma)(L_{P,K} + L_{P,A})C_S^*] \tag{9}$$
$$= \left(\frac{J_S}{\Delta C}\right)_{i=0},$$

$$\phi = \beta_K + \beta_A = \left(\frac{J_V}{i}\right)_{\Delta C=0}, \tag{10}$$

$$\rho = RT(\sigma_K L_{P,K} + \sigma_A L_{P,A}),$$
$$= RT\sigma(L_{P,K} + L_{P,A}) = \left(\frac{J_V}{\Delta C}\right)_{i=0}. \tag{11}$$

Here, σ_K and σ_A are reflection coefficients of a cation- and an anion-exchange membrane respectively ((18) and (19)). σ expresses the contributions of a cation- and an anion-exchange membrane, so it is termed the overall reflection coefficient (Equation (20)) and it is introduced as follows. From (9):

$$\sigma(1 - \sigma) = \frac{L_{P,K}\sigma_K(1 - \sigma_K) + L_{P,A}\sigma_A(1 - \sigma_A)}{L_{P,K} + L_{P,A}}. \tag{12}$$

From (11),

$$\sigma = \frac{\sigma_K L_{P,K} + \sigma_A L_{P,A}}{L_{P,K} + L_{P,A}}. \tag{13}$$

From (12) and (13),

$$\sigma = \frac{L_{P,K}\sigma_K^2 + L_{P,A}\sigma_A^2}{L_{P,K}\sigma_K + L_{P,A}\sigma_A}, \tag{14}$$

when $\sigma_K = 0$, $\sigma = \sigma_A$
when $\sigma_A = 0$, $\sigma = \sigma_K$
when $\sigma_K = \sigma_A = 1$, $\sigma = 1$.

2.3. Reflection Coefficient. Equations (8)–(11) show that σ_K and σ_A appear in μ and ρ, and disappear in λ and ϕ. These events and (5) and (6) suggest that σ_K and σ_A do not exert an influence on electric current-driven mass transport, that is, migration and electro-osmosis. Reflection coefficient is a parameter indicating the permselectivity between solutes and solvent passing through the membrane. It is essentially the concept developed in pressure-driven dialysis. This phenomenon is conceivable in electrodialysis,

but the concept developed in pressure dialysis is not directly applicable to the phenomenon in electrodialysis. In order to understand the behavior of the reflection coefficient in an electrodialysis process, it is necessary to establish a concept of zero current density. In other words, it is reasonable to image the electric current interruption (switch off) for a moment in the electrodialysis process operating under a constant electric current, and assume the disappearance of the migration and electro-osmosis in this moment. Here, we assume further that solute diffusion and hydraulic osmosis remain as they are just after the electric current interruption [10].

In order to discuss the behavior of the reflection coefficient in an electrodialysis process, we express the volume flow J_V and exchange flow J_D in an ion-exchange membrane pair by the following equations introduced by Schloegel [17]:

$$J_V = (L_{P,K} + L_{P,A})\Delta P$$
$$+ (L_{PD,K} + L_{PD,A})RT\Delta C = \left(\frac{J_S}{C_S^*}\right) + \left(\frac{J_W}{C_W^*}\right), \tag{15}$$

$$J_D = (L_{DP,K} + L_{DP,A})\Delta P$$
$$+ (L_{D,K} + L_{D,A})RT\Delta C = \left(\frac{J_S}{C_S^*}\right) - \left(\frac{J_W}{C_W^*}\right), \tag{16}$$

ΔP and ΔC are pressure difference and concentration difference across the membrane, respectively. L_P is the hydraulic conductivity and L_D is the exchange flow parameter. L_{PD} is the osmotic volume flow coefficient and L_{DP} is the ultrafiltration coefficient. J_W is the flux of water molecules. C_S^* and C_W^* are respectively, logarithmic mean concentration (3) of solutes and water (solvent) between a desalting cell and a concentrating cell. It should be noticed that (15) and (16) are originally defined in the pressure-driven transport (pressure dialysis) of neutral species with no electric current. Equation (15) presents the sum of the solute flux J_S and solvent flux J_W, while (16) shows the difference between J_S and J_W. It should be added that (6) expresses the solution flux J_V and not expresses the solvent flux J_W.

In a pressure-driven process, putting $\Delta C = 0$ in (15) and (16) introduces the following equation applicable to pressure dialysis:

$$(J_V)_{\Delta C=0} = (L_{P,K} + L_{P,A})\Delta P,$$
$$(J_D)_{\Delta C=0} = (L_{DP,K} + L_{DP,A})\Delta P, \tag{17}$$

σ_K and σ_A included in (4) are defined by Staverman [18] and Kedem-Katchalsky [19] as follows:

$$\sigma_K = -\left(\frac{J_{D,K}}{J_{V,K}}\right)_{\Delta C=0} = -\left(\frac{L_{DP,K}}{L_{P,K}}\right), \tag{18}$$

$$\sigma_A = -\left(\frac{J_{D,A}}{J_{V,A}}\right)_{\Delta C=0} = -\left(\frac{L_{DP,A}}{L_{P,A}}\right). \tag{19}$$

We define the overall reflection coefficient σ (14) using (18) and (19):

$$\sigma = -\left(\frac{J_D}{J_V}\right)_{\Delta C=0} = -\frac{L_{DP,K} + L_{DP,A}}{L_{P,K} + L_{P,A}}. \tag{20}$$

σ given in (20) is the reflection coefficient defined in the pressure-driven dialysis. Here, we term σ "overall pressure reflection coefficient," because it is the membrane pair characteristic and reflects pressure difference ΔP-driven phenomenon.

σ given in (18)–(20) defines the permselectivity between water (solvent) and solutes and it is originally established in pressure-dialysis. For understanding the permselectivity between water and solutes (ions) in electrodialysis, σ is fundamentally inapplicable. The permselectivity between water and ions in electrodialysis must be explained on the basis of σ^* (overall concentration reflection coefficient) described below.

In the electrodialysis process, ΔP is relatively low and negligible, and just after an electric current interruption (switch off), ΔC remains as it is. Putting $\Delta P = 0$ in (15) and (16) introduces the following equations applicable to electrodialysis:

$$(J_V)_{\Delta P=0} = (L_{PD,K} + L_{PD,A})RT\Delta C,$$
$$(J_D)_{\Delta P=0} = (L_{D,K} + L_{D,A})RT\Delta C. \tag{21}$$

We define here another reflection coefficient σ^* introduced from (21) and generated just after an electric current interruption as follows:

$$\sigma^* = -\left(\frac{J_D}{J_V}\right)_{\Delta P=0} = -\frac{L_{D,K} + L_{D,A}}{L_{PD,K} + L_{PD,A}}. \tag{22}$$

σ^* (22) is the reflection coefficient defined in electrodialysis, and it is termed "the overall concentration reflection coefficient" because it reflects a concentration difference ΔC-driven phenomenon.

Koter [20] suggests that the Onsager reciprocity [5, 6] is not satisfied and the relationship between L_{PD} and L_{DP} depends on the concentration difference across the membrane. One example for Nafion 417 in NaCl/H$_2$O at $C''/C' = 2$ is $L_{DP}/L_{PD} = 0.40$. Thus, the reciprocal equation is presented as follows:

$$L_{DP,K} + L_{DP,A} = k(L_{PD,K} + L_{PD,A}) \quad k \neq 1, \tag{23}$$

in which, k is defined as the Onsager reciprocity coefficient. From (20), (22), and (23):

$$\sigma\sigma^* = k\frac{L_{D,K} + L_{D,A}}{L_{P,K} + L_{P,A}}. \tag{24}$$

Cancelling, J_W/C_W^* in (15) and (16):

$$J_D = 2\frac{J_S}{C_S^*} - J_V. \tag{25}$$

From (22) and (25),

$$\sigma^* = 1 - 2\left(\frac{1}{C_S^*}\right)\left(\frac{J_S}{J_V}\right)_{\Delta P=0}. \tag{26}$$

σ^* presented in (26) gives the permselectivity between ions and water molecules just after an electric current interruption. J_S and J_V are expressed as the following equations

just after an electric current interruption by putting $i = 0$ in (5) and (6):

$$J_S = -\mu(C'' - C'),$$
$$J_V = \rho(C'' - C'). \tag{27}$$

Substituting (27) into (26):

$$\sigma^* = 1 + 2\left(\frac{\mu}{\rho}\right)\left(\frac{1}{C_S^*}\right). \tag{28}$$

2.4. Membrane Characteristics. Yamauchi and Yasuko Tanaka [21] measured σ_K, $L_{P,K}$, and ω_K of a cation-exchange membrane (Neosepta CL-25T, Tokuyama Soda Co. Ltd.) by means of pressure-driven dialysis of a KCl solution and found σ_K to be unity. This phenomenon means that K$^+$ ions do not pass through the cation-exchange membrane due to the Donnan exclusion of co-ions (Cl$^-$ ions). σ_K and σ_A included in (9), (11) and (14) are equivalent to σ_K measured in the above experiment. So, (8)–(11) are simplified as follows by substituting $\sigma_K = \sigma_A = 1$ in (9), (11), and (14):

$$t_K + t_A = \lambda F + 1, \tag{29}$$

$$\omega_K + \omega_A = \frac{\mu}{RT}, \tag{30}$$

$$\beta_K + \beta_A = \phi, \tag{31}$$

$$L_{P,K} + L_{P,A} = \frac{\rho}{RT}. \tag{32}$$

Putting $\sigma = 1$ in (20),

$$L_{DP,K} + L_{DP,A} = L_{P,K} + L_{P,A}. \tag{33}$$

From (24) and (32),

$$k(K_{D,K} + L_{D,A}) = \frac{\rho\sigma^*}{RT} = (L_{P,K} + L_{P,A})\sigma^*. \tag{34}$$

3. Experimental

3.1. Electrodialysis 1. The following commercially available ion-exchange membranes were integrated into an electrodialyzer ED and formed an electrodialysis system [22] as illustrated in Figure 2.

Aciplex K172/A172, Asahi Chemical Co.
Selemion CMV/ASR, Asahi Glass Co.
Neosepta CIMS/ACS3. Tokuyama Inco.

The specifications of the membranes are listed in Table 1. The effective membrane area was maintained to 1.72 dm^2(18 cm length, 10 cm width). The flow-pass thickness in a desalting and a concentrating cell was 0.075 cm. Number of cell pairs was 9. Seawater was supplied into the diluate tank DT, then it was supplied to desalting cells De at a linear velocity of 5 cm/s keeping temperature to 25, 35, 50, or 60°C. Passing a constant electric current through the Ti/Pt-anode

Measurement of Membrane Characteristics Using the Phenomenological Equation and the Overall Mass Transport
Equation in Ion-Exchange Membrane Electrodialysis of Saline Water

55

TABLE 1: Specifications of ion-exchange membranes.

		Thickness mm	Electric resistance $\Omega\,cm^2$	Transpor number	Exchange capacity meq/dry memb.	Water content %	Intensity kg/cm^2
Aciplex	K172	0.11–0.13	1.9–2.2	> 0.99	1.5-1.6	20–30	2.6–3.3
	A172	0.11–0.15	1.7–2.1	>0.99	1.8-1.9	24–25	2.2–3.0
	CK2	0.23	3.3	0.91			
	CA3	0.09–0.12	1.5–2.0				1.3–2.0
Selemion	CMR	0.11	2.36	0.94	3.7	34	2.0
	ASR	0.11	1.80	0.96	3.5	33	2.0
	CMV2	0.11–0.15	2.0–3.5	>0.91	1.5–1.8	18–20	3–5
	AST	0.11–0.13	1.5–2.5	>0.95			1.5–2.5
Neosepta	CIMS	0.14–0.17	1.5-1.6	>0.98	2.2–2.5	30–35	3.1–4.1
	ACS3	0.09–0.12	1.5–2.0	>0.98	2.0–2.4	20–30	1.3–2.0
	CL25T	0.15–0.17	2.2–3.0	>0.98	1.5–1.8	25–35	3–5
	AVS4T	0.15–0.17	3.7–4.7	>0.98	1.5–2.0	25–30	4–6

and stainless (SUS 304) cathode, concentrate was extracted from concentrating cells. Confirming the salt concentration of the concentrate to be stable, diluate and concentrate were sampled. Electrodialysis was repeated changing current density incrementally.

In Figure 2, partition cells were incorporated between the desalting cells and electrode cells and seawater was supplied to the partition cells for preventing the influence of electrode reactions to the performance of electrodialysis. Seawater was also supplied to the electrode cells. The electrode reactions are as follows:

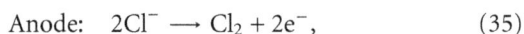

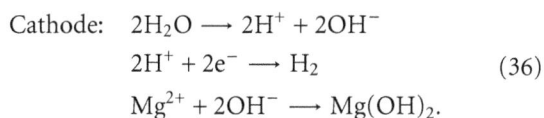

$$\text{Anode:} \quad 2Cl^- \longrightarrow Cl_2 + 2e^-, \tag{35}$$

$$\text{Cathode:} \quad 2H_2O \longrightarrow 2H^+ + 2OH^-$$
$$2H^+ + 2e^- \longrightarrow H_2 \tag{36}$$
$$Mg^{2+} + 2OH^- \longrightarrow Mg(OH)_2.$$

An HCl solution was supplied to the cathode cell to dissolve $Mg(OH)_2$.

3.2. Electrodialysis 2. The following commercially available membranes were integrated into an electrodialyzer [10] similar to the unit in Figure 2: Aciplex CK2/CA3, Selemion CMV2/AST, Neosepta CL25T/AVS4T.

The specifications of the membranes are listed in Table 1. The effective membrane area was 5 dm^2 (25 cm length, 20 cm width). The flow-pass thickness in a desalting cell and concentrating cell was 0.12 cm. Number of cell pairs was 10. Diluted seawater (0.294 eq/dm^3), seawater (0.577 eq/dm^3), and concentrated seawater (1.131 and 1.920 eq/dm^3) were supplied into the diluate tank DT, then it was supplied to desalting cells De at a linear velocity of 5 cm/s keeping temperature to 29°C. Passing a constant electric current through graphite-stainless electrodes, concentrate was extracted from concentrating cells. Diluate and concentrate were sampled and electrodialysis was repeated changing current density step by step as described in the electrodialysis 1.

3.3. Chemical Analysis. Concentration (eq/dm^3) of K$^+$, Mg^{2+}, Ca^{2+}, Cl$^-$, and SO$_4^{2-}$ ions in diluate and concentrate were analyzed. In seawater electrodialysis, total salt concentration C is given as $C = C_{Na} + C_K + C_{Mg} + C_{Ca} = C_{Cl} + C_{SO_4}$. So, sodium ion concentration is given by $C_{Na} = C_{Cl} + C_{SO_4} - C_K - C_{Ca} - C_{Mg}$.

4. Results and Discussion

4.1. Overall Mass Transport Equation and Overall Membrane Pair Characteristics; λ, μ, ϕ and ρ. J_S/i and J_V/i are plotted against $\Delta C/i$ and exemplified in Figures 3 and 4. From these plots, overall transport number λ, overall solute permeability μ, overall electro-osmotic permeability ϕ, and overall hydraulic permeability ρ are calculated as shown in the figures. The gradient of J_S/i versus $\Delta C/i$ is usually negative, so μ is positive (Figure 3). However, the gradient occasionally gives positive, so μ becomes negative (Figure 4). μ values of the commercially available membranes are generally very small due to the Donnan exclusion of the membranes against co-ions and their dense structure. Using the overall mass transport equation, μ can be detected from the above gradients. Overall membrane pair characteristics observed in this investigation are listed in Table 2. μ values of the Neosepta CIMS/ACS3 membranes (Electrodialysis 1) and the Selemion CMV2/AST membranes (Electrodialysis 2) are found to be negative.

4.2. Membrane Characteristics t, ω, β, and L$_P$. Substituting λ, μ, ϕ, and ρ into (29)–(32), membrane characteristics (phenomenological coefficients) such as transport number $t_K + t_A$, solute permeability $\omega_K + \omega_A$, electro-osmotic permeability $\beta_K + \beta_A$, and hydraulic conductivity $L_{P,K} + L_{P,A}$ are calculated. $L_{DP,K} + L_{DP,A}$ is equivalent to $L_{P,K} + L_{P,A}$ as defined in (33).

Figures 5–8 show the relationship between temperature T and the membrane characteristics obtained in Section 2.1 Electrodialysis 1. It is seen that $L_{P,K} + L_{P,A}$ increases

TABLE 2: Overall membrane pair characteristics.

(a) Electrodialysis 1.

Ion-exchange membrane	Temp. °C	$\lambda \times 10^6$ eq/C	$\mu \times 10^6$ cm/s	$\phi \times 10^3$ cm/C	$\rho \times 10^2$ cm^4/eq s
Aciplex K172/A172	25	9.724	1.434	1.403	1.218
	35	9.736	1.289	1.370	1.626
	50	9.972	4.515	1.542	1.996
	60	9.692	2.123	1.405	2.765
Selemion CMR/ASR	25	9.389	1.482	1.409	1.116
	35	9.422	2.421	1.501	1.377
	50	9.475	3.614	1.563	1.937
	60	9.583	5.103	1.621	2.380
Neosepta CIMS/ACS3	25	9.349	−1.055	1.004	1.254
	35	9.373	−1.679	1.067	1.538
	50	9.459	−1.239	1.067	1.835
	60	9.547	−0.561	1.405	2.059

(b) Electrodialysis 2.

Ion-exchange membrane	Conc. eq/dm^3	$\lambda \times 10^6$ eq/C	$\mu \times 10^6$ cm/s	$\phi \times 10^3$ cm^3/C	$\rho \times 10^2$ cm^4/eq s
Aciplex CK2/CA3	0.294	9.482	6.028	1.699	1.226
	0.576	9.407	8.570	1.619	1.234
	1.131	8.970	6.977	1.514	1.280
	1.920	8.953	9.690	1.367	1.387
Selemion CMV2/AST	0.294	9.055	−1.443	1.294	5.945
	0.577	9.371	−0.731	1.316	5.787
	1.131	9.389	−0.424	1.296	5.893
	1.921	9.139	0.850	1.193	6.371
Neosepta CL25T/AVS4T	0.294	9.584	1.458	1.513	6.354
	0.577	9.633	3.208	1.394	6.788
	1.132	9.228	0.908	1.304	7.566
	1.920	9.269	3.575	1.254	7.691

apparently with T. $\omega_K + \omega_A$ and $\beta_K + \beta_A$ increase slightly with T. $t_K + t_A$ is not influenced by T. $\omega_K + \omega_A$ for Neosepta CIMS/ACS3 membranes is negative (Figure 6) because $\mu < 0$ (Table 2).

Figures 9–12 show the relationships between salt concentrations in desalting cells C' and membrane characteristics obtained in Section 2.2 Electrodialysis 2. $\beta_K + \beta_A$ decreases and $L_{P,K} + L_{P,A}$ increases to some extent with the increase of C'. $t_K + t_A$ is not influenced by C'. $\omega_K + \omega_A$ increases slightly with C' and it takes negative for Selemion CMV2/AST membranes (Figure 10) due to $\mu < 0$ (Table 2).

4.3. Overall Concentration Reflection Coefficient σ^*. σ^* is defined by the functions of current density because σ^* is given by the function of logarithmic mean concentration C_S^* (Equation (28)). However, the experimental results show that σ^* is not influenced by current density. Thus, the observed σ^* is averaged for each current density. Figures (13) and (14) show the σ^* versus T and C'. σ^* is generally larger than 1, however, less than 1 for Neosepta CIMS/ACS3 membranes

and for Selemion CMV2/AST membranes. The phenomena exhibited in Figures (13) and (14) are understandable as follows.

σ^* defined in (26) gives the permselectivity between ions and water molecules just after an electric current interruption. Physical rationale of σ^* is given by the following relationships realized in an electrodialysis system (Figure 15) incorporated with a cation- and an anion-exchange membrane and cells I, II and III in which electrolyte solutions are filled:

$$J_{\text{solute}} > 0, \qquad \sigma^* < 1 \qquad \mu < 0, \tag{37}$$

$$J_{\text{solute}} = 0, \qquad \sigma^* = 1 \qquad \mu = 0, \tag{38}$$

$$J_{\text{solute}} < 0, \qquad \sigma^* > 1 \qquad \mu > 0. \tag{39}$$

In Figure 15, an electric current is passed through reversible electrodes and electrolyte concentration in cells I

Measurement of Membrane Characteristics Using the Phenomenological Equation and the Overall Mass Transport Equation in Ion-Exchange Membrane Electrodialysis of Saline Water

57

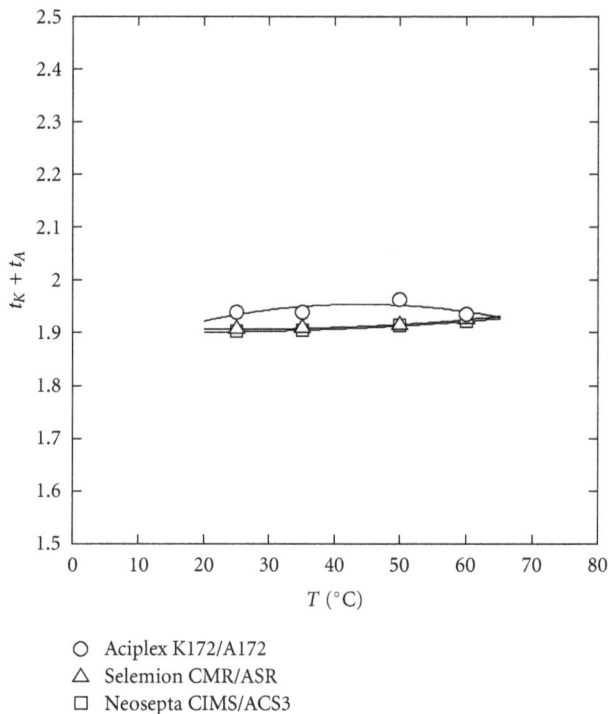

FIGURE 5: Relationship between temperature and transport number.

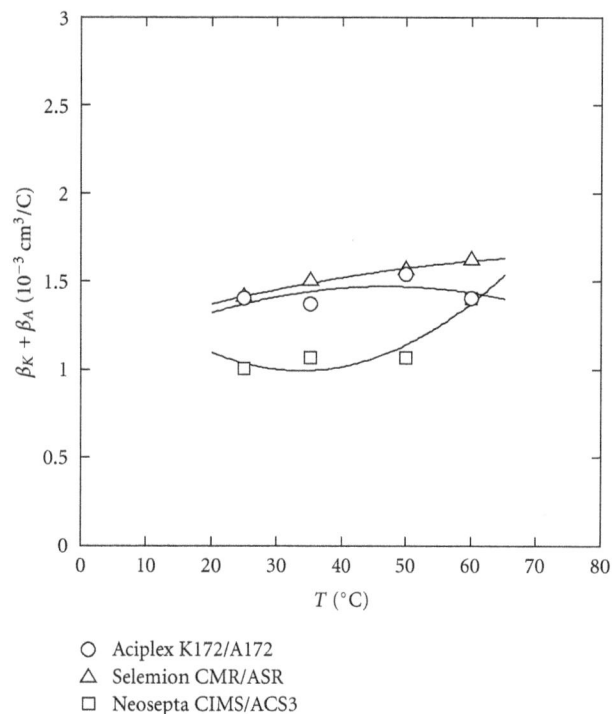

FIGURE 7: Relationship between temperature and electro-osmotic permeability.

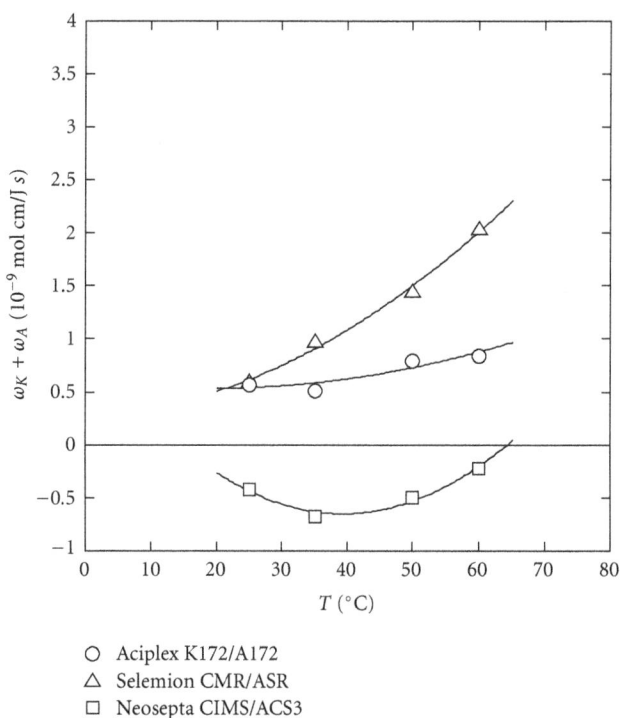

FIGURE 6: Relationship between temperature and solute permeability.

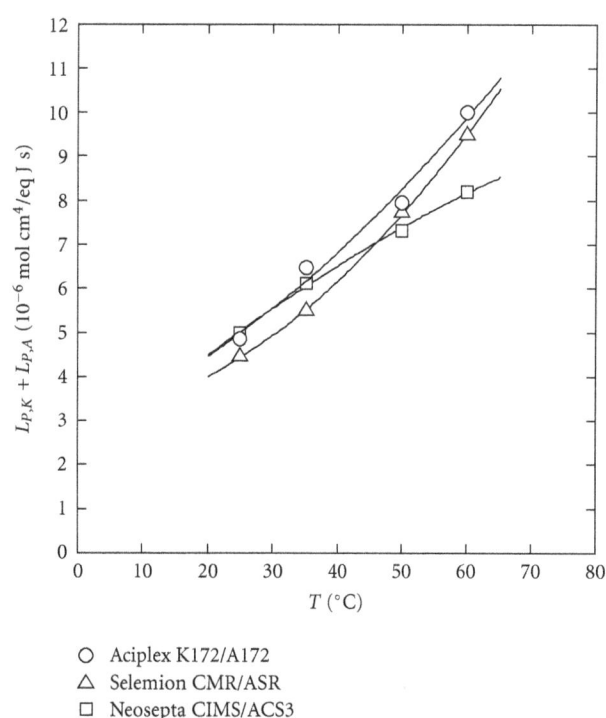

FIGURE 8: Relationship between temperature and hydraulic permeability.

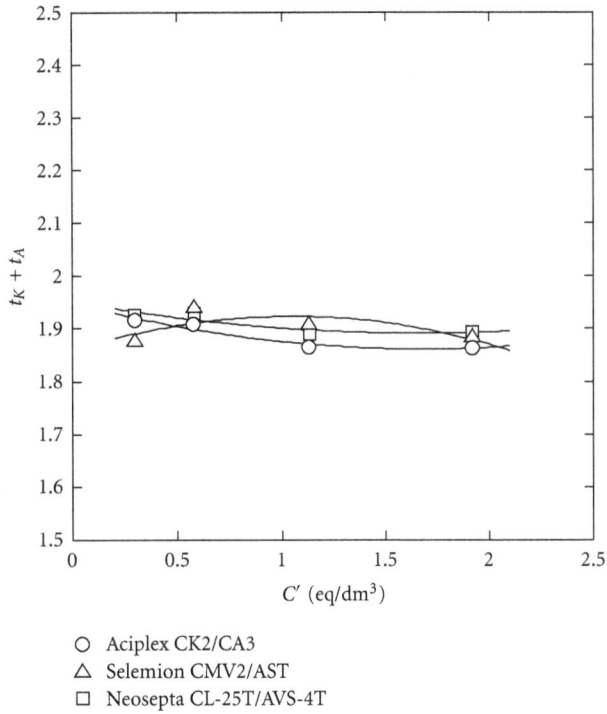

FIGURE 9: Relationship between salt concentration in a feeding solution and transport number.

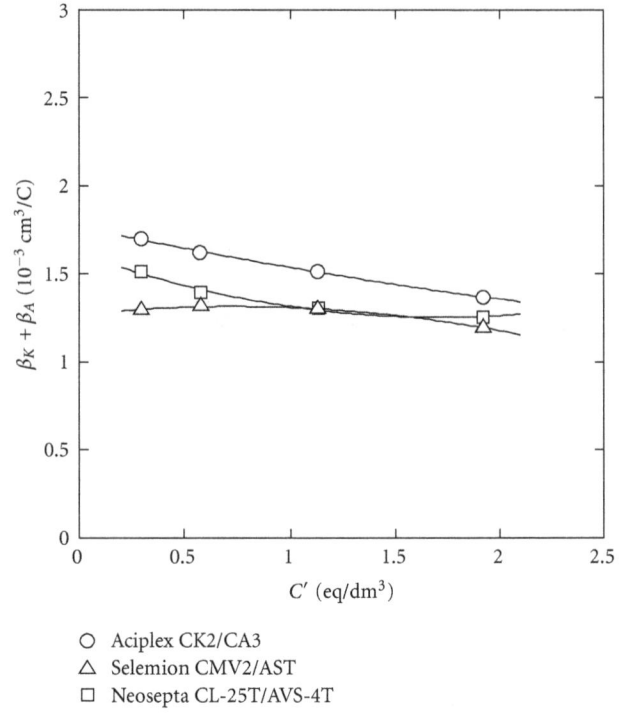

FIGURE 11: Relationship between salt concentration in a feeding solution and electro-osmotic permeability.

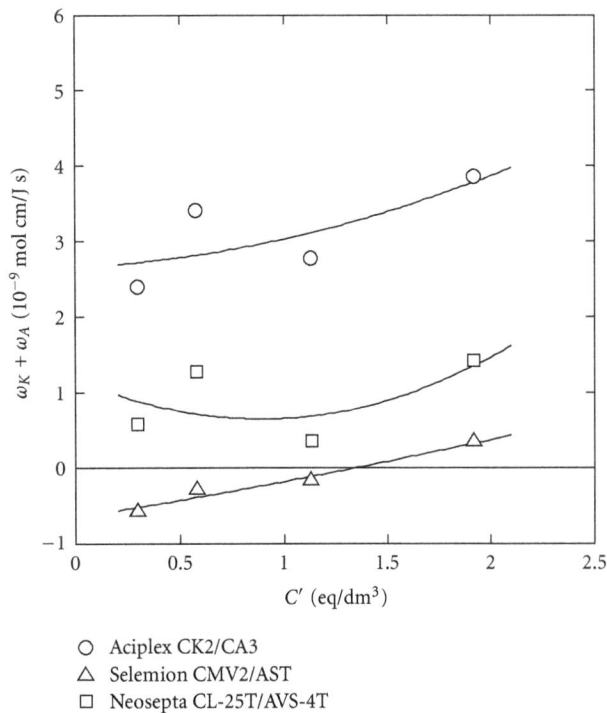

FIGURE 10: Relationship between salt concentration in a feeding solution and solute permeability.

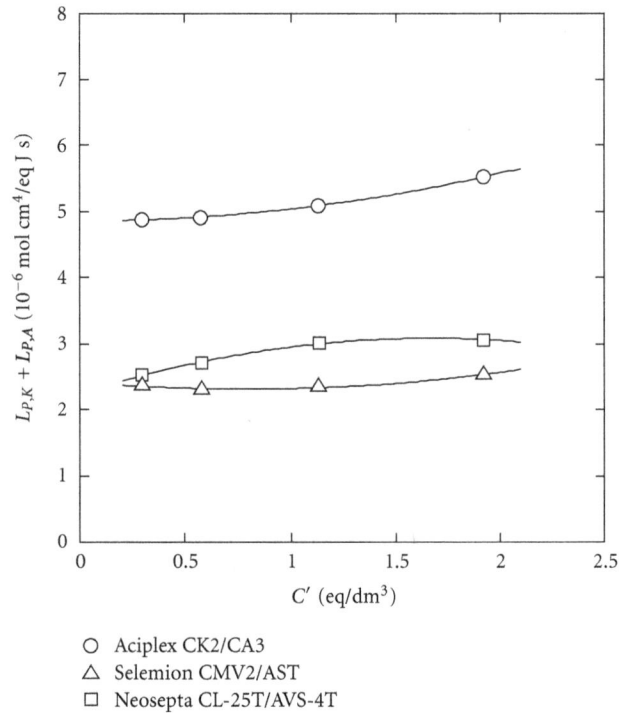

FIGURE 12: Relationship between salt concentration in a feeding solution and hydraulic conductivity.

Measurement of Membrane Characteristics Using the Phenomenological Equation and the Overall Mass Transport Equation in Ion-Exchange Membrane Electrodialysis of Saline Water

59

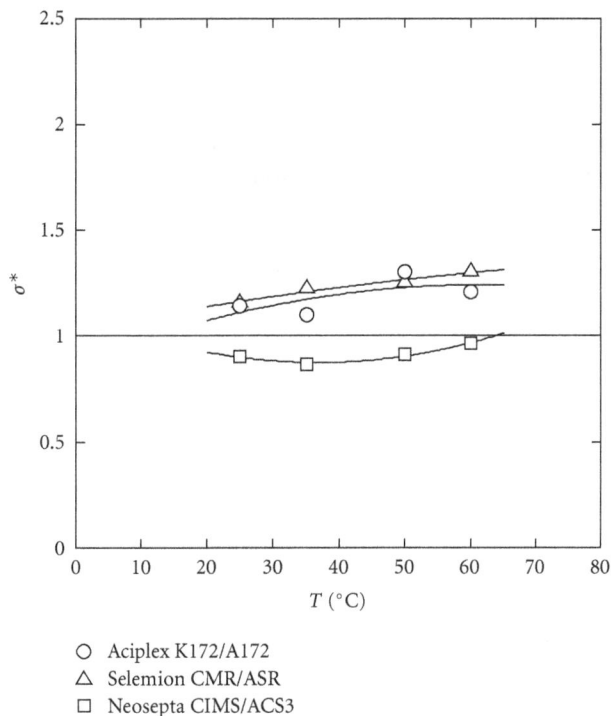

FIGURE 13: Relationship between temperature and overfull concentration reflection coefficient.

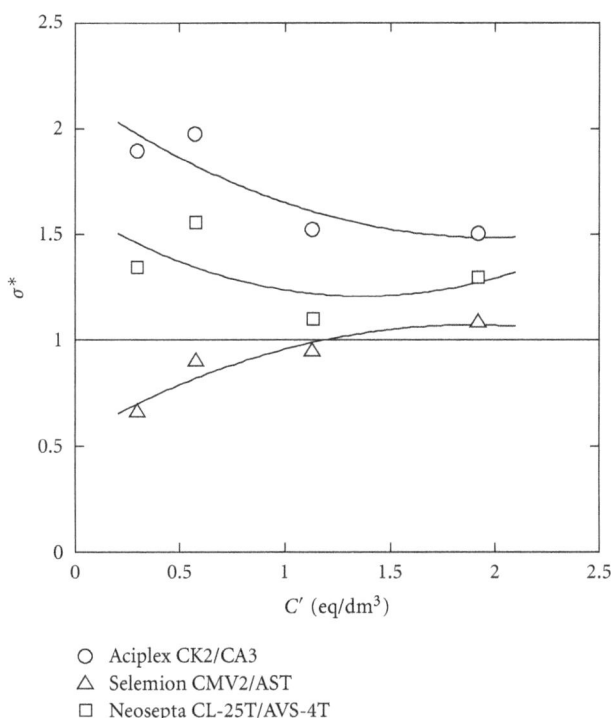

FIGURE 14: Relationship between salt concentration in a feeding solution and overfull concentration reflection coefficient.

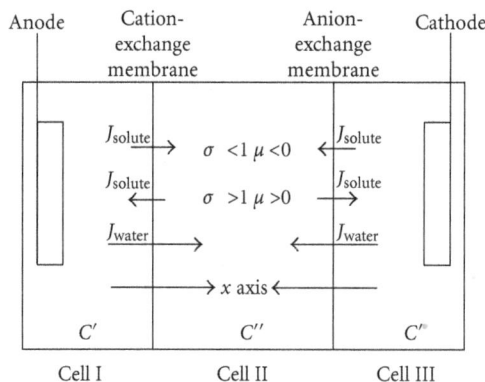

FIGURE 15: Electrodialysis system.

and III is assumed to be decreased to C' and that in cell II is increased to C'' ($C' < C''$). The x axis is drawn across each ion-exchange membrane as shown in the figure. Direction of water flux J_{water} is positive ($J_{water}0$); J_V is positive ($J_V > 0$) in every situation.

Solute flux across the membrane is generally negative ($J_{solute} < 0$) and it moves from the concentrating cell (cell II) toward desalting cells (cells I and III) due to solute diffusion just after an electric current interruption. In this situation, the concentration reflection coefficient is larger than 1 (Equation (39), $\sigma^* > 1$, Figures 13 and 14). This phenomenon is ordinary in electrodialysis, but extraordinary in pressure dialysis because the direction of the solute due to solute diffusion is against water flux due to hydraulic osmosis. In this case, the overall solute permeability becomes positive, that is, $\mu > 0$ (Table 2) and $\omega_K + \omega_A > 0$ (Figures 6 and 10).

In Neosepta CIMS/ACS3 membranes and Selemion CMV2/AST membranes, solutes move from the desalting cells (cells I and III) toward concentrating cell (cell II) ($J_{solute} > 0$) just after an electric current interruption. In this situation, the concentration reflection coefficient is less than 1 (Equation (37), $\sigma^* < 1$, Figures 13 and 14). This phenomenon is ordinary in pressure dialysis because the direction of the solute is the same to the water flux due to hydraulic osmosis. The overall solute permeability in this situation becomes negative ($\mu < 0$) (Table 2). The negative μ is indicative of coupled transport between water and solute, which is demonstrated also by negative $\omega_K + \omega_A$ (Figures 6 and 10). However, this phenomenon is extraordinary in electrodialysis because it means uphill transport of solutes against the concentration difference ($C'' - C'$).

In repeating words, ions (solutes) transfer (diffuse) generally from concentrating cells toward desalting cells ($\mu > 0$, $\sigma^* > 1$, $\omega_K + \omega_A > 0$). This direction of solute flux ($J_{solute} < 0$) is ordinary in electrodialysis but extraordinary in pressure dialysis because the direction of solute flux is against water flux ($J_{water} > 0$). In some cases, ions transfer from desalting cells toward concentrating cells ($\mu < 0$, $\sigma^* < 1$, $\omega_K + \omega_A < 0$). This direction of solute flux ($J_{solute} > 0$) is extra-ordinary in electrodialysis, but ordinary in pressure dialysis because the direction of solute flux is the same to water flux ($J_{water} >$

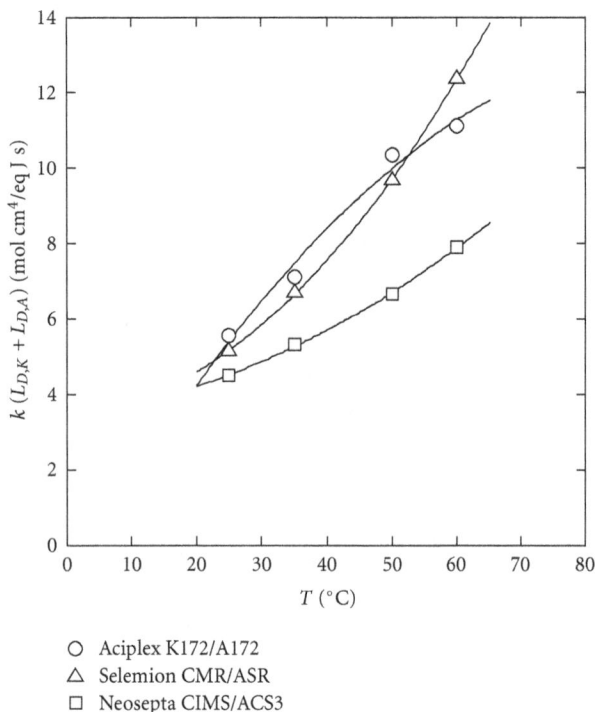

FIGURE 16: Relationship between temperature and exchange flow parameter.

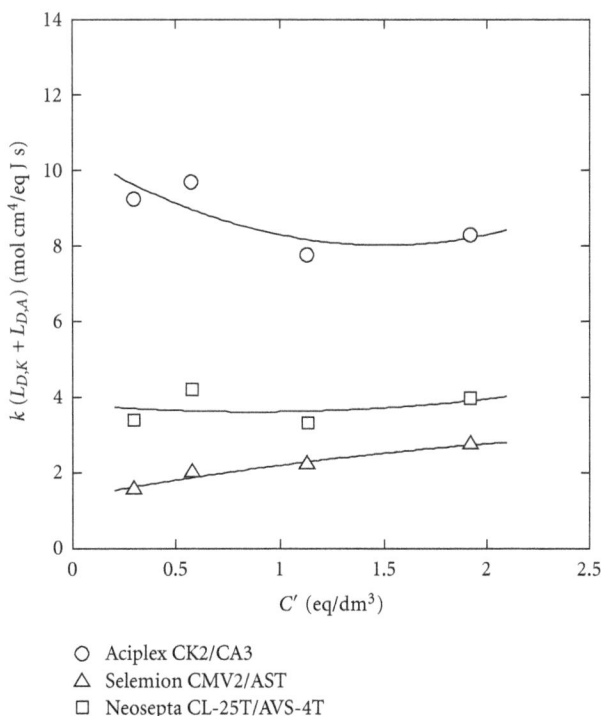

FIGURE 17: Relationship between salt concentration in a feeding solution and exchange flow parameter.

0), which indicates up-hill transport or coupled transport between water and solutes.

4.4. Exchange Flow Parameter L_D. $k(L_{D,K} + L_{D,A})$ is calculated from ρ and σ^* (Equation (34)). ρ is the leading parameter and the other membrane characteristics are governed by ρ. [10]. $k(L_{D,K} + L_{D,A})$ is increased with T (Figure 16) and hardly influenced by C' (Figure 17). These phenomena are caused by that ρ is increased with T and hardly influenced by C'.

5. Conclusion

Phenomenological coefficients such as transport number, solute permeability, electro-osmotic permeability, hydraulic permeability, and exchange flow parameter included in the phenomenological equations are evaluated by electrodialysis based on the overall mass transport equation. The overall reflection coefficient σ is essentially the concept applied in pressure-driven dialysis. In order to understand the behavior of the overall reflection coefficient in an electrodialysis process, the concept of the overall concentration reflection coefficient σ^* is applied. σ^* presents the permselectivity between ions and water molecules across a membrane pair just after an electric current interruption in an electrodialysis process. σ^* is generally larger than 1 and at the same time the overall solute permeability μ is positive in usual commercially available ion-exchange membranes. However, occasionally σ^* becomes less than 1 and μ becomes negative, which indicates up-hill transport or coupled transport between water and solutes.

Nomenclature

C: Concentration (equiv. cm^{-3})
C^*:   Mean concentration (equiv. cm^{-3})
F: Faraday constant (C equiv.$^{-1}$)
i: Current density (A cm^{-2})
I: Electric current (A)
J: Volume flow (cm s^{-1})
J_D: Exchange flow (cm s^{-1})
J_S: Salt flux (equiv.cm^{-2}s^{-1})
J_V: Volume flux (cm^{3}cm^{-2}s^{-1})
J_W: Flux of water molecules (mol cm^{-2}s^{-1})
k: Onsager reciprocity coefficient
L_D: Exchange flow parameter (mol cm^4 equiv.$^{-1}$J^{-1}s^{-1})
L_{DP}: Ultrafiltration coefficient (mol cm^{4}equiv.$^{-1}$J^{-1}s^{-1})
L_P: Hydraulic conductivity (mol cm^{4}equiv.$^{-1}$J^{-1}s^{-1})
L_{PD}: Osmotic volume flow coefficient (mol cm^{4}equiv.$^{-1}$J^{-1}s^{-1})
R: Gas constant (JK^{-1}mol^{-1})
t: Transport number
T: Temperature (°C, K).

Measurement of Membrane Characteristics Using the Phenomenological Equation and the Overall Mass Transport
Equation in Ion-Exchange Membrane Electrodialysis of Saline Water

61

Greek symbols

β: Electro-osmotic permeability (cm^3C^{-1})

ΔC: Concentration difference (equiv. cm^{-3})

ΔP: Hydraulic pressure difference (dyn cm^{-2})

$\Delta\psi$: Potential difference (V)

$\Delta\mu$: Chemical potential difference (J mol^{-1})

λ: Overall transport number (equiv. C^{-1})

μ: Overall solute permeability (cm s^{-1})

ρ: Overall hydraulic permeability (cm^4equiv.$^{-1}s^{-1}$)

σ: Reflection coefficient, overall pressure reflection coefficient

σ^*: Overall concentration reflection coefficient

ϕ: Overall electro-osmotic permeability (cm^3C^{-1})

ω: Solute permeability (mol cm $J^{-1}s^{-1}$).

Subscript

A: Anion-exchange membrane

i: Component

K: Cation-exchange membrane

S: Solute

W: Water molecule (solvent).

Superscript

$'$: Desalting cell

$''$: Concentrating cell.

Acknowledgment

The authors are grateful to Mr. S. Nagatsuka and Mr. M. Akiyama, Odawara Salt Experimental Station, Japan Tobacco and Salt Public Corp. for their cooperation in the experimental work.

References

[1] I. Progogine, *Introduction to Thermodynamics of Irreversible Processes*, John Wiley, New York, NY, USA, USA, 2nd edition, 1961.

[2] P. Glansdorff and I. Prigogine, *Thermodynamic Theory Structure, Stability and Fluctuations*, Wiley-Interscience, London, UK, 1971.

[3] S. R. de Groot and P. Mazur, *Nonequilibrium Thermodynamics*, North-Holland, New York, NY, USA, 1962.

[4] D. D. Fitts, *Nonequilibrium Thermodynamics*, Mc-Graw Hill, New York, NY, USA, 1962.

[5] L. Onsager, "Reciprocal relations in irreversible processes. I.," *Physical Review*, vol. 37, no. 4, pp. 405–426, 1931.

[6] L. Onsager, "Reciprocal relations in irreversible processes. II.," *Physical Review*, vol. 38, no. 12, pp. 2265–2279, 1931.

[7] H. B. G. Casimir, "On Onsager's principle of microscopic reversibility," *Reviews of Modern Physics*, vol. 17, no. 2-3, pp. 343–350, 1945.

[8] P. J. Dunlop, "A study of interacting flows in diffusion of the system raffinose-KCl-H_2 at 25°C," *Journal of Physical Chemistry*, vol. 61, no. 7, pp. 994–1000, 1957.

[9] P. J. Dunlop and L. J. Gosting, "Use of diffusion and thermodynamic data to test the Onsager reciprocal relation for isothermal diffusion in the system NaCl-KCl-H_2O at 25°C," *Journal of Physical Chemistry*, vol. 63, no. 1, pp. 86–93, 1959.

[10] Y. Tanaka, "Irreversible thermodynamics and overall mass transport in ion-exchange membrane electrodialysis," *Journal of Membrane Science*, vol. 281, no. 1-2, pp. 517–531, 2006.

[11] Y. Tanaka, "A computer simulation of continuous ion exchange membrane electrodialysis for desalination of saline water," *Desalination*, vol. 249, no. 2, pp. 809–821, 2009.

[12] Y. Tanaka, "A computer simulation of batch ion exchange membrane electrodialysis for desalination of saline water," *Desalination*, vol. 249, no. 3, pp. 1039–1047, 2009.

[13] Y. Tanaka, "A computer simulation of feed and bleed ion exchange membrane electrodialysis for desalination of saline water," *Desalination*, vol. 254, no. 1–3, pp. 99–107, 2010.

[14] O. Kedem and A. Katchalsky, "Permeability of composite membranes Part 1. Electric current, volume flow and flow of solute through membranes," *Transactions of the Faraday Society*, vol. 59, pp. 1918–1930, 1963.

[15] C. R. House, *Water Transport in Cells and Tissues*, Edward Arnold, London, UK, 1974.

[16] S. G. Schultz, *Basic Principles of Membrane Transport*, Cambridge University Press, Cambridge, UK, 1980.

[17] R. Z. Schloegel, *Fortschritte der physikalischen Chemie. Band 9*, 1964.

[18] A. J. Staverman, "The theory of measurement of osmotic pressure," *Recueil des Travaux Chimiques des Pays-Bas*, vol. 70, no. 4, pp. 344–352, 1951.

[19] O. Kedem and A. Katchalsky, "Thermodynamic analysis of the permeability of biological membranes to non-electrolytes," *Biochimica et Biophysica Acta*, vol. 27, no. 2, pp. 229–246, 1958.

[20] S. Koter, "Transport of electrolytes across cation-exchange membranes. Test of Onsager reciprocity in zero-current processes," *Journal of Membrane Science*, vol. 78, no. 1-2, pp. 155–162, 1993.

[21] A. Yamauchi and Y. Tanaka, "Salt transport phenomena across charged membrane driven by pressure difference," in *Effective Membrane Process—New Prospective*, R. Paterson, Ed., pp. 179–185, BHR Group; Information Press, Oxford, UK, 1993.

[22] Y. Tanaka, "Ion exchange membrane electrodialysis of saline water desalination and its application to seawater concentration," *Industrial & Engineering Chemistry Research*, vol. 50, no. 12, pp. 7494–7503, 2011.

7

Design Mixers to Minimize Effects of Erosion and Corrosion Erosion

Julian Fasano,[1] **Eric E. Janz,**[2] **and Kevin Myers**[3]

[1] *Mixer Engineering Co., 2673 Stonebridge Drive, Troy, OH 45373, USA*
[2] *Chemineer, Inc., 5870 Poe Avenue, Dayton, OH 45414, USA*
[3] *Department of Chemical & Materials Engineering, University of Dayton, 300 College Park, Dayton, OH 45469-0246, USA*

Correspondence should be addressed to Julian Fasano, j.fasano@mixerengineering.com

Academic Editor: Shunsuke Hashimoto

A thorough review of the major parameters that affect solid-liquid slurry wear on impellers and techniques for minimizing wear is presented. These major parameters include (i) chemical environment, (ii) hardness of solids, (iii) density of solids, (iv) percent solids, (v) shape of solids, (vi) fluid regime (turbulent, transitional, or laminar), (vii) hardness of the mixer's wetted parts, (viii) hydraulic efficiency of the impeller (kinetic energy dissipation rates near the impeller blades), (ix) impact velocity, and (x) impact frequency. Techniques for minimizing the wear on impellers cover the choice of impeller, size and speed of the impeller, alloy selection, and surface coating or coverings. An example is provided as well as an assessment of the approximate life improvement.

1. Introduction

There are numerous applications of mixers that deal with erosive solids, especially in the minerals processing and power industries. In many of these applications, there is an erosion-corrosion synergistic effect on the wear of a mixer's wetted parts, particularly the impeller. This paper pulls together the authors' research with numerous articles on erosion and erosion corrosion to permit a designer to optimize the cost-based life of eroding mixer parts before replacement is required.

There are a large number of factors that can affect the rate of erosion. Many of these factors have been known and studied to some extent:

(i) chemical environment,

(ii) hardness of solids,

(iii) density of solids,

(iv) difference in liquid and solid density,

(v) percent solids,

(vi) shape of solids,

(vii) fluid regime (turbulent, transitional, or laminar),

(viii) fluid rheology (e.g., pseudoplasticity),

(ix) hardness of the mixer's wetted parts,

(x) young's modulus of the mixer's wetted parts,

(xi) hydraulic efficiency of the impeller (kinetic energy dissipation rates near the impeller blades),

(xii) impact velocity,

(xiii) impact frequency,

(xiv) angle of impact.

Theoretically the rate of volume loss of material is due to the kinetic energy lost when a particle impacts a material [1]. This would suggest a velocity exponent of 2. However, presented below, experimental velocity exponents have ranged from 1.5 to 4.0. The general form of the equation relating erosion rate to velocity is given by

$$E = K \cdot V^n f(\theta), \qquad (1)$$

where E = volumetric erosion rate, K = constant (function of all parameters other than V or θ), V = particle velocity or relative velocity for rotating systems (impellers), n = velocity exponent (can also be a function of other parameters), and θ = impingement angle.

Most investigators have used this general equation form.

Sapate and RamaRao [2] used a power law correlation between volumetric erosion rate and impingement velocity in a nonrotating system. They observed exponents on velocity of 1.91 to 2.52. The velocity exponent showed an increasing trend with increasing hardness of the alloys irrespective of the hardness of the erodent particles and the impingement angle of the alloys investigated.

Stack [3] and others investigated the effect that corrosion plays in an erosion environment. These investigators studied various parameters of the corrosion-erosion environment in a nonrotating system. They observed velocity exponents that ranged from 1.4 to 3.5. They concluded that exponents derived for erosion of alloys under erosion-dominated conditions can be correlated to those derived for the strictly ductile erosion process. These are typically very near the theoretical "2" for the strictly ductile erosion process. However, those for the erosion corrosion dominated regime are higher than for the erosion-dominated regime and were in the 2.5 to 3.5 range. A publication by the Hydraulics Institute [4] suggests that the erosion velocity exponent for pumps in slurry transport is on the order of 2.5–3.0.

Fort [5] and others studied pitched blade impellers 100 mm in diameter with a blade width of 20 mm in water-solid slurries under turbulent conditions. These impellers were studied at pitch angles of 20°, 35°, and 45°. These studies included a slurry of 18.3% by volume of gypsum having a mean particle diameter of 0.1 mm and a 10% by volume slurry of 0.4 mm mean diameter sand particles. From their studies, they concluded the following.

(i) Particles of the lower hardness gypsum generated uniform sheet erosion over the entire surface of the impeller, while the particles of sand, having a higher hardness, generated predominately erosion of the leading edges of the impeller blade.

(ii) The higher the hardness of the blade material, the lower the wear rate of the blade.

(iii) The wear rate of the leading edge was not a function of pitch angle.

(iv) Sheet erosion of the blades exhibits a maximum erosion rate between 20° and 45°.

Zheng [6] and others studied the erosion-corrosion synergistic effect in an acidic slurry. The slurry was 10% by weight H_2SO_4 and 15% by weight −60 mesh (<0.251 mm) corundum sand. Their apparatus was a rotating disk with four specimen holders on its edge. They determined the rate of erosion by making electrochemical measurements during rotation. All of their studies were done under turbulent flow conditions. Erosion rate velocity exponents ranged from 1.9 to 4.0. A model was proposed and used which divided the overall erosion rate into an erosion rate via corrosion, an erosion rate via erosion, and an erosion rate due to synergism. The synergism rate was very large and varied between 32 and 99% of the total. The percent contributed by synergism diminished as the alloy became more statically corrosion resistant.

Amelyushkin and Agafonov [7] studied the erosion of cogeneration steam turbine blades caused by water droplets. If kinetic energy is high enough, even water droplets can cause erosion. They found that the toroidal and near root vortices were very intense and caused enhanced wear of the rotor blades. Also they found that they were able to eliminate erosion by making the water droplets small enough. It is expected that these effects are related to grain size. In ductile erosion, plastic deformation may occur first, before metal is removed. If erosion is due to intergranular grain fracture, then if particles are significantly smaller than the metal's grain size, erosion should be minimized. As ductile alloy grain sizes are on the order of $20\,\mu m$, particles smaller than this should have little erosive effect.

Khalid and Sapuan [8] studied wear for a centrifugal pump impeller in a slurry application. Weight and diameter losses were very nearly linear with time over 480 hours of operation. Blade height and depth loss did exhibit some nonlinearity but were modeled as linear. Typical of rotating devices in slurries, more material was lost near the periphery of the impeller than in the center because linear velocities increase with radial distance from the shaft.

López [9] and others studied the effect of corrosion erosion at relatively high velocities on 304 and 420 stainless steel. Velocities ranged from 4.5 to 8.5 m/s. Such velocities are not common in mixing equipment except in high-shear devices. They used a rotating disc device with erosion samples attached to the periphery of the disc. The aqueous liquid for the slurry was composed of 70% by weight H_2SO_4 and 3.5% NaCl. The slurry solids were 30% by weight SiO_2 particles with a mean diameter about 0.25 mm. They found that high-velocity impacts were beneficial. The combined action of erosive and corrosive mechanisms did not lead to a significant increase in mass loss if compared to corrosion tests. They suggested that pores, small cracks, and fresh pits can be covered by the prows and lips that are formed as a consequence of the wedge action of round particles which produces a smoother, uniformly corroded surface. Thus, even though the transport mechanisms which remove corrosion products are greater due to higher velocities, the surface area exposed is smaller.

Corpstein and Fasano [10] studied slurry wear through multilayer paint modeling. The paint layers are on the order of 0.0381 mm thick. The three layers of paint used had an overall thickness of approximately 0.114 mm. This is only about 7% of the blade thickness and did not change the fluid hydraulics over the blade. Erosion was studied using 8.3-inch diameter scaled down axial flow mixing impellers in a sand water slurry. These studies pointed out the strong effects that blade-shedding vortices, could have on erosive wear. Impellers, such as the four-bladed-pitched impeller that generated stronger vortices, suffered the highest degree of localized erosion. The effects of these vortices can completely wear through an impeller blade, leaving holes where the vortices contacted the surface. High-efficiency impeller blades created significantly smaller vortices and as a consequence exhibited much lower localized wear. Vortex

FIGURE 1: Backside of pitched blade impeller.

FIGURE 2: Backside of HE-3 high-efficiency impeller.

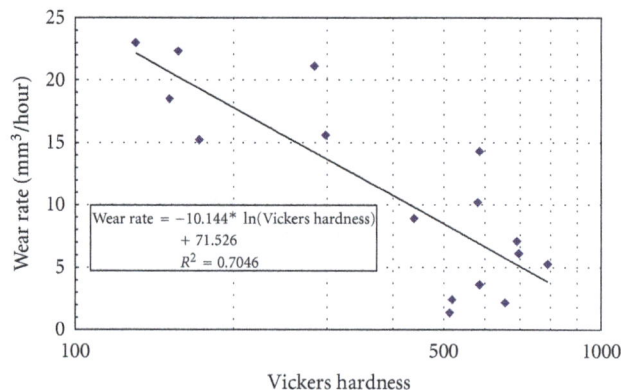

Wear rate $= -10.144 *$ ln(Vickers hardness)
$+ 71.526$
$R^2 = 0.7046$

FIGURE 3: Wear rate data of Miller and Schmidt.

erosion can be severe and occurs on the backside (low-pressure side) of the impeller blade. Comparisons of the backside wear pattern for a Chemineer HE-3 impeller and a standard generic 45° pitched four-bladed impeller are shown in Figures 1 and 2. The impellers were each 211 mm diameter and operated at 870 rpm in a 10% by weight sand slurry in water. The weight mean particle size of the sand was 360 μm. These backside erosion patterns were made after only 30 minutes of operation, and it is obvious that the erosion on the backside of the 45° four-bladed-pitched impeller was much more severe than the erosion for the Chemineer HE-3 high-efficiency impeller.

Wu [11] and others were also successful in using this technique to study five different style radial flow impellers and a low attack angle ($\sim$15°) (6-bladed) pitched impeller. As expected, the hydraulically more efficient six-bladed-pitched impeller experienced the least erosion.

Increased hardness of metals will generally provide an increased life. Miller and Schmidt [12] compared the erosion rates of 16 metals in a recycled slurry test system using 2% by weight silica sand in water. The impeller tip velocity was 15.7 m/s, and the temperature was 16°C. In addition to the erosion rate for each metal, they included the metal's hardness. The best fit for their data was logarithmic. However, probably due to the synergistic corrosion effects, the data was fairly dispersed. A plot of this data is provided in Figure 3. The effect of particle hardness depends on whether erosion is ductile or brittle. For brittle erosion, the effect of particle hardness is much more pronounced than for ductile erosion.

Changes in particle size can change the erosion mechanism. Stachowiak and Batchelor [13] reported that as the particle size was increased from 8.75 μm to 127 μm, the mode of erosion changed from ductile to brittle. The erosion study

was for silicon carbide particles impinging on glass, steel, graphite, and ceramics. The particle velocity was 152 m/s.

2. Design for Erosion Minimization

Because maximum velocities in mixing processes seldom exceed 6 m/s, erosion and corrosion erosion of materials are fatigue processes for most mixing processes. There is generally not enough particle kinetic energy to cause ductile erosion where there is some plastic flow of material. The fatigue process occurs on a micro- or localized scale, and, as with macroscale fatigue, two stages of the erosion process have been observed. There is an incubation period followed by the formation and growth of pits involving the removal of the metal or material. Refer to a materials behavior text such as that by Armstrong and Zerilli [14] for a more in-depth discussion on material behavior.

Due to the vast number of parameters that can affect erosion or erosion-corrosion processes, and the fact that this area of mixer service has not been widely studied, it is very difficult to predict a priori what the rate of erosion will be for any given liquid-solid application. However, there are certain factors within the control of the designer that can be used to optimize the life of the mixer's wetted parts.

Most mixer designers will not have control over the type of slurry, the percent solids, the hardness of the solids, the shape of the solids, the liquid, the pH, and so forth. However designers will generally have control over

(i) the mixer's wetted parts materials, coating, or lining,

(ii) the impeller style,

(iii) the impeller horsepower and speed combination.

2.1. Material Selection. The choice in selecting a material is to either go hard or soft and elastic. All else being equal, when selecting a metal alloy, a higher hardness will lead to a longer life. Thus, when selecting a metal alloy material, select a hard material which will also provide good corrosion resistance.

There are a number of hard surface ceramic coatings such as tungsten carbide or silicon carbide, which could be applied to the high-wear areas such as impeller blades. Ceramics are the most wear resistant but are low in toughness and impact strength. Ceramic coatings as well must be corrosion resistant to the liquid medium. Ceramics also do not have the ability to absorb much strain. These high strains on flexing blades may allow cracking of the ceramic coating. Ceramic coating applicators should be able to provide the maximum allowable strain for the ceramic coating under consideration. Coatings of the more common ceramic materials tend to be more costly than high-hardness metals, or elastomeric coverings [15].

Glass-lined equipment has a glass hardness of 5 to 6 on the Moh scale. For the great majority of solids, this hardness would be very acceptable. However, there are numerous materials and minerals including, Al_2O_3, SiO_2, WC, SiC, and ZiO_2 that have higher harnesses and would tend to wear away the glass lining. Glass linings have very many of the same limitations as ceramic coatings. They tend to be brittle and cannot tolerate much strain.

Elastomeric coverings on the order of $3/8''$ thick for industrial scale impellers have a long history of providing longer life in slurry applications. Instead of having to absorb most of the particle's impact energy, an elastomer releases most of the energy back to the particle after impact. Elastomeric lining manufacturers and applicators will generally recommend an elastomeric hardness of 40–60 Durometer A for optimum life. As with metals or hard surface coatings, the lining must also be compatible with the fluid medium. An elastomer's hardness is directly related to its corrosion resistance. However, as an elastomer's hardness increases above a 40 A Durometer, its erosion resistance decreases. A Durometer selection of 40–60 A is somewhat of a compromise between erosion and corrosion resistance. Elastomers should not be used when large particles are present. The term "large particles" is relative to the impinging velocity and mass of the particle, as well as the thickness of the elastomeric covering. If the impinging particle can bottom out against the metallic substrate, elastomeric coverings should not be used. Even if most of the slurry might be suitable, a small percentage of tramp particles can do significant damage to the elastomeric covering. Since impact energy is a function of the impingement angle, the leading edges of impeller blades are almost always double wrapped. The most popular elastomeric coverings are natural rubber, neoprene, butyl, chlorobutyl, and hypalon. Improperly applied linings on high-efficiency impellers that significantly change the profile of the blade can cause increased erosion problems. Typically linings are double layered on the leading, trailing, and outside edges of impeller blades. These linings must be adequately feathered such that the transition from the double layer to the single layer is smooth to avoid generation of additional vortices.

Both thermoplastic and thermoset polymers do not have the ability to restore back to the particle most of the kinetic energy and are generally not as good in mixing slurry service. Hercules 1900 UHMWPE, touted as being a very abrasion-resistant polymer, was tested by the authors against polymeric protective coatings elastomers 2001-B (natural rubber Durometer A 30–40) and 1054-B (chlorobutyl rubber Durometer A 35–45). The testing procedure was identical to that specified in the Hercules 1900 UHMWPE bulletin. A

TABLE 1

Material	Rate of weight loss, g/hr
Hercules 1900 UHMWPE	0.0975
Polymeric protective coatings 2001-B (natural rubber)	0.0104
Polymeric protective coatings 1054-B (chlorobutyl)	0.0124

50% sand-water slurry was used with a sand weight mean particle size of 53 μm. The specimen tip speed was 2.22 m/s. Weight loss was determined at various time intervals over an 8-hour period. The weight loss versus time was found to be linear with R^2 values for all three falling between 0.93 and 0.94. The rates of erosion were shown in Table 1.

As can be observed, the rate of weight loss for the thermoplastic polymer is 7 to 10 times greater than that for the elastomers tested.

Dickey and Fasano have provided a general reference on materials selection considerations [16].

2.2. Impeller Selection. There are many different impeller styles available to the designer. Selecting the correct impeller can often make a difference in impeller life of two or three times. Erosion of impeller blades can depend heavily on the flow regime, with flow regime being determined by the impeller Reynolds number

$$N_{Re} = \frac{\rho N D^2}{\mu}, \qquad (2)$$

where ρ = density, N = impeller rotational speed, D = impeller diameter, μ = viscosity, and N_{Re} = Reynolds number, dimensionless.

Impellers in turbulent flow create shedding vortices that attach themselves to the back of impeller blades. There are a number of techniques that can be used to visualize these vortices. In Figures 4 and 5, telltales attached to the blade are used to visualize these vortices [17]. They are shown for both a relatively inefficient 45° pitched four-blade impeller and the Chemineer HE-3 high-efficiency impeller.

In these figures, the impellers are rotating clockwise when viewed from above. Thus, the blades are moving into the page, and the view is of the backside of the blades. These vortices cause very localized wear emanating from the backside or low-pressure side of impeller blades. These impeller vortices begin to diminish at Reynolds numbers below 10,000, become very weak below a Reynolds number of 500, and have completely disappeared below a Reynolds number of 10. As most slurry particles are heavier than the fluid, the centrifugal effect caused by the vortex will cause particles caught in the vortex to migrate to the OD of the vortex. Thus, the concentration of solids at the periphery of the vortex is much higher, and the rate of solid particle to surface impacts is much greater, increasing locally the rate of erosion. Of course, particles must be small enough to be captured in

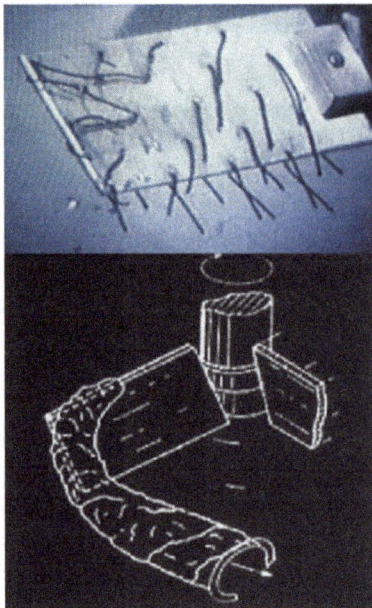

FIGURE 4: Pitched blade impeller vortex.

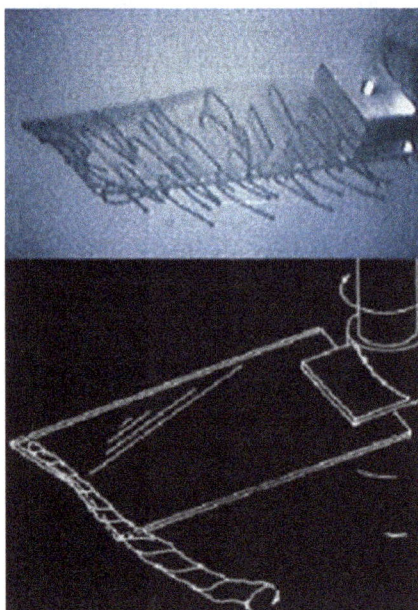

FIGURE 5: HE-3 impeller vortex.

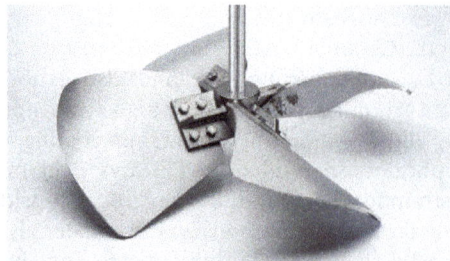

FIGURE 6: Chemineer Maxflo W impeller.

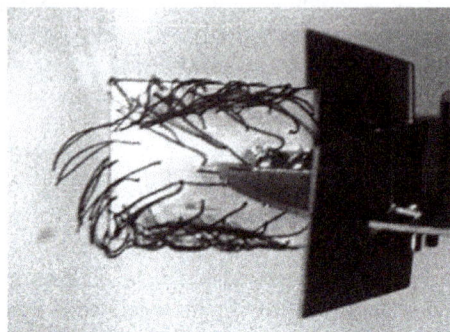

FIGURE 7: Telltales on D-6 impeller blade.

these vortices before this effect would be observed. On an industrial scale, however, the greatest majority of slurry applications would have particles sufficiently small to be captured by these vortices.

There are a number of relatively efficient wide blade impellers used in solids suspension service. Fasano and Reeder [18] compared the erosion rate between a Chemineer Maxflo W impeller, Figure 6, and a standard 45° four-bladed-pitched impeller (refer to Figure 4).

For the same level of solid suspension, these impellers utilize the same impeller diameter at the same speed. Therefore, velocities at the impeller are the same. The rate of erosion however for the pitched-blade impeller was, on a percentage basis, 59% greater than the erosion rate of the Maxflo W impeller.

Radial flow impellers are not very efficient in suspending solids. However, radial flow impellers are efficient in dispersing gasses. In applications where solids are present and a gas must be dispersed, they are often used. As with axial flow impellers, impeller efficiency changes with design. The Rushton or D-6 impeller, introduced in the late 1940s, creates a pair of vortices behind each blade in turbulent flow. These vortices, as we have observed with the pitched-blade impeller, can cause severe erosion from the backside of the blade. Figure 7 photo of the intertwined telltales on the backside of the blade demonstrates the size and nature of these vortices.

There exist today highly efficient radial flow impellers. The Chemineer CD-6 impeller was introduced into the marketplace in 1988, and the even more efficient Chemineer BT-6 was introduced in 1998. Both of these impellers exhibit very little tendency for vortex formation on the backside of the blade. This is demonstrated in telltale photos, in Figures 8 and 9. In each of these photos, the blades are rotating into the plane of the paper.

Under turbulent flow conditions, it is usually very beneficial, both from an erosion standpoint as well as a process efficiency standpoint, to select a high-efficiency style impeller. In transitional flow, there are still benefits to be achieved by the use of a high-efficiency impeller, but not as profound as in turbulent flow. In laminar flow, there is

TABLE 2: Possible process design selections for example.

Impeller diameter, m	Impeller dia./tank dia.	Impeller speed, rpm	Impeller power, kW	Impeller torque, kNm	Impeller tip speed, m/min	Relative wear life*	Approx. relative cap. cost
0.889	0.243	159	3.49	0.210	444	1.00	1.00
1.016	0.278	123	2.94	0.228	392	1.54	1.06
1.143	0.313	102	3.15	0.294	366	1.96	1.25
1.220	0.347	92	3.59	0.372	367	1.95	1.45
1.397	0.382	84	4.05	0.460	369	1.92	1.67
1.524	0.417	75	4.61	0.587	359	2.10	1.95
1.651	0.451	70	5.23	0.713	363	2.02	2.22
1.778	0.486	67	6.32	0.972	374	1.82	2.71
1.905	0.521	65	9.16	1.345	389	1.59	3.35

*Relative wear based on assumed velocity exponent of 3.5.

FIGURE 8: Telltales on CD-6 impeller blade.

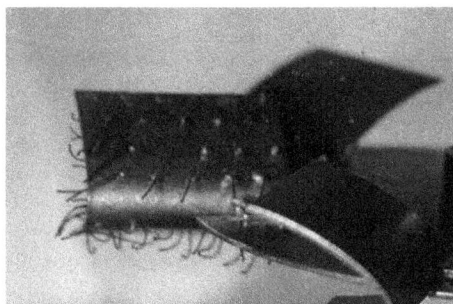

FIGURE 9: Telltales on BT-6 impeller blade.

no advantage in using a high-efficiency impeller either for erosion or solids suspension efficiency.

2.3. *Horsepower and Speed Selection.* As discussed earlier, erosion is very dependent on velocity, and typically in erosion-corrosion environments, the velocity exponent for the volumetric rate or weight rate of erosion is typically observed to be 2.5 to 4.0. In designing an agitator for the suspension of solids, the designer has a choice of selecting a number of power and speed combinations. Because solid suspension impeller efficiencies change with impeller style, impeller to tank diameter, and off-bottom clearance to

tank diameter [19], a number of possible horsepower and speed choices can meet process objectives. The selection of a specific agitator design in the end should come down to economics. There are capital costs and operating costs. Capital costs are largely associated with the general size of the machine. The torque can generally best characterize the capital cost. Operating costs include the energy costs to operate the machine plus any maintenance costs. Maintenance costs include the costs of oil changes, new bearings, new gears, and new seal components, and in the case of erosion applications, the replacement of in-tank wear components, typically impeller blades. A close examination of the various horsepower and speed options should be examined closely in order to make an economic selection.

2.3.1. *Example.* For the sake of demonstration, let us assume that we need to design an agitator for suspending a 10% solution of sand in water. The sand will be assumed to have a weight mean particle size of $360\,\mu\text{m}$. The tank is 3.66 m diameter with a 2 : 1 elliptical dished bottom, and the water-sand slurry will have a liquid volume such that the depth of liquid in the tank is 3.66 m. The sand will be assumed to have a specific gravity of 2.4 and the water a specific gravity of 1.0. The viscosity of water will be assumed to be 1 mPa-s. The process solution requires that the solids be suspended to the "just suspended" condition such that no particles settle on the bottom of the tank for more than 2 seconds. All designs are to utilize a single HE-3 impeller one-third the tank diameter off-bottom. The horsepower, speed, and impeller diameter combinations that satisfy the process objective can be determined using solids suspension design procedures such as the article presented by Corpstein and others [19]. The following designs shown in Table 2 below all satisfy the off-bottom solids suspension process requirement.

Even though using small impellers operating at high speeds reduces capital costs and power, the high tip speeds of these designs lead to short wear lives. Using an intermediate size impeller that minimizes tip speed at the cost of higher capital and power costs maximizes the wear life. As can be observed, the torque and consequently the capital cost

TABLE 3: Comparison of equal suspension to equal tip speed.

Condition	Impeller dia., cm	Impeller speed, rpm	Impeller tip speed, m/min
Full scale design, 144 in. dia. tank	139.7	84	369
Scale-down design based on equal suspension	17.46	456	250
Scale-down design based on equal tip speed	17.46	672	369

increase dramatically with impeller to tank diameter ratios greater than 0.45 due to changes in the flow pattern generated by the impeller.

2.4. Scale-Down Studies. Since there is often an erosion-corrosion synergistic effect that cannot be predicted a priori with today's current data, any material selection should be studied on a smaller scale, before making a final choice. Since it is important to model the same hydraulic behavior over the blade, the authors recommend that the ratio of mean particle diameter to impeller diameter not exceed 0.008. For example, a mean particle diameter of 1 mm would suggest a small-scale impeller no less than 125 mm (4.9 in) diameter. It is also important to ensure that fluid regimes have not changed. If the impeller operation is turbulent on the large scale, it should also be turbulent on the small scale.

An optimum agitator horsepower and speed selection can be made as described above for the full scale. However, in order to determine the rate of erosion, scale-down studies should be made. A geometric scale-down for an equal level of solid suspension will result in a tip speed that will always be lower on the smaller scale, except when scaling down geometrically for very slow settling solids (<0.1 m/min). Wear rate, as previously demonstrated, is a strong function of velocity. Therefore, all scale-down tests should be made at equal tip speed. As an example, if we examine the 1.397 m impeller solution for the above described problem and scale this down to a 0.4572 m diameter tank, we would have the comparison provided in Table 3.

3. Conclusions and Summary

The rate of erosion is dependent on the following major static environmental factors: chemical environment, hardness of the solid particles, density of the solid particles, percent solids, the shape of the solids, the size of the solids including whether or not tramp solids are present, and type of impeller. The dynamic factors affecting erosion rate are fluid regime, impact velocity, impact frequency, and angle of impact. As there are no good means currently of predicting erosion rate, small-scale studies should be conducted emulating as much of the total environment as possible. These small-scale studies should be conducted using equal tip speed to mimic the full-scale rate of erosion.

Erosion in most mixing processes is a fatigue process normally accelerated by a liquid corrosive environment. The fatigue process occurs on a micro- or localized scale, and, as with macroscale fatigue, two stages of the erosion process have been observed. There is an incubation period followed by the formation and growth of pits involving the removal of the metal or material. One of two routes is generally utilized in dealing with an erosion application. The high-velocity areas such as the blades are either made from hard materials or coated with hard ceramic materials such as tungsten carbide or silicon carbide. Alternatively, the blades are covered with some type of elastomeric covering.

Impeller selection is important especially in turbulent flow conditions. High-efficiency impellers will generally erode at a slower rate because the backside of the blades has minimized shedding vortices. In laminar flow, from an erosion standpoint, most impellers behave similarly due to a lack of vortices. Thus, the selection of the impeller should be based primarily on what is required to accomplish the desired process result.

A number of horsepower and speed selections that satisfy the process requirement should be examined to conduct an economic analysis. The lowest possible speed may not be the most economical. It is best to first design the most cost optimum agitator for the full scale. Then in order to estimate the corrosion rate, scale down on the basis of equal tip speed.

References

[1] R. Chattopadhyay, *Surface Wear, Analysis, Treatment and Prevention*, ASM International, Metals Park, Ohio, USA, 2001.

[2] S. G. Sapate and A. V. RamaRao, "Effect of erodent particle hardness on velocity exponent in erosion of steels and cast irons," *Materials and Manufacturing Processes*, vol. 18, no. 5, pp. 783–802, 2003.

[3] M. M. Stack, F. H. Stott, and G. C. Wood, "The significance of velocity exponents in identifying erosion-corrosion mechanisms," *Journal de Physique IV, Colloque C9, Supplement au Journal de Physique III*, vol. 3, 1993.

[4] *A New Slurry Pump Standard, Pumps and Systems*, The Hydraulic Institute, 2006.

[5] I. Fort, J. Medek, and F. Ambros, "Erosion wear of axial flow impellers in a solid-liquid suspension," *Acta Polytechnica*, vol. 41, no. 1, pp. 23–28, 2001.

[6] Y. Zheng, Z. Yao, X. Wei, and W. Ke, "The synergistic effect between erosion and corrosion in acidic slurry medium," *Wear*, vol. 186-187, no. 2, pp. 555–561, 1995.

[7] V. N. Amelyushkin and B. N. Agafonov, "Special features of erosion wear of rotor blades of cogeneration steam turbines," *Power Technology and Engineering*, vol. 36, no. 6, pp. 359–362, 2002.

[8] Y. A. Khalid and S. M. Sapuan, "Wear analysis of centrifugal slurry pump impellers," *Industrial Lubrication and Tribology*, vol. 59, no. 1, pp. 18–28, 2007.

[9] D. López, J. P. Congote, J. R. Cano, A. Toro, and A. P. Tschiptschin, "Effect of particle velocity and impact angle on

the corrosion-erosion of AISI 304 and AISI 420 stainless steels," *Wear*, vol. 259, no. 1-6, pp. 118–124, 2005.

[10] R. C. Corpstein and J. B. Fasano, "Erosion of rubber covered impeller blades in an abrasive service," *The Indian Chemical Engineer*, vol. 36, no. 1, 1990.

[11] J. Wu, B. Ngyuen, L. Graham, Y. Zhu, T. Kilpatrick, and J. Davis, "Minimizing impeller slurry wear through multilayer paint modelling," *Canadian Journal of Chemical Engineering*, vol. 83, no. 5, pp. 835–842, 2005.

[12] J. E. Miller and F. Schmidt, *Slurry Erosion: Uses, Applications and Test Methods*, ASTM, 1987.

[13] G. W. Stachowiak and A. W. Batchelor, *Engineering Tribology*, Butterworth-Heinemann, 3rd edition, 2005.

[14] R. W. Armstrong and F. J. Zerilli, "Dislocation mechanics based viscoplasticity description of FCC, BCC and HCP metal deformation and fracturing behaviors," in *Proceedings of ASME International Mechanical Congress and Exposition*, pp. 417–428, November 1995.

[15] K. C. Wilson, G. R. Addie, A. Sellgren, and R. Clift, *Slurry Transport Using Centrifugal Pumps*, Springer, 2nd edition, 2005.

[16] D. S. Dickey and J. B. Fasano, *Handbook of Industrial Mixing*, chapter 21, section 9, John Wiley and Sons, New Jersey, NJ, USA, 2004, Edited By Paul, Atiemo-Obeng and Kresta.

[17] J. B. Fasano, "Flow visualization techniques on rotating impellers," in *Proceedings of the Engineering Foundation Mixing Conference XII*, Pitosi, MO, USA, 1989.

[18] J. B. Fasano and M. F. Reeder, "An improved maxflo impeller," in *Proceedings of the North American Mixing Forum, Mixing Conference XVII*, Banff, Canada, August 1999, Paper 2.4.

[19] K. J. Myers, R. R. Corpstein, A. Bakker, and J. B. Fasano, "Solid suspension agitator design with pitched blade and high efficiency impellers," in *Proceedings of the AIChE Annual Meeting*, St. Louis, MO, USA, November 1993.

8

Study of Chromium Removal by the Electrodialysis of Tannery and Metal-Finishing Effluents

Ruan C. A. Moura,[1] **Daniel A. Bertuol,**[2] **Carlos A. Ferreira,**[3] **and Franco D. R. Amado**[1]

[1] *PROCIMM, State University of Santa Cruz, Road Ilhéus-Itabuna km 16, 45662-000 Ilhéus, BA, Brazil*
[2] *DEQ, Federal University of Santa Maria, 97105-900 Santa Maria, RS, Brazil*
[3] *PPGEM, Federal University of Rio Grande do Sul, 91501-970 Porto Alegre, RS, Brazil*

Correspondence should be addressed to Ruan C. A. Moura, ruan_moura@yahoo.com.br

Academic Editor: Yoshinobu Tanaka

The metal-finishing and tannery industries have been under strong pressure to replace their current wastewater treatment based on a physicochemical process. The electrodialysis process is becoming an interesting alternative for wastewater treatment. Electrodialysis is a membrane separation technique, in which ions are transported from one solution to another through ion-exchange membranes, using an electric field as the driving force. Blends of polystyrene and polyaniline were obtained in order to produce membranes for electrodialysis. The produced membranes were applied in the recovery of baths from the metal-finishing and tannery industries. The parameter for electrodialysis evaluation was the percentage of chromium extraction. The results obtained using these membranes were compared to those obtained with the commercial membrane Nafion 450.

1. Introduction

Over the past few decades, there has been increased concern for the preservation of water resources. Industrial activities have led to widespread heavy metal contamination of soils and natural waters. Among the various sources of water contamination, the electroplating industry stands out as one of the most important, because it generates a considerable volume of effluents containing high concentrations of metal ions and, often, high concentrations of organic matter [1]. Another aggravating factor is that the traditional process for the treatment of these effluents, not very efficient and in some cases totally inefficient, produces dangerous solid waste (electroplating sludge), which should, therefore, be disposed of in appropriate landfills.

The most commonly used technology for the treatment of effluents is the physicochemical one, followed by units of biological treatment, usually consisting of activated sludge or aerated lagoon systems [2]. These conventional treatments are generally not able to reduce all the polluting parameters. Chemical Oxygen Demand (COD), chlorides, sulfates, and chromium often do not reach the required limits [3].

In this context, the leather and metal-finishing industries urge researchers to investigate new technologies for the recovery or recycling of chemical wastewater [4]. Because of their toxicity, these effluents cannot be rejected without pretreatment in the environment [5, 6].

Membrane technology has become increasingly attractive for wastewater treatment and recycling [7]. The main advantage of a membrane process is that concentration and separation are achieved without changing the physical state or using chemical products. Because of their modularity, membrane techniques in general and electromembrane techniques in particular are very well adapted to pollution treatment at its source; within this process, the electrodialysis process is becoming a good alternative when compared to the traditional methods of wastewater treatment [8, 9].

Electrodialysis (ED) is a membrane separation process based on the selective migration of aqueous ions through an ion-exchange membrane as a result of an electrical driving force. The transport direction and rate for each ion depend on a number of conditions, such as, its charge, mobility, relative concentrations, and applied voltage. Ion separation is closely associated with the characteristics of the ion-exchange membrane, especially its permselectivity. ED was first used for the desalination of saline solutions, but other applications, such as, the treatment of industrial effluents, have gained importance [10, 11].

The purpose of this study is the investigation of the transport of some ions through synthesized membranes and a commercial one by electrodialysis. For the tannery effluents, photoelectrochemical oxidation (PEO) processes were previously used to degrade organic matter [12–14].

2. Experimental

For this study, two different real effluents were collected at two industries in the Southern Brazil. One effluent was collected at the discharge point of the conventional effluent treatment plant (CET) of a tannery plant. This plant carries out all the industrial processes from raw hides to finished leather. This effluent was then photoelectrooxidized for 24 hours and then treated by ED. A scheme of the PEO system used in this work is shown in Figure 1. It is made up of two serial, one liter PVC electrolytic reactors.

A 400 W high-pressure mercury-vapor lamp was used as a light source. Before each experiment, the UV light was turned on for 15 min to allow the UV energy to become stable. Two pairs of electrodes were used. The cathode and anode were DSA ($70TiO_2/30RuO_2$). The electrode area inside the cell was 118 cm^2. During the experiments, the reactor was operated in a batch recirculation mode. The effluent was recirculated at a flow rate of 4 $L \cdot h^{-1}$, and 50 L effluent was treated by PEO for each experiment. The photoelectrochemical oxidation experiments were carried out using a DC power supply with an applied current density of 20 $mA\,cm^{-2}$.

In the metal-finishing plant, the effluent was also collected at the discharge point of the conventional effluent treatment plant (CET). The chromium concentrations were 0.5 ppm for the tannery effluent and 60 ppm for the metal-finishing effluent.

2.1. Membranes.

The membranes were prepared by mixing conventional polymer (HIPS) with conducting polymers polyaniline (PAni). Two different mixing methods were tested to evaluate the effect of the production method. Dopants for polyaniline (PAni), camphorsulfonic acid (CSA), and p-toluenesulfonic acid (p-TSA) were also used.

HIPS and PAni were dissolved in 20 mL of tetrachloroethylene. After dissolution, PAni was dispersed in an HIPS polymeric matrix for 30 minutes. This dispersion was performed at 1,000 rpm in a mixer (Fisaton). The membranes were molded on glass plates using a laminator to keep thickness constant, and the solvent evaporated slowly for 24 hours under room temperature. The membranes were referred to as MCS and MTS.

2.2. Membrane Characterization

2.2.1. Infrared Spectroscopy.

The samples were prepared with potassium bromide (KBr) powder. All of the samples were

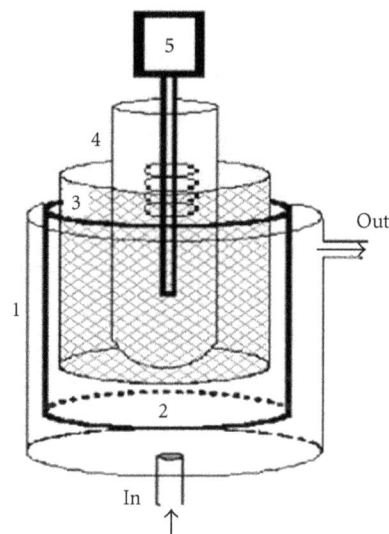

FIGURE 1: Schematic representation of PEO system: (1) PVC reservoir; (2) titanium oxide cathode; (3) titanium oxide recovered with TiO_2/RuO_2 anode; (4) quartz tube; (5) mercury steam lamp.

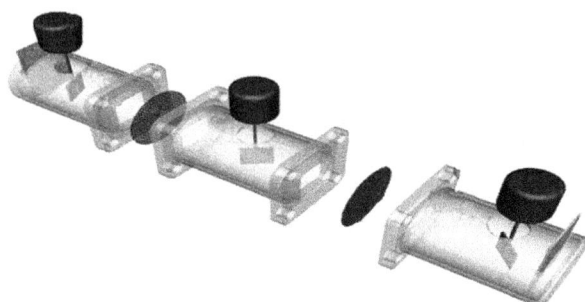

FIGURE 2: Three-compartment cell used for electrodialysis.

analyzed using an FTIR Perkin Elmer spectrometer model Spectrum 1000. The spectra were recorded in the spectral range of 400–4,000 cm^{-1}.

2.2.2. Swelling.

Excess water was removed with a paper filter, and the membranes were weighed and kept in the oven at 80°C for 10 hours and then weighed again. The uptake of water was determined by the mass difference between the wet and the dried membranes (after heating at 80°C). Water absorption is expressed in percentage.

2.2.3. Morphology.

Scanning electron micrographs of the membranes' surface were obtained using a microscope (Philips XL20) after the samples were sputter-coated with gold.

2.2.4. Electrodialysis.

The membranes were synthesized (MTS and MCS) [15–17] and the commercial membrane (Nafion 450) was used as a cation selective membrane. Selemion AMV was used as an anion selective membrane. All of the membranes were maintained in contact with the solutions for 48 h in order to achieve equilibrium.

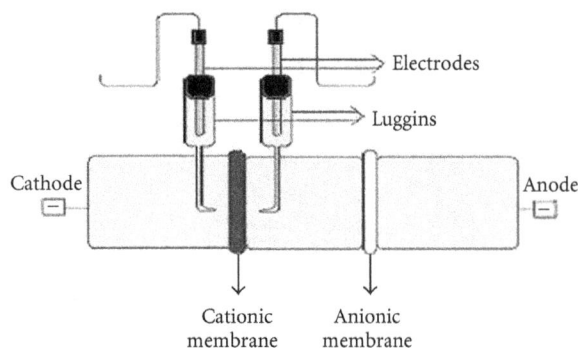

FIGURE 3: Three-compartment cell used to determine polarizations curves.

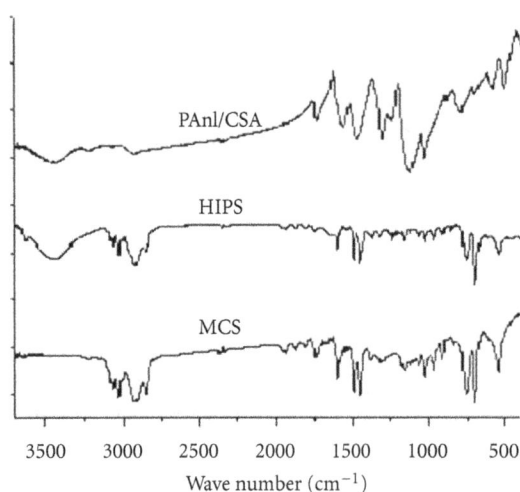

FIGURE 4: FTIR spectrum of PAni/CSA, HIPS sample, and MCS membrane.

The membranes were also equilibrated in deionized water at room temperature for 24 hours, with the aim of testing the hydrophilic behavior of $-SO_3^-$ from the doping acid that was used in polyaniline.

The electrodialysis experiments were performed using a three-compartment cell with a capacity of 200 mL each, as shown in Figure 2. A platinized titanium electrode was used as the anode and cathode. A Selemion AMT anionic membrane was utilized and the cationic membranes were the synthesized membranes (MCS and MTS) and Nafion 450. The area of the membranes was 16 cm^2 and all the experiments were galvanostatic, with a current density of $10 \, \text{mA} \cdot \text{cm}^{-2}$. All of the electrodialysis experiments were carried out during 180 minutes.

The evaluation of the electrodialysis process was expressed in percent extraction, that is, how much of the ion in question was transferred from the diluted to the concentrated compartment:

$$\text{Pe} \, \% = \frac{M_i - M_f}{M_i} \times 100, \qquad (1)$$

where Pe is the percent extraction (%), M_i is the ion concentration considered in the diluted compartment in

TABLE 1: Solutions used in electrodialysis tests for recovery of metals tannery and metal-finishing effluents.

Experiment	Cathodic compartment	Intermediary compartment	Anodic compartment
1	0.1 M Na$_2$SO$_4$	Tannery effluent photoelectro-oxidized with 20 A and 5 h	0.1 M Na$_2$SO$_4$
2	0.1 M Na$_2$SO$_4$	Metal-finishing effluent	0.1 M Na$_2$SO$_4$

TABLE 2: Thickness and swelling of the membranes.

Membrane	Thickness (mm)	Swelling (%)
MCS	0.20–0.25	13.6
MTS	0.20–0.25	12.9
Nafion 450	0.40	28.6

time zero, and M_f is the ion concentration considered in the diluted compartment at the final time.

Table 1 shows the solution's distributions that were used in the experiments.

2.2.5. Polarization Curves. Current-voltage curves (CVCS) were obtained in galvanostatic mode using a classical three compartment cell [18, 19]. This cell was composed of three symmetrical 200 cm^3 half cells. These compartments were separated by gaskets, which clamp the membrane. In the geometrical center of the gaskets there was a cylindrical hole. The working area of the AMV membrane was 16 cm^2. Two Ag/AgCl electrodes, immersed in Luggin's capillaries, allowed the measurement of the potential difference between the two sides of the membrane. Mechanical stirrers were placed in each compartment. The same solutions were used on both sides of the membrane. The electrical current was supplied with two platinum electrodes (Figure 3). Electric current was applied using a DC power source for 120 seconds. The curves were obtained by potential measurements through the membrane corresponding to the applied current.

3. Results and Discussion

3.1. Thickness and Swelling. Table 2 shows the thickness and swelling of the membranes produced for this study and of the Nafion 450 membrane.

The swelling capacity of the membrane affects not only its dimensional stability but also its selectivity, electric resistance, and hydraulic permeability. Dimensional stability increases as the polymer affinity for water decreases. Conversely, as the polymer affinity for water increases, ionic transport resistance decreases [20].

Membranes that use CSA as a doping acid show slightly greater swelling than other membranes. However, the Nafion 450 membrane showed much greater swelling than the synthesized membranes. This difference may be associated with the fact that Nafion is a supported membrane and is thicker than the membranes under study.

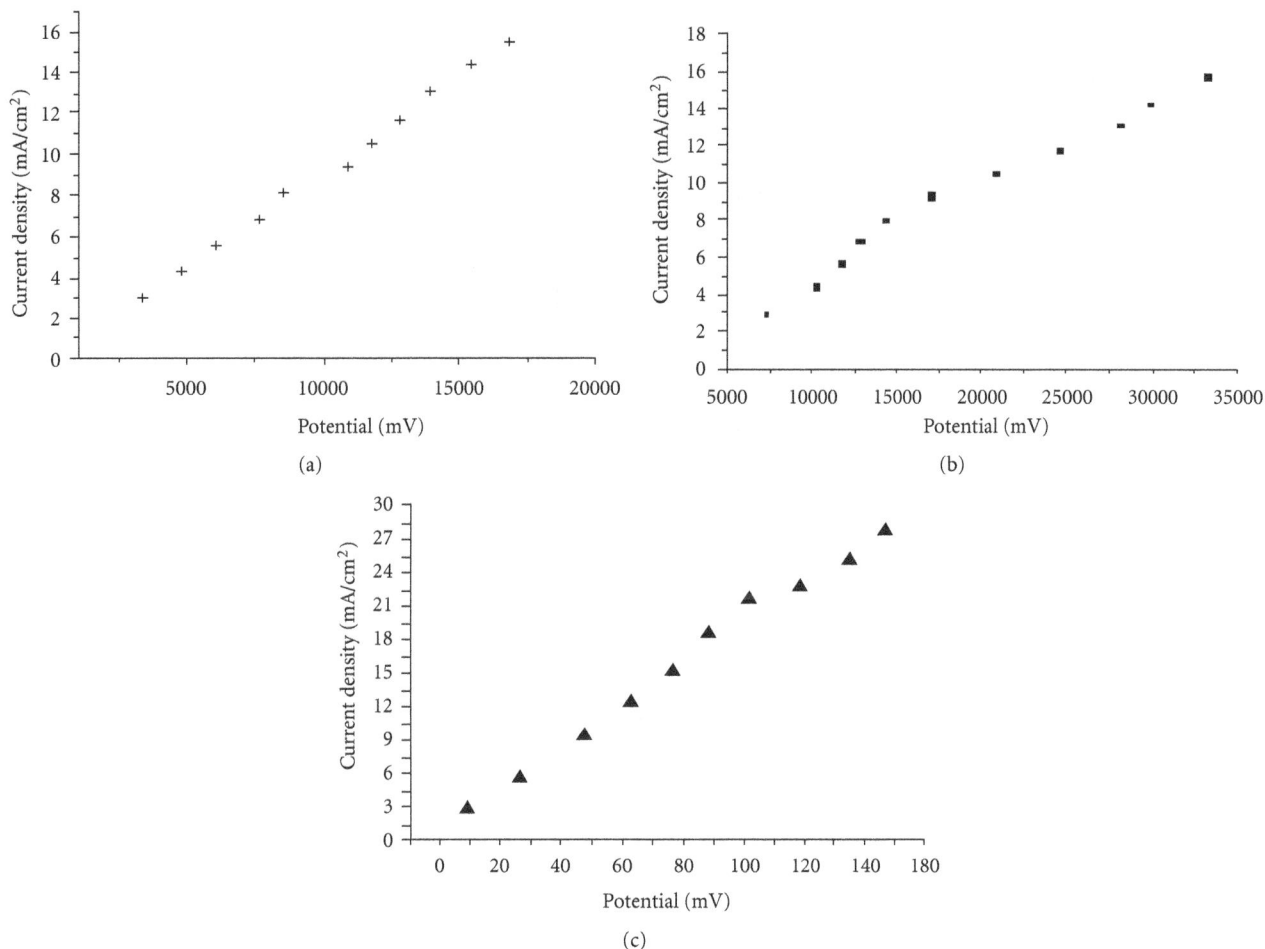

FIGURE 5: Polarization curves of the synthesized membranes and commercial membrane using the metal-finishing effluents. (a) MCS, (b) MTS, and (c) Nafion 450.

MPC for HIPS membrane with polyaniline doped with camphorsulfonic acid (CSA), MPT for HIPS membrane with polyaniline doped with p-toluenesulfonic acid (TSA). This difference between the transport numbers may be related to the structure of polyaniline dopant acid because its dopant (CSA) is more hydrophilic than the other one (p-TSA).

3.2. Infrared Spectroscopy.
To ensure incorporation into the polymeric matrix, samples of PAni/CSA, HIPS, and MCS membrane were analyzed. Figure 4 shows the FTIR spectra of these samples.

In Figure 4, HIPS spectrum and different peaks were observed. The peak at $2,948 \, cm^{-1}$ corresponds to an angular deformation of CH_3. At $1,731 \, cm^{-1}$, there is a peak attributed to the stretching of C=O groups. The peaks at $1,645 \, cm^{-1}$ and $1,554 \, cm^{-1}$ correspond to N_2H stretching. The peaks at $1,075 \, cm^{-1}$ and $1,140 \, cm^{-1}$ are associated to the stretching of C–O–C groups [17].

The MCS membrane spectrum displays peaks of PAni and HIPS spectra, thus showing the incorporation of PAni into the plastic matrix. Some of the peaks are overlapped, as seen in the stretching of N–H groups, in approximately $3,430 \, cm^{-1}$.

3.3. Polarization Curves.
Figure 5 presents the polarization curves of the synthesized membranes and commercial membrane Nafion 450 using the metal-finishing effluents.

According to the classical theory [21, 22] of concentration polarization for ion-exchange membranes, the current-voltage response shows three regions. The shape of current-voltage curves can be distinguished. In the first region, a linear relationship is obtained between the current and voltage drop that is referred to as the ohmic region. In the second region, the current varies very slightly with voltage, denoting an almost unrelated current applied voltage, corresponding to the so-called limiting current. In the region III is an over-limiting current region, and then current intensity increases again with the applied voltage.

The MCS and MTS membranes presented a higher electric resistance than the Nafion 450 membrane and the limit current density was around $11 \, mA \cdot cm^{-2}$. For the Nafion 450 membrane, the limit current density was around $20 \, mA \cdot cm^{-2}$, thus showing that electric resistance is lower.

Figure 6 presents the polarization curves of the synthesized membranes and commercial membrane Nafion 450, using the tannery effluents.

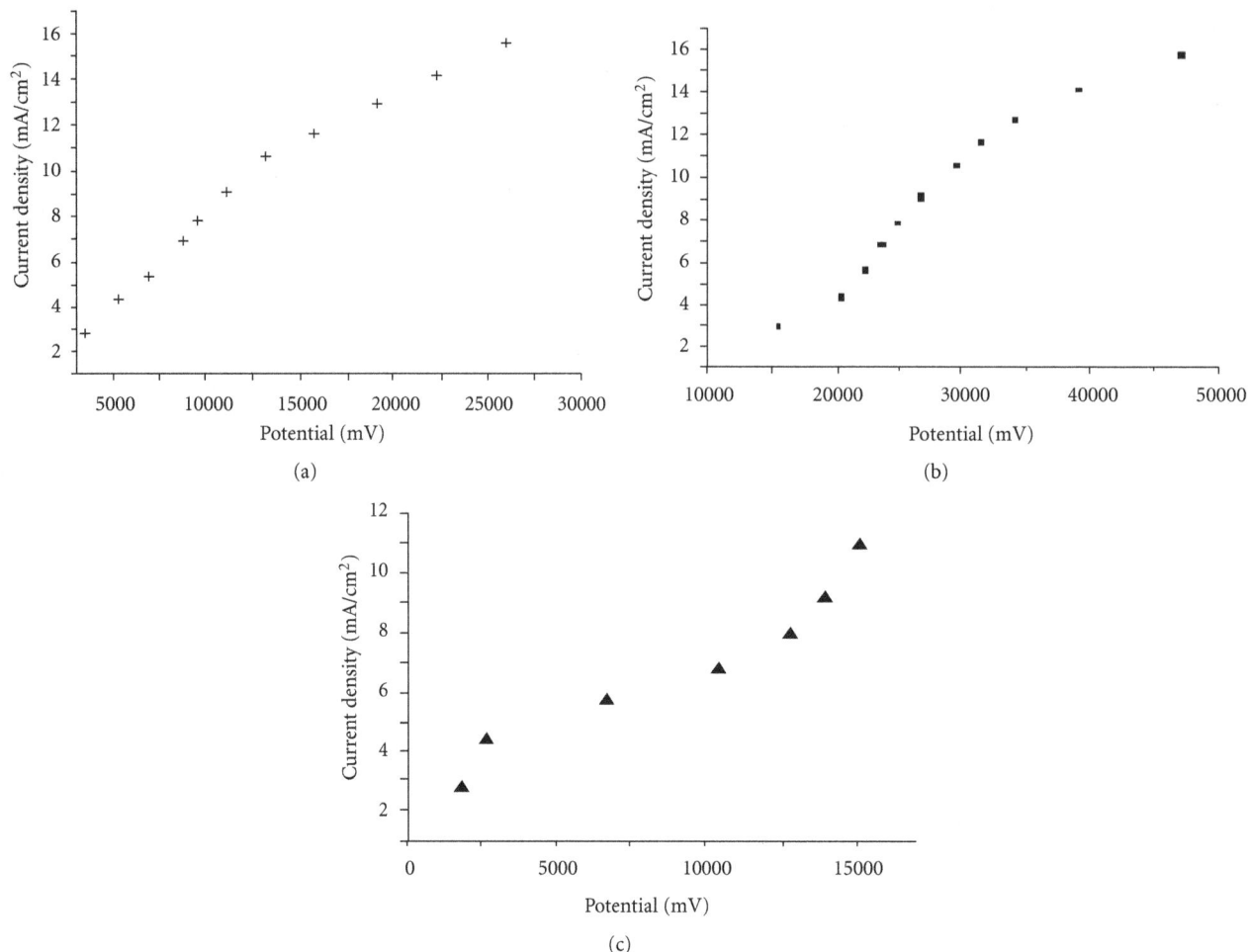

FIGURE 6: Polarization curves of the synthesized membranes and commercial membrane using the tannery effluents (a) MCS, (b) MTS, and (c) Nafion 450.

In Figure 6, it is possible to observe the current-potential curves of the membranes used for the treatment of the tannery effluents after photoelectrochemical oxidation. It is verified that the curves present the same behavior as the curves obtained with the metal-finishing effluent.

The membranes had higher resistance due to the residual organic matter present in the effluent, which might have caused the membranes fouling, hindering the transport, and consequently increasing electric resistance. This phenomenon was also observed for the commercial membrane Nafion 450.

3.4. Electrodialysis. Table 3 shows the chromium transport from metal-finishing effluents through the synthesized membranes and the commercial membrane. It is possible to verify that the Nafion 450 membrane presented better results when compared to the synthesized membranes (MTS and MCS).

The analysis of chromium in the tannery effluent is shown in Table 4. The MCS membrane had better chromium transport than the MTS membrane. The commercial Nafion membrane showed a better result, once again. Transport results confirmed the effect of the acid structure used

TABLE 3: Chromium percent extraction through membranes using the metal-finishing effluents.

Membrane	Pe Cr^{3+} (%)
MCS	18.7
MTS	19.3
Nafion 450	37.9

as polyaniline dopant. TSA (toluenesulfonic acid) is an aromatic acid and CSA (camphorsulfonic acid) is a cyclic acid; this difference may affect the interactions between the (HSO_3^-) groups from the dopant acid and nitrogen from polyaniline, which may, in turn, affect ionic transport through the membrane.

3.5. Morphology. Regarding the morphology of membranes (Figure 7), it is possible to observe the MTS (A) and MCS (B) surfaces. The addition of polyaniline clearly promoted changes in the morphology of the HIPS polymeric matrix. The main differences between the MCS and MTS membranes were observed in the polyaniline structure. The morphology

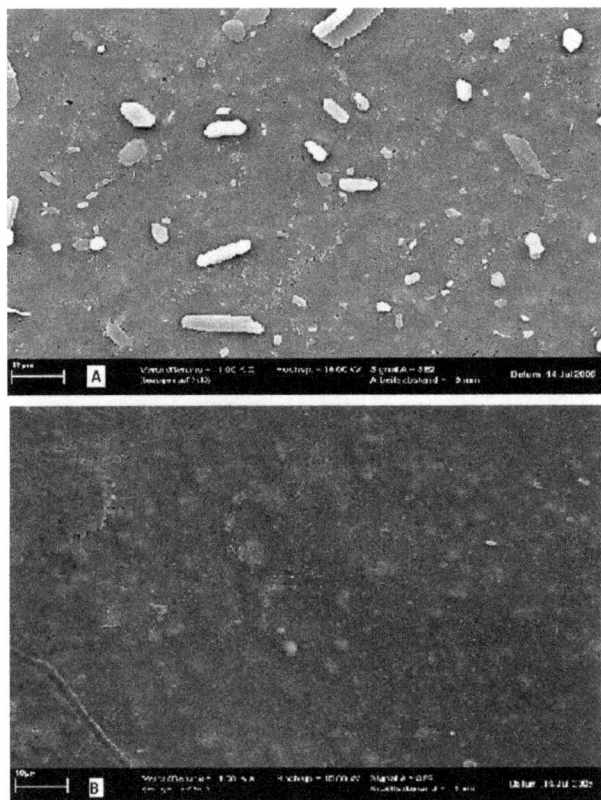

FIGURE 7: Microscopy (A) MTS and (B) MCS membrane surface.

TABLE 4: Chromium percent extraction through membranes using tannery effluents.

Membrane	Pe Cr^{3+}
MTS	95
MCS	96
Nafion 450	100

of the MTS membrane resembles needles; such a difference is due to the fact that during the synthesis of polyaniline doped with p-toluenesulfonic acid (p-TSA), the complete oxidation reaction of the aniline took place.

4. Conclusions

Infrared spectroscopy showed characteristic bands of PAni in the spectra of the membranes, especially the peak at $1{,}034\,\text{cm}^{-1}$ regarding the S=O group. The synthesized membranes presented similar chromium transport to that observed in the Nafion 450 membrane using the tannery effluent. Electric resistance was higher in the synthesized membrane than in the commercial membrane.

Using the metal-finishing effluent, it was possible to verify that the MCS and MTS membranes presented similar results in chromium transport. The Nafion 450 membrane, however, presented better results, because its electric resistance is lower.

The study proved the feasibility of using an alternative technology in the treatment of tannery and metal-finishing effluents, bringing great advantages to water reuse.

References

[1] M. Haroun, A. Idris, and S. R. Syed Omar, "A study of heavy metals and their fate in the composting of tannery sludge," *Waste Management*, vol. 27, no. 11, pp. 1541–1550, 2007.

[2] A. A. Dantas Neto, T. N. de Castro Dantas, and M. C. P. Alencar Moura, "Evaluation and optimization of chromium removal from tannery effluent by microemulsion in the Morris extractor," *Journal of Hazardous Materials*, vol. 114, no. 1–3, pp. 115–122, 2004.

[3] A. Cassano, R. Molinari, M. Romano, and E. Drioli, "Treatment of aqueous effluents of the leather industry by membrane processes: a review," *Journal of Membrane Science*, vol. 181, no. 1, pp. 111–126, 2001.

[4] V. Sivakumar, V. J. Sundar, T. Rangasamy, C. Muralidharan, and G. Swaminathan, "Management of total dissolved solids in tanning process through improved techniques," *Journal of Cleaner Production*, vol. 13, no. 7, pp. 699–703, 2005.

[5] G. Tiravanti, D. Petruzzelli, and R. Passino, "Low and non waste technologies for metals recovery by reactive polymers," *Waste Management*, vol. 16, no. 7, pp. 597–605, 1996.

[6] Z. Bajza and I. V. Vrcek, "Water quality analysis of mixtures obtained from tannery waste effluents," *Ecotoxicology and Environmental Safety*, vol. 50, no. 1, pp. 15–18, 2001.

[7] J.-H. Tay and S. Jeyaseelan, "Membrane filtration for reuse of wastewater from beverage industry," *Resources, Conservation and Recycling*, vol. 15, no. 1, pp. 33–40, 1995.

[8] A. Cassano, J. Adzet, R. Molinari, M. G. Buonomenna, J. Roig, and E. Drioli, "Membrane treatment by nanofiltration of exhausted vegetable tannin liquors from the leather industry," *Water Research*, vol. 37, no. 10, pp. 2426–2434, 2003.

[9] S. M. Kulikov, O. M. Kulikova, O. V. Scharkova, R. I. Maximovskaya, and I. V. Kozhevnikov, "Use of electromembrane technology for waste water treatment and modern acidic catalyst synthesis," *Desalination*, vol. 104, no. 1-2, pp. 107–111, 1996.

[10] X. Tongwen, "Electrodialysis processes with bipolar membranes (EDBM) in environmental protection—a review," *Resources, Conservation and Recycling*, vol. 37, no. 1, pp. 1–22, 2002.

[11] H. Strathmann, "Electrodialysis and related processes," in *Membrane Separations Technology—Principles and Applications*, R. D. Noble and S. Stern, Eds., p. 213, Elsevier, New York, NY, USA, 1995.

[12] L. Pinhedo, R. Pelegrini, R. Bertazzoli, and A. J. Motheo, "Photoelectrochemical degradation of humic acid on a (TiO_2) $0.7(RuO_2)0.3$ dimensionally stable anode," *Applied Catalysis B*, vol. 57, no. 2, pp. 75–81, 2005.

[13] L. Szpyrkowicz, G. H. Kelsall, S. N. Kaul, and M. De Faveri, "Performance of electrochemical reactor for treatment of tannery wastewaters," *Chemical Engineering Science*, vol. 56, no. 4, pp. 1579–1586, 2001.

[14] M. A. S. Rodrigues, F. D. R. Amado, J. L. N. Xavier, K. F. Streit, A. M. Bernardes, and J. Z. Ferreira, "Application of photoelectrochemical-electrodialysis treatment for the recovery and reuse of water from tannery effluents," *Journal of Cleaner Production*, vol. 16, no. 5, pp. 605–611, 2008.

[15] F. D. R. Amado, E. Gondran, J. Z. Ferreira, M. A. S. Rodrigues, and C. A. Ferreira, "Synthesis and characterisation

of high impact polystyrene/polyaniline composite membranes for electrodialysis," *Journal of Membrane Science*, vol. 234, no. 1-2, pp. 139–145, 2004.

[16] F. D. R. Amado, L. F. Rodrigues Jr., M. A. S. Rodrigues, A. M. Bernardes, J. Z. Ferreira, and C. A. Ferreira, "Development of polyurethane/polyaniline membranes for zinc recovery through electrodialysis," *Desalination*, vol. 186, no. 1–3, pp. 199–206, 2005.

[17] F. D. R. Amado, M. A. S. Rodrigues, F. D. P. Morisso, A. M. Bernardes, J. Z. Ferreira, and C. A. Ferreira, "High-impact polystyrene/polyaniline membranes for acid solution treatment by electrodialysis: preparation, evaluation, and chemical calculation," *Journal of Colloid and Interface Science*, vol. 320, no. 1, pp. 52–61, 2008.

[18] P. Sistat and G. Pourcelly, "Chronopotentiometric response of an ion-exchange membrane in the underlimiting current-range. Transport phenomena within the diffusion layers," *Journal of Membrane Science*, vol. 123, no. 1, pp. 121–131, 1997.

[19] M. Taky, G. Pourcelly, F. Lebon, and C. Gavach, "Polarization phenomena at the interfaces between an electrolyte solution and an ion exchange membrane. Part I. Ion transfer with a cation exchange membrane," *Journal of Electroanalytical Chemistry*, vol. 336, no. 1-2, pp. 171–194, 1992.

[20] R. F. D. Costa, M. A. S. Rodrigues, and J. Z. Ferreira, "Transport of trivalent and hexavalent chromium through different ion-selective membranes in acidic aqueous media," *Separation Science and Technology*, vol. 33, no. 8, pp. 1135–1143, 1998.

[21] R. Valerdi-Pérez and J. A. Ibáñez-Mengual, "Current-voltage curves for an electrodialysis reversal pilot plant: determination of limiting currents," *Desalination*, vol. 141, no. 1, pp. 23–37, 2001.

[22] N. Pismenskaya, V. Nikonenko, B. Auclair, and G. Pourcelly, "Transport of weak-electrolyte anions through anion exchange membranes: current-voltage characteristics," *Journal of Membrane Science*, vol. 189, no. 1, pp. 129–140, 2001.

9

Optimization of Wind Turbine Airfoil Using Nondominated Sorting Genetic Algorithm and Pareto Optimal Front

Ziaul Huque,[1, 2] Ghizlane Zemmouri,[1, 2] Donald Harby,[1, 2] and Raghava Kommalapati[2, 3]

[1] Department of Mechanical Engineering, Prairie View A&M University, P.O. Box 519, Mail Stop 2525, Prairie View, TX 77446, USA
[2] Center for Energy and Environmental Sustainability, Prairie View A&M University, P.O. Box 519, Mail Stop 2500, Prairie View, TX 77446, USA
[3] Department of Civil and Environmental Engineering, Prairie View A&M University, P.O. Box 519, Mail Stop 2510, Prairie View, TX 77446, USA

Correspondence should be addressed to Ziaul Huque, zihuque@pvamu.edu

Academic Editor: Mahesh T. Dhotre

A Computational Fluid Dynamics (CFD) and response surface-based multiobjective design optimization were performed for six different 2D airfoil profiles, and the Pareto optimal front of each airfoil is presented. FLUENT, which is a commercial CFD simulation code, was used to determine the relevant aerodynamic loads. The Lift Coefficient (C_L) and Drag Coefficient (C_D) data at a range of 0° to 12° angles of attack (α) and at three different Reynolds numbers (Re = 68,459, 479,210, and 958,422) for all the six airfoils were obtained. Realizable k-ε turbulence model with a second-order upwind solution method was used in the simulations. The standard least square method was used to generate response surface by the statistical code JMP. Elitist Nondominated Sorting Genetic Algorithm (NSGA-II) was used to determine the Pareto optimal set based on the response surfaces. Each Pareto optimal solution represents a different compromise between design objectives. This gives the designer a choice to select a design compromise that best suits the requirements from a set of optimal solutions. The Pareto solution set is presented in the form of a Pareto optimal front.

1. Introduction

According to the US Department of energy, the combustion of fossil fuels results a net increase of 10.65 billion tones of atmospheric carbon dioxide every year [1] which deteriorates the environmental balance. Fossil fuels also give out sulfur dioxide into the air, which, after reacting with the moisture in air, produces sulfuric acid and leads to acid rain. Furthermore, depletion of these nonrenewable sources of energy is taking place at a rapid pace because of the increasing demands of energy, with the modernization of our society. Estimates from the US Department of Energy predict that the years of production left in the ground for oil are 43 years, gas 167 years, and coal 417 years [2]. Therefore, it is critical that we start looking for some renewable sources of energy that can be used as alternatives to fossil fuels. Renewable energy resources can play a key role in producing local, clean,

and inexhaustible energy to supply the growing demand for electricity, heat, and transportation fuel. Wind energy as a source of energy to produce electricity is favoured widely as an alternative to fossil fuels. It is plentiful, renewable, widely distributed, and clean and produces no greenhouse gas emissions. Wind turbines convert kinetic energy from the wind into mechanical energy which can be used to generate electricity. When the wind blows and flows around the blades of the wind turbines, which have essentially airfoil cross sections, it generates lift forces which makes the blades spin. These blades are connected to a drive shaft that turns an electric generator to produce electricity which therefore can be sent through a cable down the turbine tower to a transmission line. The blades of a wind turbine rotor are generally regarded as the most critical component of the wind turbine system [3]. The aerodynamic profiles of wind turbine blades have crucial influence on aerodynamic efficiency of wind turbines. Even minor alterations in the shape of the profile can greatly alter the power curve and noise level. Some of the important design parameters include the number of blades, blade solidity, blade taper, and twist as well as tip-speed ratio. The aerodynamic theory of the wind turbines gradually developed, starting with the simple one-dimensional momentum analysis [4] of the actuator disc to the more commonly used BEM theory.

The BEM theory is based on the assumption that the flow at a given annulus does not affect the flow at adjacent annuli [5]. This allows the rotor blade to be analysed in sections, where the resulting forces are summed over all sections to get the overall forces of the rotor. The theory uses both axial and angular momentum balances to determine the flow and the resulting forces at the blade. BEM methods are very fast and reliable in the design process, nevertheless these are limited due to their two dimensional nature. These codes require tabulated data for the lift, drag and moment distributions versus the angle of attack to calculate the blade aerodynamic loads. Furthermore, empirical corrections are necessary to account for rotational effects near the root and three dimensional flows around the tip region [6].

Over the last several years, the wind turbine community has started to look at CFD methods to complement wind tunnel [7] and in field tests on the understanding of the complex flow physics around rotating wind turbine blades. CFD codes can be very useful on the calculation of aerodynamic coefficients required by engineering methods and on the explicit determination of loads since no corrections are necessary as in BEM method.

The overall goal of this study is to perform a response surface-based multiobjective optimization of selected 2D airfoil profiles using Elitist Nondominated Sorting Genetic Algorithm (NSGA). In order to achieve this overall goal, several specific objectives were determined. The first specific objective was to identify several airfoil profiles with their geometric coordinates. The second objective was to perform CFD simulations around the airfoils. Simulations for each airfoil were performed for several values of Re and α. The third objective was to determine response surfaces for lift and drag coefficients as a function of Re and α. The fourth and final objective was to perform the optimization using genetic algorithm to determine a set of nondominated solution for each airfoil. Based on the optimization results, designers can opt choose multiple airfoils for a single blade depending on Re and α variation along the length of the blade after appropriate twist and taper is applied. It is also worth to mention other interesting works [8, 9] that are reported where numerical models are used to design turbines specifically in the wave energy area.

2. Approaches

2.1. CFD Modelling. In this work, we consider the flow around six different airfoil shapes (NACA 63-218, E387, FX63-137, NACA 63-421, NACA 64-421, and NACA 65-421) at 3 different Reynolds number (Re = 68, 459, Re = 479, 210, and Re = 958, 422) for a range of 0° to 12° angles of attack. These airfoils are created from a set of vertices generated

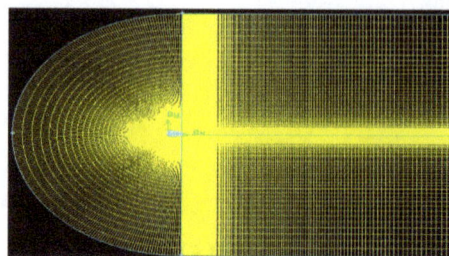

FIGURE 1: 2D mesh of the entire domain using a map scheme with around 50,000 quadrilateral elements.

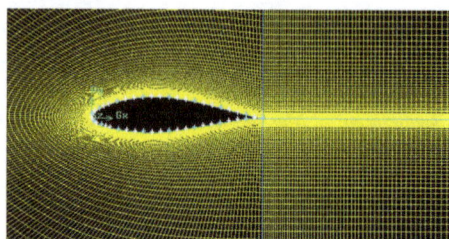

FIGURE 2: Mesh generated around the airfoil.

from the University of Illinois at Urbana Champagne (UIUC) airfoil database [10]. These vertices are connected using a smooth curve, creating the surface of the airfoil. A flow domain is created surrounding the airfoil and this domain is split for meshing purposes.

The CFD data of the 15 simulated cases for each airfoil were used to generate a response surface. The response surfaces were fit using standard least-square regression with quadratic polynomial using JMP. These response surfaces are obtained between design variable (Re and α) and objective functions (C_L and C_D) for each airfoil profile. All the design variables and the objective functions are normalized between 0 and 1 based on their maximum and minimum values in order to determine the response surface.

2.2. Grid Description. Grid generation is the most important step in the CFD simulations. The quality of the grid plays a direct role on the quality of the analysis, regardless of the flow solver used. Additionally, the solver will be more robust and efficient when using a well-constructed mesh. In this work, structured grids were generated using the commercial code GAMBIT. Figure 1 shows a 2D mesh of the entire domain using a map scheme with around 50,000 quadrilateral elements (the number varies slightly for different airfoils) while Figure 2 shows the blowup of the mesh generated around the airfoil.

In order to have a stable solution, the generated grids had the least number of elements with high aspect ratios. To be able to resolve adequately the boundary layer along the airfoil wall, grid points were clustered near the wall. The grids were also clustered near the trailing edge in order to catch the flow separation.

2.3. Boundary Conditions. Boundary conditions specify the flow and thermal variables on the boundaries of the physical model. They are, therefore, a critical component of the CFD simulations, and it is important that they are specified appropriately. In this work, 3 different types of boundary conditions were used: no-slip boundary condition over the airfoil surface, inlet boundary condition for free stream flow, and pressure outlet. The outlet boundary of the domain was set to a constant pressure value. It was set to be atmospheric pressure. The object in the computational domain (i.e., the airfoil surface), around which the flow was simulated, was set to be no-slip boundary (wall). The no-slip boundary condition sets the stream wise velocity to zero. The velocity Inlet boundary condition was used to define the flow velocity at the flow inlet. Figure 3 shows the different assigned boundary conditions for all the CFD simulations as follows.

In Gambit, the boundary conditions were declared (i.e., wall, velocity inlet, and pressure outlet), but actual values for these boundaries were defined in fluent. For velocity inlet, we used 3 different velocities for each airfoil at every angle of attack. We set $v = 1$ m/s (for Re = 68,459), $v = 7$ m/s (for Re = 479,210), and $v = 14$ m/s (for Re = 958,422). The Realizable k-epsilon turbulence model and a second-order upwind solution method were used to get more accurate results.

2.4. Response Surface Methodology. The response surface method fits an approximate function to a set of experimentally or numerically evaluated design data points [11]. There are various response surface approximation methods available in the literature [11, 12], with the polynomial-based approximations being the most popular. In this technique, an appropriate ordered polynomial is fitted to a set of data points, such that the adjusted RMS error σ_a is minimized and quality parameter R^2_{adj} is made as close as possible to one [12]. The σ_a and R^2_{adj} are defined as follows.

Let N be the number of data points and let N_p be the number of coefficients, and error e_i at any point i is defined:

$$e_i = f_i^a - f_i^p, \tag{1}$$

where f_i^p is the actual value of the function at the design point and f_i^p is the predicted value. Hence,

$$\sigma_a = \sqrt{\frac{\sum_{i=1}^N e_i^2}{(N - N_P)}}, \tag{2}$$

$$R^2_{adj} = 1 - \frac{\sigma_a^2 (N_P - 1)}{\sum_{i=1}^N (y_i - \overline{y})^2}, \tag{3}$$

where

$$\overline{y} = \frac{\sum_{i=1}^N y_i}{N_p}. \tag{4}$$

The number of data N has to be greater than the number of coefficients N_p so that the denominator of (2) is always positive and well posed. Since R^2_{adj} needs to be as close as possible to 1 to represent a good fit, the terms in the numerator of (3) $(\sigma_a)^2(N_P - 1)$ should be less than or equal

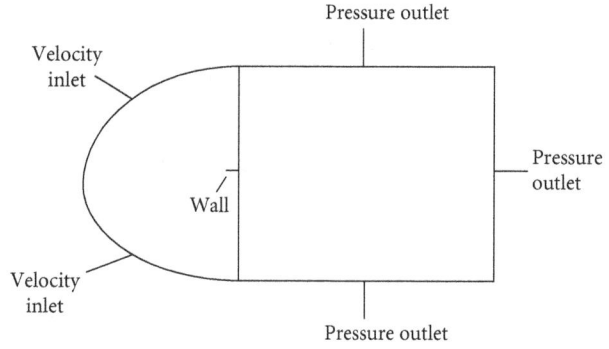

FIGURE 3: The flow domain with the boundary conditions.

to the denominator $\sum_1^N (y_i - \overline{y})^2$ so that R^2_{adj} will always be positive. In this study, the response surface method is applied with two objectives, namely, to generate response surface from the CFD simulation results and to approximate the global Pareto optimal front by representing one objective in terms of others. Both aspects will be discussed in upcoming sections.

2.5. Optimization Approach. The methodology used for generating Pareto optimal front is a multiobjective evolutionary algorithm (MOEA). The specific algorithm used is the Elitist Nondominated Sorting Genetic Algorithm (NSGA-II) [13]. All genetic algorithm codes use some form of sorting scheme to get nondominated solutions. Nondominated solutions are the best solutions. Among the nondominated solutions one cannot be said to be better than the other. NSGA-II uses an explicit diversity-preserving mechanism. The starting point is the identification of constraints, performance criteria, design variables, and allowable range of design variables. The response surfaces generated from the results of the CFD simulations are incorporated into the NSGA-II code. In running the code input values for several parameters must be provided. These parameters and their best values, as suggested by Deb [11, 13] after their extensive parametric study, are as follows:

(i) Population size: 100,

(ii) Generations: 250,

(iii) Crossover probability (P_{cross}): 1.00,

(iv) Distribution parameter (for crossover): 20,

(v) Mutation probability (P_{mut}): 0.250,

(vi) Distribution parameter (for mutation): 200.

Where the population size is the size of the non-dominated solutions and the generations are equivalent to the number of iterations. Crossover probability, mutation probability, distribution parameter for crossover, and distribution parameter for mutation are used to create the offspring population from the parent population. The crossover probability is mainly responsible for the search aspect of the genetic algorithm while mutation probability keeps the diversity in the population. The distribution parameter for crossover controls the diversity of the children solutions obtained after crossover while distribution parameter for mutation controls the spread of the solutions after mutation.

In NSGA-II algorithm, the code first creates a parent population of P_t of size N. From the parent set it then creates an offspring population Q_t of size N. The NSGA-II algorithm, instead of finding the non-dominated front of Q_t only, uses the combined population of P_t and Q_t to form R_t of size $2N$. Then, a nondominated sorting is used to classify the entire population R_t. Although this requires more effort compared to performing a non-dominated sorting on Q_t alone, it allows a global non-domination check among the offspring and parent solutions. Once the non-dominated sorting is over, the new population is filled by solutions of different non-dominated fronts, F_i, one at a time. The filling starts with the best non-dominated front and continues with solutions of the second non-dominated front and so on. Since the overall population size of R_t is $2N$, not all fronts may be accommodated in N slots available in the new population. All fronts which could not be accommodated are simply deleted. When the last allowed front is being considered, there may exist more solutions in the last front than the remaining slots in the new population. Instead of arbitrarily discarding some members from the last front, a niching strategy [13] is used to choose the members of the last front based on crowding distance. The solutions kept in the population are those which have the largest crowding distance thus keeping the diversity of the solution. This new set of solutions is now the parent set for the next generation. The procedure is then repeated till the best non-dominated set is obtained.

3. Results and Discussions

3.1. Flow Field Overview.
The CFD validation is a very important part of computational fluid dynamics. It is used to evaluate the accuracy of CFD results. We compared the CFD results of Lift Coefficient of E387 airfoil for 5 different angles of attack with National Renewable Energy Laboratory (NREL) experimental data [14] for Re = 479, 210 as shown in Figure 4. The CFD results of Lift Coefficient show good agreement with NREL experimental results.

Colour contours of static pressure and velocity vectors by velocity magnitude of NACA 64-421 airfoil at an angle of attack of 6° and at Re = 479, 210 are shown in Figures 5 and 6. The static pressure plot clearly shows the higher pressure (indicated by red, yellow, and green colours) on the bottom surface and lower pressure (indicated by blue colour) on the top surface. In the velocity vector plot, higher velocity corresponding to lower pressure and lower velocity corresponding to higher pressure can be clearly observed.

3.2. Performance Trend

3.2.1. C_p Distribution.
Figures 7(a)–7(c) show the C_p distribution for NACA 63-421 for the three Reynolds numbers. In each figure and for each angle of attack (uniform colour), the bottom line represents the C_p distribution at the top surface of the airfoil, indicating lower pressure, and the top line represents the C_p distribution on the bottom surface of

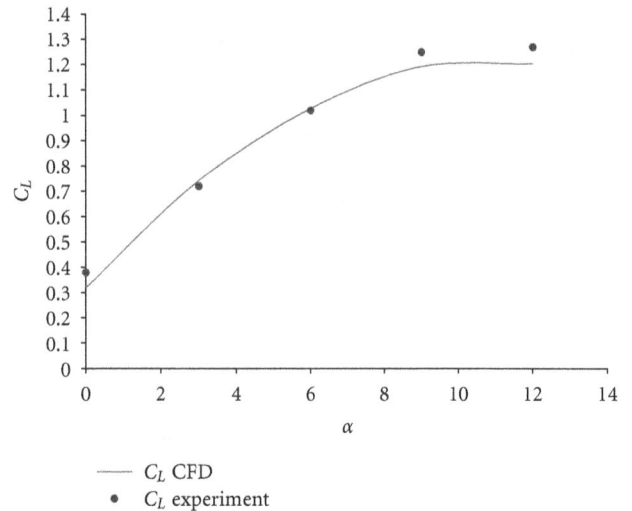

FIGURE 4: Comparison of C_L Versus α of CFD simulation results with NREL experimental data.

the airfoil indicating higher pressure. As the angle of attack increases from 0 to 12 for any Re, the area under the C_p curve increases indicating larger pressure difference between the bottom and the top surfaces. Similar trend is observed for different Re with the same angle of attack. These are expected trends for any airfoil.

3.2.2. Integrated Pressure Coefficient, C_p.
Figure 8 represents the overall integrated pressure coefficient (C_p) as a function of angle of attack (α) of NACA 63-421 airfoil at the three different Reynolds numbers. As expected, as we increase the angle of attack, the overall pressure coefficient increases for all six airfoils. However, within the same airfoil, C_p has little change as we move from a lower Reynolds number (Re = 68, 459) to a higher Reynolds number (Re = 958, 422). The C_p of NACA 63-218 airfoil increases continuously as we increase the angle of attack which indicates that it has not reached the stall condition yet, while the C_p plot of the other airfoils starts to flatten at around 11° to 12° of angle of attack which indicates that it is close to its stall condition. In addition, NACA 63-218, NACA 63-421, NACA 64-421, and NACA 65-421 airfoils have small integrated C_p (C_p around 1.3 or 1.4) at stall condition which are much smaller than FX 63137 and E 387 airfoils (C_p around 1.6 or 1.8). Thus, we can conclude that the stall conditions could vary significantly between various airfoil profiles.

3.2.3. Coefficient of Lift, C_L.
Figure 9 shows the plot of lift coefficient for NACA 63-421 airfoil. C_L is plotted as functions of angles of attack and Reynolds number. The general trends of all the plots are similar as expected; that is, C_L increases with increasing α and Re. Some of the observations from the plots are as follows.

(i) The variations of C_L between different Re are not significant.

(ii) The differences are more significant at higher α for NACA 65-421 and E 387 airfoil.

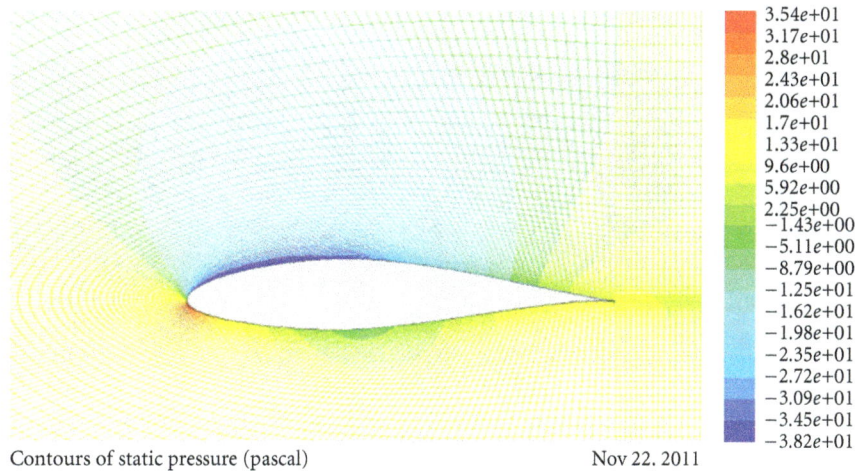

Contours of static pressure (pascal)   Nov 22. 2011

FIGURE 5: Colour contour of static pressure around NACA 64-421 airfoil at $\alpha = 6$ and Re $= 479,210$.

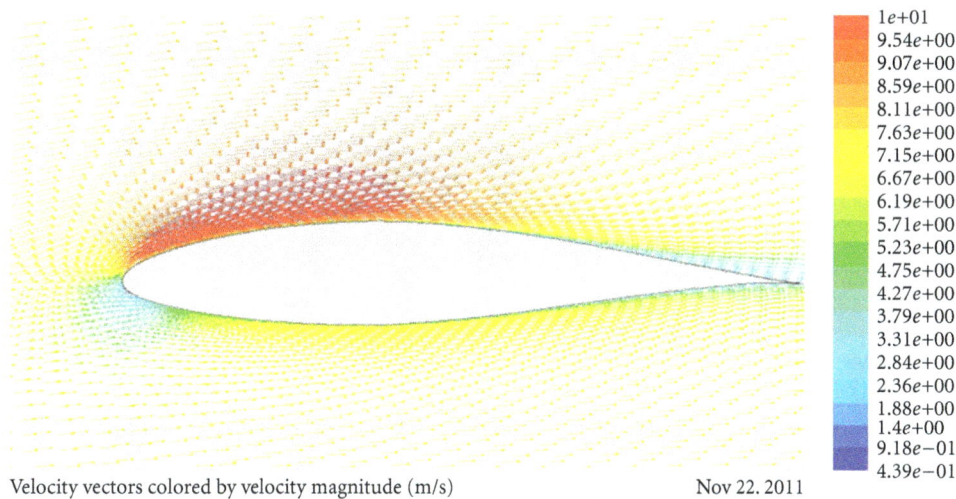

Velocity vectors colored by velocity magnitude (m/s)   Nov 22. 2011

FIGURE 6: Colour velocity vectors by velocity magnitude around NACA 64-421 airfoil at $\alpha = 6$ and Re $= 479,210$.

(iii) C_L for NACA 63-218 at all Re did not reach the stall conditions; that is, the stall condition will be reached at much higher than $\alpha = 12°$.

(iv) Both FX 63-137 and E 387 indicate reaching stall condition at around $\alpha = 12°$.

(v) Both FX 63-137 and E 387 show smaller variation with Re and reach higher values of $C_L =$ around 1.8 and 1.6 at $\alpha = 12°$.

It is obvious from the previous observations that different airfoils behave differently with angle of attack and Reynolds numbers.

3.2.4. Coefficient of Drag, C_D. Drag Coefficient as a function of angle of attack of NACA 63-421 at the three different Reynolds numbers is shown in Figure 10. As the velocity goes down from $v = 7$ m/s (Re $= 479,210$) to $v = 1$ m/s (Re $= 68,459$), the C_D curves increase drastically for all six airfoils, while if we increase the velocity from $v = 7$ m/s to

$v = 14$ m/s (Re $= 958,422$), the C_D curves do not change a lot. As expected, for lower Re and larger α, the higher is the Drag Coefficient. For instance, NACA 64-421 airfoil has the highest C_D ($C_D = 0.5259$) at Re $= 68,459$ and $\alpha = 12$. Hence there is an optimum combination of α and Re for the maximum ratio of C_L by C_D for each airfoil. These optimum conditions are presented in the next sections.

The CFD simulation results for the 15 cases are shown in Table 1 of NACA 63-421 airfoil where the design variables and objective functions are given in normalized form.

3.3. Response Surface Approximation. The CFD data of 15 cases were used to generate a response surface for each of the two objective functions for each airfoil shape. The response surfaces were fit using standard least-square regression with quadratic polynomial using JMP [15]. The following response surfaces for each of the objective functions were obtained as functions of the two design variables of NACA 63-421 airfoil:

(a)

(b)

(c)

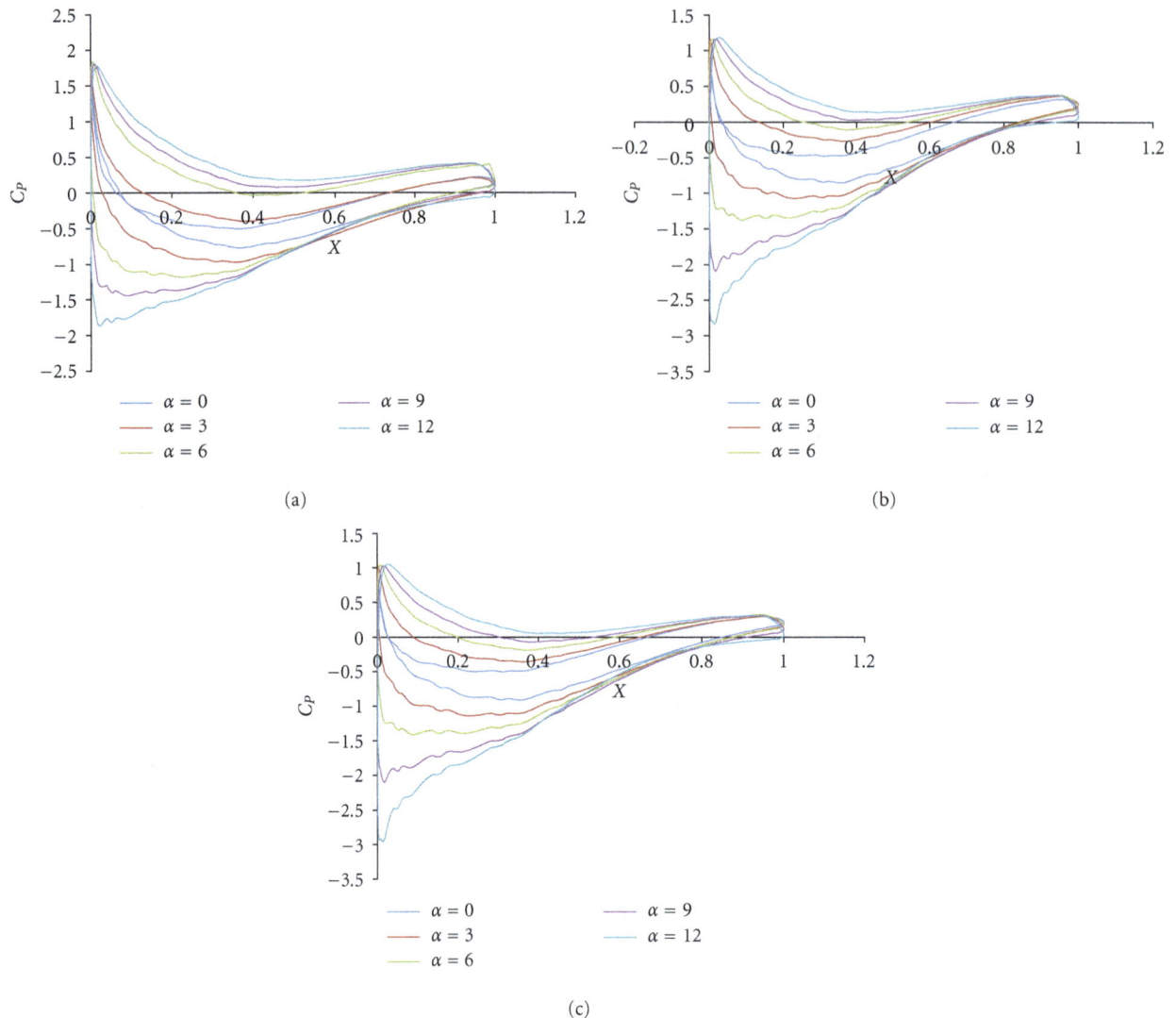

FIGURE 7: (a) C_p distribution of NACA 63-421 airfoil at Re = 68,459. (b) C_p distribution of NACA 63-421 airfoil at Re = 479,210. (c) C_p distribution of NACA 63-421 airfoil at Re = 958,422.

Lift Coefficient

$$(C_L) = 0.1306 + (0.1646 * \text{Re}) + (1.2649 * \alpha)$$
$$- (0.0766 * \text{Re} * \text{Re}) - (0.1732 * \text{Re} * \alpha) \qquad (5)$$
$$- (0.3843 * \alpha * \alpha).$$

Drag Coefficient

$$(C_D) = 0.6795 - (1.6805 * \text{Re}) + (0.1358 * \alpha)$$
$$+ (1.1322 * \text{Re} * \text{Re}) - (0.2220 * \text{Re} * \alpha) \qquad (6)$$
$$+ (0.2329 * \alpha * \alpha).$$

The quality of the response surface of this airfoil is shown in Table 2. The response surface for the entire objective had very high adjusted coefficient of both C_L and C_D which indicate good capabilities for this airfoil.

3.4. Optimization Results. In order to obtain the Pareto optimal solutions, the two response surface equations were incorporated in the NSGA-II code with the input parameters as mentioned in the previous section. After the simulation, the code generates a file containing 100 non-dominated solutions created during the final iteration. Non-dominated values are the best values according to the desired maximization of the objective functions among the entire population. For better understanding 2D plots of C_L versus C_D and C_L/C_D versus C_L are depicted using only 100 nondominated solutions for NACA 63-421 airfoil in Figures 11 and 12. The C_L versus C_D plot represents 100 best combinations of C_L and C_D corresponding to the best combinations of Re and α. Therefore, the designer can choose any of these combinations and get good results. But, to be able to get one optimum result of these 100 combinations, we can plot C_L/C_D versus C_L. For example, for NACA 63-421 airfoil, the optimum point is at $C_L = 0.717$ for C_L/C_D around 40. This optimum

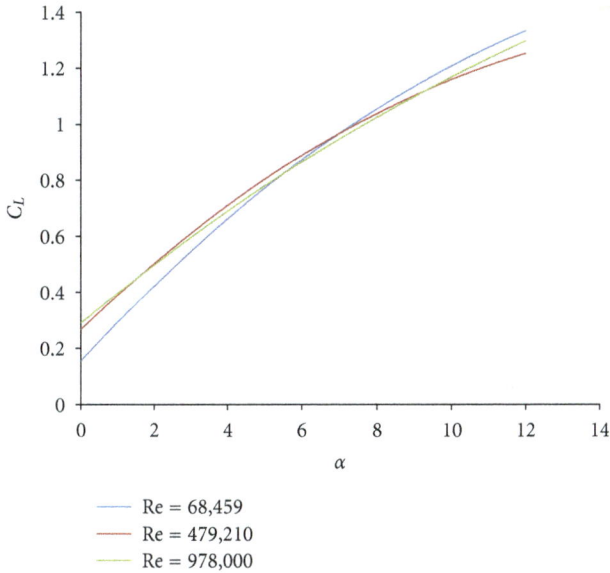

FIGURE 8: Integrated pressure coefficient of NACA 63-421 airfoil at different Re.

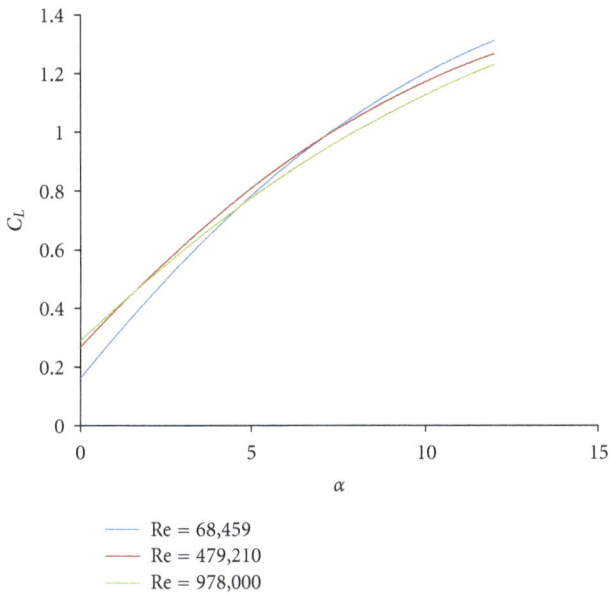

FIGURE 9: CFD simulation results of C_L versus α of NACA 63-421 airfoil.

C_L at the maximum value of C_L/C_D corresponds to an angle of attack of around 4°.

Similar optimizations were performed on all the airfoil shapes. The results obtained are as follows: for NACA 63-218 airfoil, the optimum $C_L = 0.54$ and $C_D = 0.026$ at $\alpha = 4.58$ and Re = 777, 284; for E387 airfoil, the best combination is at $\alpha = 2.90$ and Re = 721, 398 which give $C_L = 0.7$ and $C_D = 0.129$; for FX 63137 airfoil, the optimum $C_L = 0.976$ and $C_D = 0.024$ at $\alpha = 2.10$ and Re = 730, 996; for NACA 63-421 airfoil, $C_L = 0.717$ and $C_D = 0.018$ for Re = 770, 937

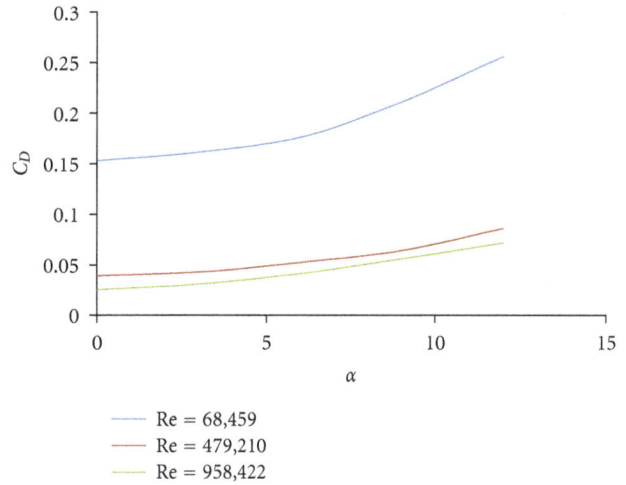

FIGURE 10: CFD simulation results of C_D versus α of NACA 63-421 airfoil.

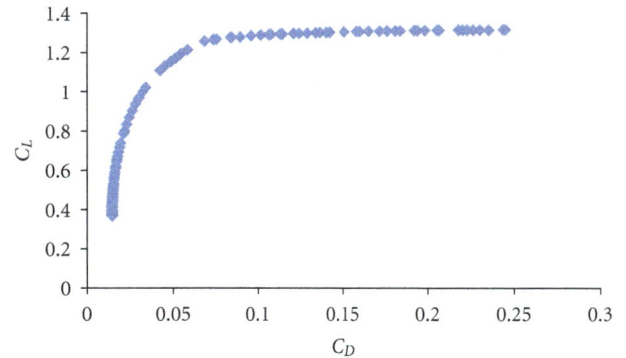

FIGURE 11: 2D presentation of non-dominated solutions of C_L versus C_D.

and $\alpha = 4.06$; for NACA 64-421 airfoil, at $\alpha = 3.12$ and Re = 725, 083 for $C_L = 0.68$ and $C_D = 0.02$; finally for NACA 65-421 airfoil, the optimum $C_L = 0.775$ and $C_D = 0.023$ corresponding to $\alpha = 3.07$ and Re = 759, 718. The optimum C_L corresponding to maximum C_L/C_D as a function of angle of attack for all the airfoils was plotted in Figure 13. From this figure, the designer can choose the airfoil shape corresponding to the angle of attack dictated by the twist angle he/she is using. For example, if the twist angle is 4° at any blade section, the designer can use NACA 63-421 airfoil shape for that section of the blade in order to get the optimum C_L.

4. Conclusions

The Pareto optimal front in multiobjective optimization problem is useful to visualize the tradeoffs among different objectives. In addition to identify compromise solutions, this also helps the designer set realistic design goals. The goal of the current research is focused on the determination and optimization of wind turbine airfoil performance. For this

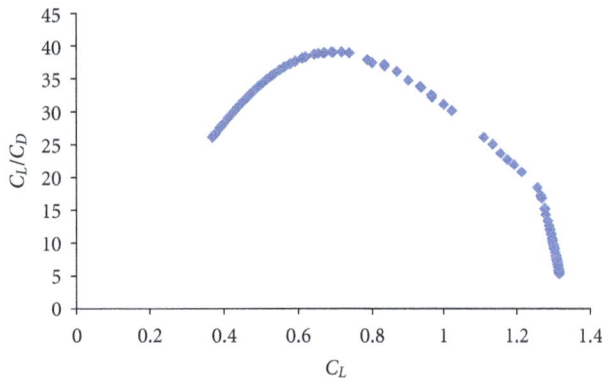

FIGURE 12: 2D presentation of non-dominated solutions of C_L/C_D versus C_L.

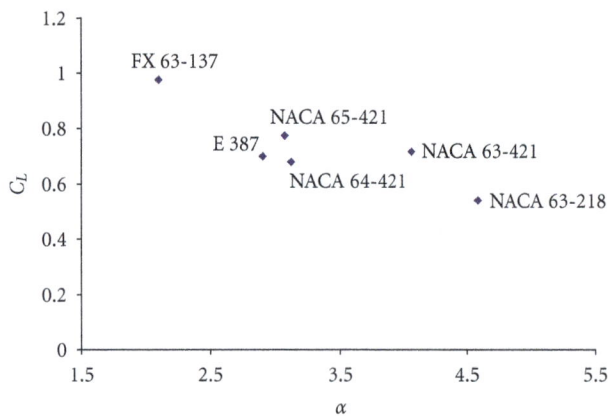

FIGURE 13: Optimum C_L versus α for all airfoils.

TABLE 2: Quality parameters of response surface of NACA 63-421 airfoil.

Observations	C_L	C_D
R^2	0.994164	0.992416
R^2 Adj	0.990921	0.988203
Root Mean Square Error	0.028065	0.031312
Mean of Response	0.627276	0.381842

purpose, six different two-dimensional airfoil profiles were studied over two important design variables. The NSGA-II approach of optimization and response surface methodology has been used to generate Pareto optimal front. The optimum C_L, corresponding to the best combination of α and Re, is different from one airfoil shape to another. In summary, the proposed systematic approach is very useful for optimization of designs with many variables and objectives and can be practically applied in many disciplines.

Acknowledgment

This work is supported by the National Science Foundation (NSF) through the Center for Energy and Environmental Sustainability (CEES), a CREST Center, award no. 1036593.

References

[1] US Government, Department of Energy, US Department of Energy on Green House Gases, 2009, http://www.eia.doe.gov/bookshelf/brochures/greenhouse/Chapter1.htm.

[2] Department of Energy, US Fossil Energy, 2009, http://www.fossil.energy.gov/index.html.

[3] C. Kong, J. Bang, and Y. Sugiyama, "Structural investigation of composite wind turbine blade considering various load cases and fatigue life," *Energy*, vol. 30, no. 11-12, pp. 2101–2114, 2005.

[4] H. Glauert, *The Elements of Aerofoil and Airscrew Theory*, Cambridge Science Classic Series, Cambridge University Press, 2nd edition, 1993.

[5] R. E. Wilson and P. B. S. Lissaman, "Applied aerodynamics of wind power machines, PB238595," Report NSF-RA-N-74-113, NTIS, Springfield, Va, USA, 1974.

[6] H. Glauert, "Airplane propellers," in *Aerodynamic Theory*, vol. 4, Dover, 1963.

[7] M. Hand, D. Simms, L. Fingersh et al., "Unsteady aerodynamics experiment phase vi: wind tunnel test configurations and available data campaigns," Tech. Rep. NREL/TP-500-29955, NREL, 2001.

[8] M. Alberdi, M. Amundarain, A. J. Garrido, I. Garrido, O. Casquero, and M. De La Sen, "Complementary control of oscillating water column-based wave energy conversion plants to improve the instantaneous power output," *IEEE Transactions on Energy Conversion*, vol. 26, no. 4, pp. 1021–1032, 2011.

[9] A. E. Marzani, F. C Ruiz, M. A. Rodriguez, and M. T. Parra Santos, "Numerical modelling in wave energy conversion systems," *Energy*, vol. 33, no. 8, pp. 1246–1253, 2008.

[10] "UIUC Airfoil Coordinates Database," http://www.ae.illinois.edu/m-selig/ads/coord_database.html.

[11] T. Goel, R. Vaidyanathan, R. T. Haftka, W. Shyy, N. V. Queipo, and K. Tucker, "Response surface approximation of pareto

TABLE 1: Normalized design variables and objective function values from CFD simulations of NACA 63-421.

CFD normalized results			
Re	A	C_L	C_D
0.499999	0	0.208173299	0.152746
0.499999	0.25	0.467045629	0.167201
0.499999	0.5	0.682900599	0.204313
0.499999	0.75	0.868028883	0.25002
0.499999	1	0.967890613	0.335964
1	0	0.227377477	0.098445
1	0.25	0.450145952	0.120713
1	0.5	0.657935167	0.16095
1	0.75	0.828468275	0.217986
1	1	0.937932094	0.279319
0.071428	0	0.145183592	0.597703
0.071428	0.25	0.372561069	0.630909
0.071428	0.5	0.718397603	0.688374
0.071428	0.75	0.877093255	0.822994
0.071428	1	1	1

optimal front in multi-objective optimization," in *Proceedings of the 10th AIAA/ISSMO Multidisciplinary Analysis and Optimization Conference*, pp. 2230–2245, Albany, NY, USA, September 2004.

[12] R. H. Myers and D. C. Montgomery, *Response Surface Methodology-Process and Product Optimization Using Designed Experiment*, Wiley-Interscience, 1995.

[13] K. Deb, *Multi-Objective Optimization using Evolutionary Algorithms*, John Wiley and Sons, New York, NY, USA, 2004.

[14] M. S. Selig and B. D. McGranahan, *Wind Tunnel Aerodynamic Tests of Six Airfoils for Use on Small Wind Turbines*, University of Illinois at Urbana Champaign, Urbana, Ill, USA, 2004.

[15] JMP, "The statistical discovery software," 1989–2002, Version 5, SAS Institute Inc., Cary, NC, USA.

10

CFD Study of Industrial FCC Risers: The Effect of Outlet Configurations on Hydrodynamics and Reactions

Gabriela C. Lopes,[1] Leonardo M. Rosa,[1] Milton Mori,[1] José R. Nunhez,[1] and Waldir P. Martignoni[2]

[1] *School of Chemical Engineering, University of Campinas, 500 Albert Einstein Avenue, 13083-970 Campinas, SP, Brazil*
[2] *PETROBRAS, 65 República do Chile Avenue, 20031-912 Rio de Janeiro, RJ, Brazil*

Correspondence should be addressed to Milton Mori, mori@feq.unicamp.br

Academic Editor: Jerzy Bałdyga

Fluid catalytic cracking (FCC) riser reactors have complex hydrodynamics, which depend not only on operating conditions, feedstock quality, and catalyst particles characteristics, but also on the geometric configurations of the reactor. This paper presents a numerical study of the influence of different riser outlet designs on the dynamic of the flow and reactor efficiency. A three-dimensional, three-phase flow model and a four-lump kinetic scheme were used to predict the performance of the reactor. The phenomenon of vaporization of the liquid oil droplets was also analyzed. Results showed that small changes in the outlet configuration had a significant effect on the flow patterns and consequently, on the reaction yields.

1. Introduction

Although commercially established for over half a century, the fluid catalytic cracking (FCC) process is still widely studied nowadays. Since it is a very profitable operation, any improvement in it can result in large savings for the refinery. In the FCC process, preheated high-boiling liquid oil is injected into the riser reactor, where it is vaporized and cracked into smaller molecules by contact and mixing with the very hot catalyst particles coming from the regenerator. These phenomena cause a gas expansion, which drags the catalyst to the top of the reactor. Since catalytic cracking reactions can only occur after the vaporization of liquid feedstock, mixing of hydrocarbon droplets with catalyst must take place in the riser as soon as possible.

It is known that riser reactors have complex hydrodynamics. They present a high solids concentration near the walls and are also axially divided into dense and dilute regions. In addition, different riser configurations such as the inlet and outlet structures can have a profound effect on the flow patterns mentioned above.

The influence of riser exit geometry on the hydrodynamics of gas-solid circulating fluidized beds (CFB) has been in-vestigated in many studies [1–6]. Although the results reported in these studies apparently conflict quantitatively concerning the influence of riser exit, some common aspects can be observed: (1) the design of the exit has a large effect upon the reflux of solids; (2) abrupt exits cause an increase in the solids holdup and a large backmixing at the top of the riser; (3) increasing the refluxing effect of the exit has proved to increase the mean particle residence time; (4) larger and denser clusters are formed at the walls in the risers with abrupt exits.

In an experimental work, Lim et al. [7] investigated a cold model of a circulating fluidized bed, in a riser with an horizontal and flat cover at the top. They limited the operating conditions, that resulted in stable operation of the circulating fluidized bed, and proposed a model to estimate the ratio of solids that exit the riser to solids that recirculate back into the riser. This model predicts the cases in which solids inertia dominates, and the cases when solids have insufficient inertia to resist the change in airflow.

The presence of reverse core-annulus profile under certain conditions in gas-solid CFBs was also observed in an experimental study by Chew et al. [8]. Although some previous works explain this behavior as a consequence of the impact of gas-phase turbulence associated with dilute flows, Chew et al. [8] suggest a dominant factor for reverse core-annulus flow: the particle Stokes number (St). According to their work, particles with large St are more likely to follow more diffuse trajectories after collision rather than following fluid streamlines, because of greater particle inertia relative to fluid viscous forces.

Van Der Meer et al. [3] defined a parameter to quantify the reflux of solids in a square cross-section riser of a laboratory CFB. They concluded that the values of this parameter obtained for the different outlet configurations vary in a factor of 25, showing that the exit design has a significant effect on riser flow regime.

Pugsley et al. [9] performed experiments with sand and FCC catalyst in two cold CFBs of 0.1 m and 0.2 m diameter in order to observe the relative influence of smooth and abrupt exit configurations on the axial pressure drop profile. They concluded that the reflux along the riser length induced by abrupt riser exits is related to the riser diameter and the particle terminal velocity. For heavier and largest particles, the influence of abrupt exits was observed to affect a longer section of the riser when compared to the effect of smaller particles. They also observed that these effects are less pronounced for smaller riser diameters.

In a recent study, Van Engelandt et al. [10] analyzed experimentally and computationally the riser outlet effects induced by an L-outlet and by abrupt T-outlets with different extension heights, outlet surface areas, and gas flow rates. They also observed that the T-outlet configuration is found to induce recirculation by vortex formation in the extension part of the riser and cause a main reflux at the wall opposite to the riser outlet. Moreover, a reduction of the outlet surface area of a T-outlet results in an increased solids holdup in the extension part of the riser. With the increase of the gas flow rate, the position of the vortex and the anisotropy of the fluctuating particle velocities are strongly affected.

A numerical study about the effects of different outlet surface area on the flow pattern was also performed by De Wilde et al. [11]. The results obtained in their simulations showed the need of performing 3D calculations in order to predict accurately the exit effects of abrupt outlet configurations. Chalermsinsuwan et al. [12] studied different designs of the riser geometries based on the improvement of main factors that have an effect on combustion, gasification, and cracking reaction characteristics, using a 2D transient Eulerian approach.

Das et al. [13] used CFD simulations to investigate the effect of different flow and design parameters on the adsorption reactions in a SO_2–NO_x riser. They showed that the solids recirculation at the top section of the riser, induced by abrupt T outlets significantly, decreases the NO and NO_2 removal, worsening the reactor efficiency. They emphasized that this analysis is just possible due to the use of 3D models, which allows predicting the effects of outlet geometries on the flow and reaction fields.

These works indicate that the hydrodynamics of CFB risers is extremely complex, showing that the distribution of axial solids concentration depends not only on operating conditions such as gas velocity, solids flux, and particle properties, but also on the geometric configuration of the riser. However, despite the considerable number of published studies on the influence of riser exit geometry on the hydrodynamics in gas-solid circulating fluidized beds, very few have addressed the importance of the riser geometry on the FCC reactor performance.

Due to the complexity of the flow and extreme operating conditions, experimental studies of industrial FCC process are rarely found. In this context, computational fluid dynamic (CFD) tools have been used as a way to better understand these phenomena and look for alternatives to improve the reactor performance.

Despite the complex hydrodynamics present in these reactors, most numerical studies in FCC risers that take into account cracking reactions and related phenomena simplify some important fluid dynamics aspects, like considering that the riser follows a plug flow model [14–18] or applying empirical radial dispersion models to consider the core-annulus patterns [19–21].

Through the simulation of FCC process using three different models (plug flow, one-dimensional, and two-dimensional dispersion models), Deng et al. [21] showed that the consideration of radial nonuniformity is necessary since it provides a better approximation, reflecting in better results for conversions and yields when compared with experimental plant data. Afterward, Lopes et al. [22] simulated an industrial riser using 3D models and showed that the flow presents nonuniformities and asymmetric patterns which are dependent upon feed flow rates. They also observed that these dynamic characteristics of the flow influence the yields of the cracking reactions. Based on this, the use of simplified models can fail to accurately predict the flow in industrial risers.

Another common simplification applied in numerical studies of FCC processes is related to the instantaneous vaporization of the feedstock, which can result in an incorrect representation of the reactor performance. This assumption is justified in many studies (e.g., [14, 23]) by the rapid vaporization of feed. However, Theologos et al. [24], Gupta and Rao [25], and Lopes et al. [26] presented significant differences in product yields when the feedstock vaporization was modeled using instantaneous or rate vaporization models.

In the present work, different designs of the riser exit are studied using CFD techniques. Sophisticated models which take into account vaporization of the liquid droplets, heterogeneous cracking reactions, and catalyst deactivation are applied to the model in order to simulate the reactive flow in industrial scale FCC riser reactors. A parametric study of the hydrodynamic behaviors induced by use of different exit configurations and their effects on the reactions yields is performed. The results obtained are then used to develop a criterion for specifying the riser exit geometries which can improve the reactor efficiency.

2. Mathematical Model

The three-phase model used in this work considers a three-dimensional gas-liquid-solid flow including heat transfer,

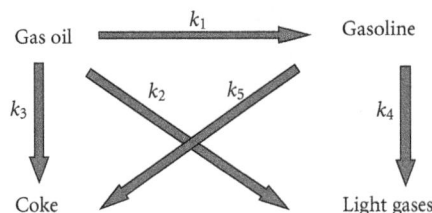

FIGURE 1: Kinetic scheme of the four-lump reaction model.

droplets vaporization, and chemical reactions. This model uses an Eulerian description of the gas and solid phases, while the Lagrangian approach was chosen to describe the liquid droplets. This phase consists of droplets represented as spherical particles and dispersed in the continuous phase. Details of these models are shown below.

2.1. Eulerian Gas-Solid Model. In the Eulerian approach both gas and solid phases are modeled as continuous phases. The characteristics of these phases are then determined by solving the mass, momentum, and energy transport equations and applying closure equations to predict the interaction between the phases.

Since the turbulence in the gas phase was modeled using the Reynolds Stress Model (RSM), one transport equation for each of the Reynolds stress tensor components should be solved. As being an anisotropic model, the RSM is able to predict the flow characteristics in regions where sudden changes in flow directions are expected.

The momentum transfer was modeled using the Gidaspow drag model, which combines the Wen Yu correlation with the Ergun equation. The heat transfer was predicted using the Ranz-Marshall correlation for the Nusselt number. Fluctuations in particle velocity were modeled using the Kinetic Theory of Granular Flow. This theory applies the concept of granular temperature, a quantity related to the kinetic energy due to particle movement, to provide closure terms for the solid-phase stress terms. An algebraic formulation, which assumes that the generation and dissipation of the granular temperature are in equilibrium, was applied to estimate this quantity.

A summary of the governing and constitutive equations can be seen in Table 1.

2.1.1. Catalytic Cracking Kinetic Model. Many complex reactions occur simultaneously inside the reactor during the FCC process. In order to simplify this kinetic net, a technique widely used is to describe the complex mixtures of hydrocarbons as a reduced number of component groups or lumps. This produces a small number of representative pseudocomponents reacting with each other. This work used a four-lump model proposed by lee et al. [29], in which species with similar properties are grouped into four different lumps: gas oil, gasoline, light gases, and coke. Each lump was defined according to the number of carbons in the molecules as presented in Table 2. The representative reactions of this kinetic scheme are shown in Figure 1.

The general rate equation for reaction r is given by

$$R_{i,r} = k_r C_i^2 \phi. \tag{1}$$

The dependence of kinetic constants on temperature is given by the Arrhenius equation:

$$k_r = k_r^0 \exp\left(-\frac{E_r}{RT}\right). \tag{2}$$

The parameter ϕ, appearing in (1), is the catalyst activity function which is related to the deposition of coke on the catalyst surface. It is expressed by

$$\phi = \exp(-K_c q_1), \tag{3}$$

where K_c is the activity constant estimated by Farag et al. [27] as a function of the catalyst type. The value obtained for FCC10 catalyst (sample free of metal traps, nickel, and vanadium) was used in this study. The specific coke concentration, q_1, is given by

$$q_1 = \frac{C_{coke}}{\rho_s \varepsilon_s}. \tag{4}$$

For simplification, it is assumed that coke is not physically deposited on the catalyst surface.

The net source of chemical species i due to reaction (see equation (G) in (Table 1)) is estimated as a sum of the Arrhenius reaction sources over the N_r reactions that the species participate in:

$$\hat{R}_i = M_{w,i} \sum_{r=1}^{N_r} (\nu_i R_{i,r}), \tag{5}$$

where ν_i is the stoichiometric coefficient of the species i, which is positive for products and negative for reagents.

The values used for the kinetic constants were those obtained by Farag et al. [27] for FCC10 catalyst at the temperature of 823 K. Farag et al. [27] concluded that the overcracking of the gasoline formed was negligible because the kinetic constant values for the cracking of the gasoline obtained in their study were very close to zero. Since they do not estimate the activation energies and the heats of reaction, the values reported by Juárez et al. [30] and Han and Chung [16] were adopted. These values are listed in Table 3.

The kinetic constants, given in Table 3, were evaluated at 823 K and are dependent on the amount of solids. In order to predict these values at any temperature and catalyst concentration, the preexponential factor was isolated from (2), applied for this temperature, and multiplied by the local concentration of solids. Then the kinetic constants were evaluated as follows:

$$k_r(T, \varepsilon_s) = k_{r,823\,K}(\rho_s \varepsilon_s) \exp\left[-\frac{E_r}{R}\left(\frac{1}{T} - \frac{1}{823\,K}\right)\right]. \tag{6}$$

2.2. Lagrangian Discrete Phase Model. Heavy oil is injected into the reactor as liquid droplets, which are modeled in this work using a Lagrangian discrete phase model. The Lagrangian approach gives more accurate results than the

TABLE 1: Summary of governing and constitutive equations for Eulerian approach.

Gas-solid Eulerian flow model		
Governing equations		

Continuity equations:

Gas-phase:
$$\frac{\partial}{\partial t}(\varepsilon_g \rho_g) + \nabla \cdot (\varepsilon_g \rho_g \mathbf{u}_g) = n_d \frac{\partial m_d}{\partial t} \tag{A}$$

Solid-phase:
$$\frac{\partial}{\partial t}(\varepsilon_s \rho_s) + \nabla \cdot (\varepsilon_s \rho_s \mathbf{u}_s) = 0 \tag{B}$$

Momentum equations:

Gas-phase:
$$\frac{\partial}{\partial t}(\varepsilon_g \rho_g \mathbf{u}_g) + \nabla \cdot (\varepsilon_g \rho_g \mathbf{u}_g \mathbf{u}_g) = \nabla \cdot \left[\varepsilon_g \mu_g \left(\nabla \mathbf{u}_g + (\nabla \mathbf{u}_g)^T \right) - \frac{2}{3} \varepsilon_g \mu_g (\nabla \cdot \mathbf{u}_g)\mathbf{I} - \varepsilon_g \rho_g \overline{\mathbf{u}'\mathbf{u}'} \right]$$
$$+ \varepsilon_g \rho_g \mathbf{g} - \varepsilon_g \nabla p + \beta(\mathbf{u}_s - \mathbf{u}_g) + \mathbf{u}_g n_d \frac{\partial m_d}{\partial t} \tag{C}$$

Solid-phase:
$$\frac{\partial}{\partial t}(\varepsilon_s \rho_s \mathbf{u}_s) + \nabla \cdot (\varepsilon_s \rho_s \mathbf{u}_s \mathbf{u}_s) = \nabla \cdot \left[\varepsilon_s \mu_s \left(\nabla \mathbf{u}_s + (\nabla \mathbf{u}_s)^T \right) - \frac{2}{3} \varepsilon_s \mu_s (\nabla \cdot \mathbf{u}_s)\mathbf{I} \right] + \varepsilon_s \rho_s \mathbf{g}$$
$$- \varepsilon_s \nabla p - \nabla p_s + \beta(\mathbf{u}_g - \mathbf{u}_s) \tag{D}$$

Energy equations:

Gas-phase:
$$\frac{\partial}{\partial t}(\varepsilon_g \rho_g H_g) + \nabla \cdot (\varepsilon_g \rho_g \mathbf{u}_g H_g) = \nabla \cdot (\varepsilon_g \lambda_g \nabla T_g) + \gamma(T_s - T_g) + \sum_r \Delta H_r \frac{\partial(\varepsilon_g \rho_g Y_{g,i})}{\partial t}$$
$$+ H_g n_d \frac{\partial m_d}{\partial t} \tag{E}$$

Solid-phase:
$$\frac{\partial}{\partial t}(\varepsilon_s \rho_s H_s) + \nabla \cdot (\varepsilon_s \rho_s \mathbf{u}_s H_s) = \nabla \cdot (\varepsilon_s \lambda_s \nabla T_s) + \gamma(T_g - T_s) \tag{F}$$

Species conservation:
$$\frac{\partial}{\partial t}(\varepsilon_g \rho_g Y_{g,i}) + \nabla \cdot (\varepsilon_g \rho_g \mathbf{u}_g Y_{g,i}) = \nabla \cdot (\varepsilon_g \rho_g \Gamma_i \nabla Y_{g,i}) + \hat{R}_i + Y_{g,i} n_d \frac{\partial m_d}{\partial t} \tag{G}$$

Reynolds stresses:
$$\frac{\partial}{\partial t}(\varepsilon_g \rho_g \overline{\mathbf{u}'\mathbf{u}'}) + \nabla \cdot (\varepsilon_g \mathbf{u} \rho_g \overline{\mathbf{u}'\mathbf{u}'}) = -\varepsilon_g \rho_g \left[\overline{\mathbf{u}'\mathbf{u}'}(\nabla \mathbf{u})^T + (\nabla \mathbf{u})\overline{\mathbf{u}'\mathbf{u}'} \right]$$
$$+ \nabla \cdot \left[\varepsilon_g \left(\mu_g + \rho_g \frac{C_\mu}{\sigma_k} \frac{k^2}{\epsilon} \right) \nabla \overline{\mathbf{u}'\mathbf{u}'} \right]$$
$$+ \varepsilon_g \Phi - \frac{2}{3} \varepsilon_g \delta \rho_g \epsilon + \Pi_{R,ij} \tag{H}$$

Constitutive equations		

Gas-solid momentum exchange
(Gidaspow drag model):

when $\varepsilon_s > 0.2$,
$$\beta = 150 \frac{\varepsilon_s^2 \mu_g}{\varepsilon_g d_s^2} + \frac{7}{4} \frac{|\mathbf{u}_s - \mathbf{u}_g| \varepsilon_s \rho_g}{d_s} \tag{I}$$

when $\varepsilon_s \leq 0.2$,
$$\beta = \frac{3}{4} C_D \frac{|\mathbf{u}_s - \mathbf{u}_g| \varepsilon_s \rho_g}{d_s} \varepsilon_s^{-2.65} \tag{J}$$

for Re > 1000,
$$C_D = 0.44 \tag{K}$$

for Re < 1000,
$$C_D = \frac{24}{Re}(1 + 0.15 Re^{0.687}) \tag{L}$$

Gas-solid heat exchange:
$$\gamma = \frac{(2 + 0.6 Re^{0.5} Pr^{0.3}) 6 \lambda \varepsilon_s \varepsilon_g}{d_s^2} \tag{M}$$

Kinetic theory of granular flow:

Solids pressure:
$$p_s = \varepsilon_s \rho_s \Theta_s + 2\rho_s(1 + e_{ss}) \varepsilon_s^2 g_{0,ss} \Theta_s \tag{N}$$

Radial distribution:
$$g_{0,ss} = \left[1 - \left(\frac{\varepsilon_s}{\varepsilon_{s,max}} \right)^{1/3} \right]^{-1} \tag{O}$$

Granular temperature:
$$\Theta_s = \left\{ \frac{-K_1 \varepsilon_s \nabla \cdot \mathbf{u}_s + \sqrt{(K_1 \varepsilon_s \nabla \cdot \mathbf{u}_s)^2 + 4 K_4 \varepsilon_s \left[K_2 (\nabla \cdot \mathbf{u}_s)^2 + 2 K_3 (\nabla \cdot \mathbf{u}_s^2) \right]}}{2 \varepsilon_s K_4} \right\}^2 \tag{P}$$

where,
$$K_1 = 2(1 + e_{ss})\rho_s g_{0,ss} \tag{Q}$$
$$K_2 = \frac{4 d_s \rho_s (1 - e_{ss}) \varepsilon_s g_{0,ss}}{3\sqrt{\pi}} - \frac{2}{3} K_3 \tag{R}$$
$$K_3 = \frac{d_s \rho_s}{2} \left\{ \frac{\sqrt{\pi}}{3(3 - e_{ss})} \left[1 + 0.4(1 - e_{ss})(3 e_{ss} - 1) \varepsilon_s g_{0,ss} \right] + \frac{8 \varepsilon_s g_{0,ss}(1 + e_{ss})}{5\sqrt{\pi}} \right\} \tag{S}$$
$$K_4 = \frac{12(1 - e_{ss}^2)\rho_s g_{0,ss}}{d_s \sqrt{\pi}} \tag{T}$$

TABLE 2: Definition of lumps [27].

Lump	Number of carbons
Gas oil	C_{13} and higher
Gasoline	C_5–C_{12}
Light gases	C_1–C_4

TABLE 3: Kinetic constants, activation energies, and heats of reaction.

Reaction r	$k_r|_{823\,K}$ (m^6 kmol^{-1} kg$_{\mathrm{cat}}^{-1}$ s^{-1})	E_r (J mol^{-1})	ΔH_r (kJ kg^{-1})
Gas oil → gasoline	20.4	57360	195
Gas oil → light gases	7.8	52750	670
Gas oil → coke	3.0	31820	745

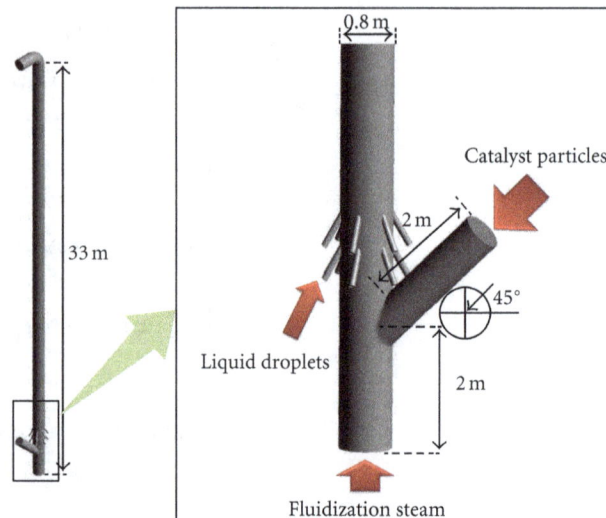

FIGURE 2: Geometry of the industrial riser.

Eulerian approach, since equations for position, velocity, temperature, and masses of species are solved individually for each discrete particle.

The trajectory of each droplet is predicted solving the forces acting on the particle: the force of gravity and the drag with the gas phase. The Morsi and Alexander drag correlation for spherical particles [31] is used to model the drag coefficient.

According to the heat balance equations, these droplets are heated by the gas phase at higher temperatures, reducing their diameters until the liquid phase is completely vaporized. While the droplet temperature is below its vaporization temperature, convective heat is transferred from the continuous phase to the discrete phase and the droplet temperature is predicted using a heat balance. In this process, the heat transfer coefficient was estimated using the Ranz-Marshall correlation.

In the temperature range between the vaporization and the boiling temperature, mass is transferred from the liquid droplet to the continuous phase. The droplet temperature is then increased according to a heat balance that relates the droplet sensible heat change to the convective and latent heat transfer between the droplet and the continuous phase. When the droplet reaches the boiling temperature, it is assumed to be constant, and the diameter of the droplet begins to decrease until complete vaporization.

Since gas oil droplets rapidly vaporize, particle-particle interactions and effects of the particle volume fraction on the gas phase are negligible. The dispersion of particles due to turbulence in the fluid phase was also neglected.

A summary of the equations applied to model the discrete droplets can be seen in Table 4.

3. Simulations

The commercial code ANSYS FLUENT 12.0 was used to solve the proposed model. Appropriate user-defined functions were developed to implement the heterogeneous kinetics and the catalyst deactivation model into the software. FLUENT applies the finite volume method, where the domain is divided into a finite number of control volumes in which discrete variables are calculated. The pressure field was de-termined using a pressure-based approach in which the continuity and momentum equations are manipulated to obtain an approximate equation for the pressure. The least squares cell-based method was applied to evaluate the diffusion and pressure gradients. The convective terms were discretized using a first-order upwind difference scheme, and transient terms were approximated using a first-order implicit scheme. The set of algebraic approximate equations was solved in a segregated way, using the simple algorithm for the pressure-velocity coupling [32].

The geometry of the simulated riser was proposed based on configurations found in the literature. The height and diameter of the column were based on an industrial riser reported by Ali et al. [14] and are shown in Table 6. Since they do not report details about the entrance and the outlet geometric configurations of that riser, a lateral entrance located 2 m from the base of the riser was adopted, for the catalyst particles feeding. Twelve 0.5 in diameter ducts at a 30° angle to the main duct were used to feed in the liquid droplets. These configurations are commonly found in studies of CFB risers [33–35] and about the vaporization of the FCC feedstock [24, 36, 37]. This riser reactor geometry is illustrated in Figure 2.

Four different kinds of configurations for the riser outlet (Figure 3) are proposed in the present study, based on the works of Cheng et al. [1], Gupta and Berruti [2], Van Der Meer et al. [3], Harris et al. [4], Chan et al. [5], Pugsley et al. [9], and Van Engelandt et al. [10]. These configurations can be classified in two main groups: abrupt exits, in which there are sharp changes in the flow direction, inducing solids recirculation in the top section of the riser, and smooth exits, in which the flow changes according to the outlet bend, following a natural path.

The meshes applied consist of approximately 1.2 million hexahedral volumes depending on the configuration used for the riser exit. In order to guarantee that smaller control volumes are present where variable gradients are steeper, nonuniform grids are used.

TABLE 4: Summary of Lagrangian equations.

Lagrangian discrete phase model		
Discrete phase trajectory:	$\dfrac{dx_d}{dt} = u_d$	(A)
Discrete phase velocity:	$\dfrac{du_d}{dt} = \dfrac{\mathbf{g}(\rho_d - \rho_g)}{\rho_d} + F_D(u_g - u_d)$	(B)
Drag force:	$F_D = \dfrac{18\mu_g}{\rho_d d_d^2} \dfrac{C_D^d \mathrm{Re}_d}{24}$	(C)
Droplet Reynolds number:	$\mathrm{Re}_d = \dfrac{\rho_g d_d \left| u_d - u_g \right|}{\mu_g}$	(D)
Drag coefficient:	$C_D^d = a_1 + \dfrac{a_2}{\mathrm{Re}} + \dfrac{a_3}{\mathrm{Re}^2}$	(E)
Heat balances:		
$\quad T_d < T_{\mathrm{vap}}$:	$m_d C_p \dfrac{dT_d}{dt} = hA_d(T_\infty - T_d)$	(F)
$\quad T_{\mathrm{vap}} \le T_d < T_{\mathrm{bp}}$:	$m_d C_p \dfrac{dT_d}{dt} = hA_d(T_\infty - T_d) + \dfrac{dm_d}{dt}L$	(G)
$\quad$ Heat transfer coefficient:	$h = \dfrac{\lambda}{d_d}(2.0 + 0.6\mathrm{Re}_d^{1/2}\mathrm{Pr}^{1/3})$	(H)
Mass transfer:	$\dfrac{dm_d}{dt} = -k_c(C_{i,\mathrm{surf}} - C_{i,\infty})A_d M_{w,i}$	(I)
$\quad$ Mass transfer coefficient:	$k_c = \dfrac{D_{m,i}}{d_d}(2.0 + 0.6\mathrm{Re}_d^{1/2}\mathrm{Sc}^{1/3})$	(J)
Droplet diameter variation:	$-\dfrac{d(d_d)}{dt} = \dfrac{4k_\infty(1 + 0.23\sqrt{\mathrm{Re}_d})}{\rho_d C_{p,\infty} d_d}\ln\left[1 + \dfrac{(T_\infty - T_d)}{L}\right]$	(K)

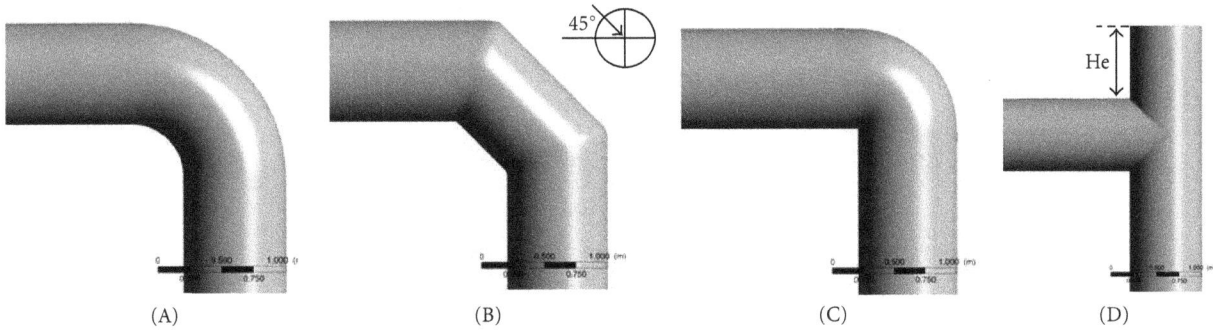

FIGURE 3: Exit configurations used in the simulations.

Since the heavy oil is not a pure component, its vaporization does not occur in a constant temperature. The properties of the liquid heavy oil and the values used for the range of temperature in which its vaporization occurs were taken from Nayak et al. [17] and are shown in Table 5.

The operating conditions used in the simulations are the same of the industrial riser reported by Ali et al. [14]. They are listed in Table 6. About 7% of the total steam is fed with the catalyst particles at the lateral entrance, while the remaining steam is injected at the base of the reactor to help fluidization inside the riser. The physical properties and characteristics of the reactive species and the catalyst were taken from Martignoni and De Lasa [38] Van Landeghem et al. [19] and are listed in Table 7. The density of each reactive species was considered constant, independent of the system pressure and temperature. However, the density of the gas phase varies as the mixture composition changes with cracking reactions.

The simulations were carried out using a parallel code with 16 partitions on computers provided with Intel Xeon 3 GHz quad-core processors. The time step used for the solution was 10^{-3} s. The convergence criterion for advancing in time was that the RMS residuals were less than 10^{-4}. Initially 2 s of transient flow, in which just the steam and the catalyst particles were present, were calculated. After this time, the feed oil droplets started to be injected into the reactor and 8 s more of flow were simulated. Then the time average values of the flow variables were obtained for five additional seconds. About one day of calculation was necessary to predict each second of reactive flow for each simulation.

4. Results and Discussion

In the FCC process, heavy oil is injected into the riser reactor as liquid droplets. The cracking reactions are just initiated when these droplets vaporize and the reactive species are transferred from the liquid to the gas phase. Many studies about the FCC reactors, assume instantaneous vaporization of the oil. Lopes et al. [26], however, observed a significant

TABLE 5: Physical properties of the liquid feed oil.

Density	Vaporization temperature	Boiling temperature	Specific heat	Thermal conductivity	Latent heat
$870 \, \text{kg m}^{-3}$	530 K	560 K	$1040 \, \text{J kg}^{-1} \, \text{K}^{-1}$	$0.2 \, \text{W m}^{-1} \, \text{K}^{-1}$	$4 \times 10^5 \, \text{J kg}^{-1}$

TABLE 6: Geometric details and operating conditions of the industrial riser.

Riser height (m)	33
Riser diameter (m)	0.8
Mass flux of feed oil ($\text{kg m}^{-2} \, \text{s}^{-1}$)	40
Catalyst-to-oil ratio ($\text{kg}_{cat}/\text{kg}_{gasoil}$)	7
Steam (wt.%)	7
Feed oil temperature (K)	500
Catalyst inlet temperature (K)	960
Droplet inlet diameter (μm)	100

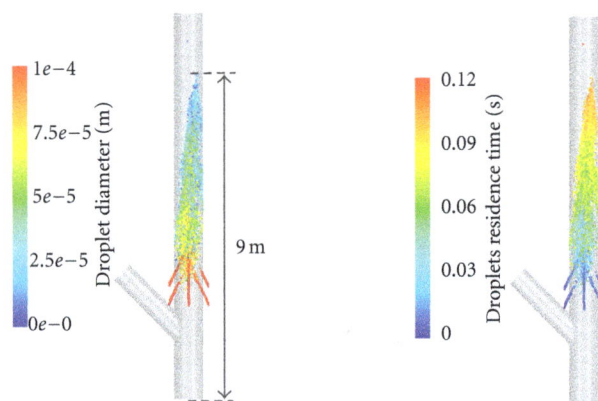

FIGURE 4: Droplets' diameter and residence time inside the reactor.

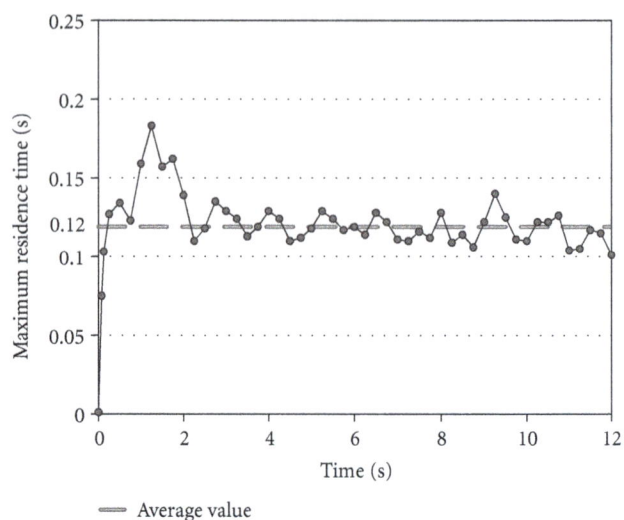

FIGURE 5: Maximum residence time of the droplets.

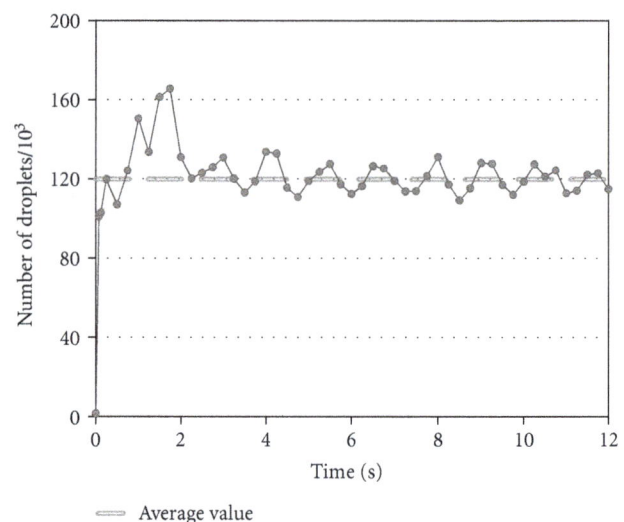

FIGURE 6: Instantaneous number of droplets.

difference in product yields when the hypothesis of instantaneous vaporization was adopted and when this phenomenon is modeled. Based on these results, the vaporization of the droplets was taken into account in the present work. The diameter of the droplets and their residence time inside the reactor are shown in Figure 4. These results were collected 10 s after the start of the liquid injection. At this moment, the complete vaporization of the liquid droplets takes place at a height of approximately 9 m.

The maximum residence time of the droplets before they are completely vaporized was monitored along the time. As can be seen in Figure 5, the maximum value observed was 0.18 s, obtained 1.25 s after the start of the liquid droplets injection. After this time, the maximum residence time decreases, reaching an average value of 0.12 s. According to Liu and Han [39] (cited in [37]) the complete vaporization of the charge is completed within 0.2 s, in risers of typical FCC units. This shows that the model used in the simulations is appropriate to describe the phenomenon of vaporization. The number of liquid droplets present in the system follows the same trend (Figure 6), reaching its maximum value about two seconds after the beginning of the charge injection into the reactor. The average number of liquid droplets inside the

riser reactor is about 120 thousand. These results emphasize the importance of simulating the vaporization of the liquid oil charge stock in the FCC process.

The inlet zone of the riser is a very complex part of the reactor, where intense turbulence and flow inhomogeneities are observed. A detailed study about this region was performed by Lopes et al. [22], where it is shown that, depending on the operating conditions applied, the inlets arrangement can affect the flow patterns along all riser height. Extremely complex hydrodynamics has been also observed in the outlet

TABLE 7: Physical properties of reactive species and catalyst.

Species	Vaporized gas oil	Gasoline	Light gases	Coke	Catalyst
Density (kg m^{-3})	6	1.5	0.8	1400	1400
Specific heat (J kg^{-1} K^{-1})	2420	2420	2420	1090	1090
Thermal conductivity (W m^{-1} K^{-1})	0.025	0.025	0.025	0.045	0.045
Molecular weight (kg kmol^{-1})	400	100	50	400	—
Particle diameter (μm)	—	—	—	—	65
Viscosity (kg m^{-1} s^{-1})	5.0×10^{-5}	1.6×10^{-5}	1.6×10^{-5}	1.7×10^{-5}	1.7×10^{-5}

TABLE 8: Definition of simulated cases.

Case	Configuration[1]	He (m)[1]
1	A	—
2	B	—
3	C	—
4	D	0
5	D	0.8

[1] According to Figure 3.

region of CFB risers [3, 9, 10], which can affect the reactor efficiency.

In order to verify the influence of different riser exit configurations on reactor performance, five cases were initially proposed. The four geometric configurations shown in Figure 3 were tested. Two variations of the abrupt T exit (Figure 3(d)) with different projected heights (He) were also simulated. The definition of each case according to the configuration applied in the simulations can be seen in Table 8.

The effect of the exits on the flow was evaluated using parameters reported by Van Der Meer et al. [3] and Schut et al. [28]. Their definitions are listed in Table 9. These parameters help to quantify some important characteristics of the flow in the riser, such as core-annulus patterns, the increase in the downflux in abrupt exit risers, and the asymmetry of the flow.

The values of parameter H_{bf} obtained for each of the five cases are shown in Table 10. For Cases 1 and 2 these values are zero, indicating that under the conditions applied, there is no solids reflux when configurations A and B (Figure 3) are used. These results help to characterize these configurations as smooth exits, since they do not restrict the solids flux. Despite the similarity between configurations A and C (Figure 3), the case simulated using the latter (Case 3) provided some solids backmixing ($H_{bf} = 0.036$) induced by the sharp right angle in this configuration. The maximum reflux of particles was obtained when the T-shape exit with a projected height of 0.8 m was used (Case 5). In this case, solids flowing in counter current to the main flow reached a distance of 9.75 m from the top of the riser, corresponding to about 28% of its total height.

The movement of solid particles in center planes at the outlet bends is represented as velocity vectors in Figure 7. As shown previously, no solids reflux can be observed in Cases 1 and 2, justifying the zero value for parameter H_{bf} obtained in these cases. The internal sharp right angle seen in Cases 3, 4,

and 5 induces the formation of recirculation areas just after the bend. A second vortex is formed at the riser wall opposite the outlet opening in Cases 4 and 5, in which the exit is at sharp right angles also at the outer side of the bend. The higher the downward flow of the particles is, the more intense the restriction on the direct flow is, causing an increase in its velocity in the central region of the riser.

Harris et al. [40] described a physical mechanism to help to explain the motion of solids at a riser exit. They postulate that when solids follow a curved path of mean radius R at the exit, the centrifugal acceleration is balanced against the average component of the acceleration due to gravity acting toward the center of the bend. In regions where the solids slow down due to geometric accidents at the exit bend, gravity overlaps centrifugal acceleration, causing a change in trajectory and creating recirculation zones, as observed in the simulated cases.

In Figure 8, solids volume fraction fields are shown for the five simulated cases, where an accumulation of particles at the top of the riser can be seen. As the particle density is much higher than the gas mixture density, the particle inertia is also higher, which means that particles are easily separated from the gas stream near the exit, even at outlets considered smooth. In addition, the abrupt exits create an extra resistance, causing an increase in the concentration of this phase at the outlet bend. As a result, the average solids fraction and the annulus thickness near the top of the riser are much higher in Cases 4 and 5, in which the solids recirculation is also higher. As observed in other studies [10, 11], for risers with abrupt exits the accumulation of solids is more pronounced at the side opposite to the outlet than the side of the riser at which the outlet bend is positioned.

The accumulation of solids observed in the cases with abrupt exits results in a densification of the annulus structure near the outlet bends. This is confirmed in Figure 9, which shows the solids volume fraction in radial planes located at heights of 20 and 32 m in Cases 1 and 5. At a height of 20 m the core-annulus patterns are very similar in both cases. However, when approaching the outlet bend there is a coarsening of the annular structure in Case 5 resulting from the solids recirculation observed in this case.

In order to analyze the solids accumulation near the exit of the risers with different configurations, average values of the solids volume fraction in transversal planes were taken along the riser height. The shape of the average solids hold-up profile is a good indicator of how the exit effects propagate down the riser. As shown in Figure 10, in all cases studied

TABLE 9: Parameters applied to evaluate the influence of the exits on the flow.

Parameter	Definition
H_{bf}	Backflux relative height: distance from the top of the riser traveled by the particulate phase in downward motion divided by riser height.
k_a	Exit reflection coefficient: fraction of the riser transversal area within which solids move downwards.
k_a^*	Asymmetry of k_a: defined by the difference between the fraction of the area with negative velocity of solids located at the positive and the negative side of the x-axis.
δ	Film thickness of downflow: given by $D(1 - \sqrt{1 - k_a})/2$ [28].

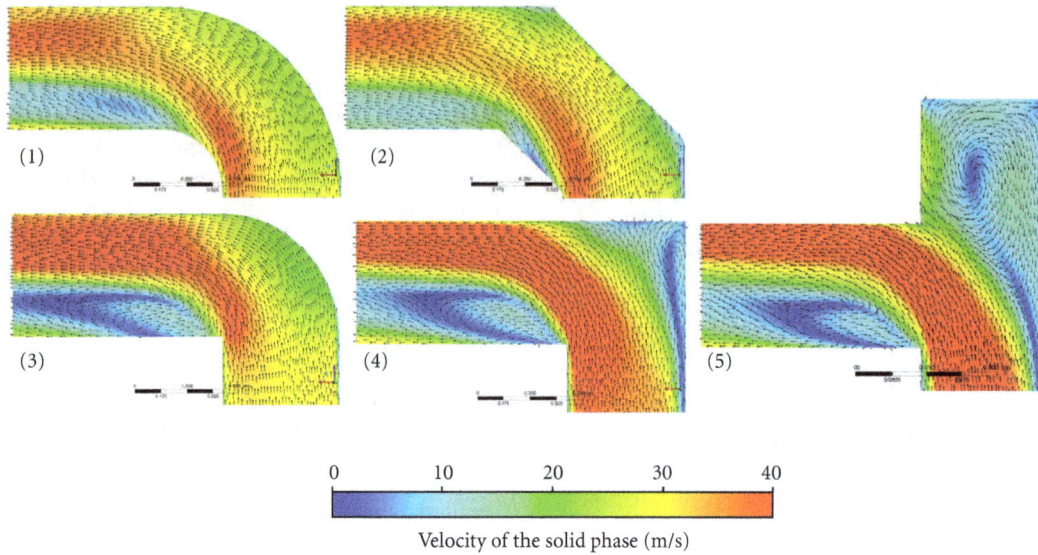

Velocity of the solid phase (m/s)

FIGURE 7: Solid phase velocity and movement.

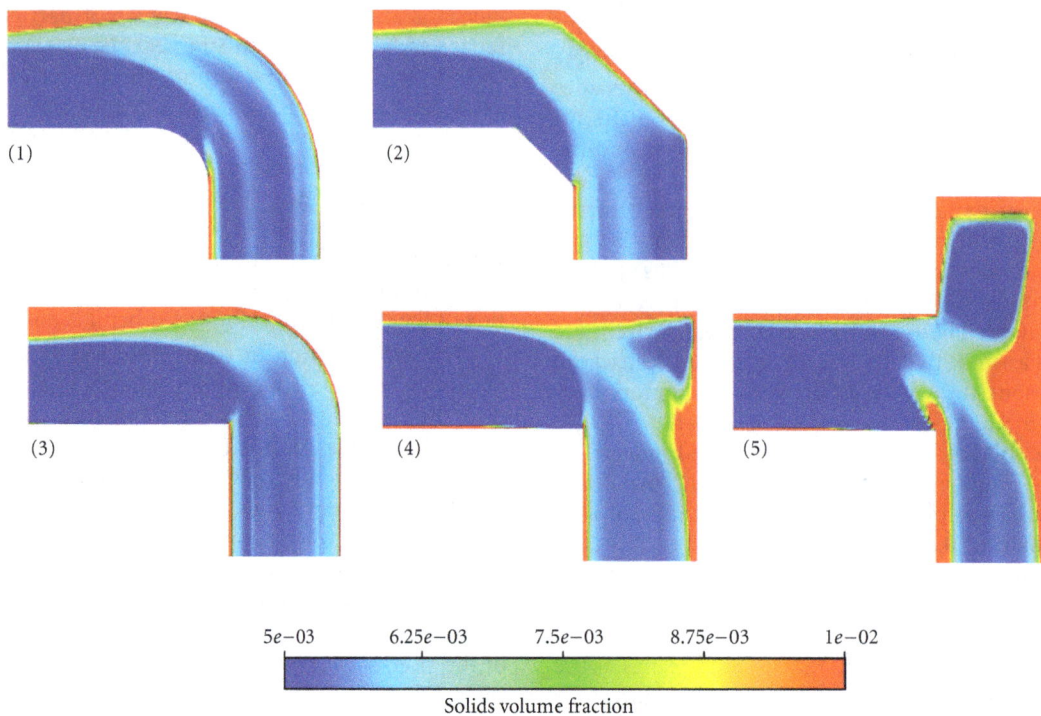

Solids volume fraction

FIGURE 8: Solids volume fraction profiles at the exits.

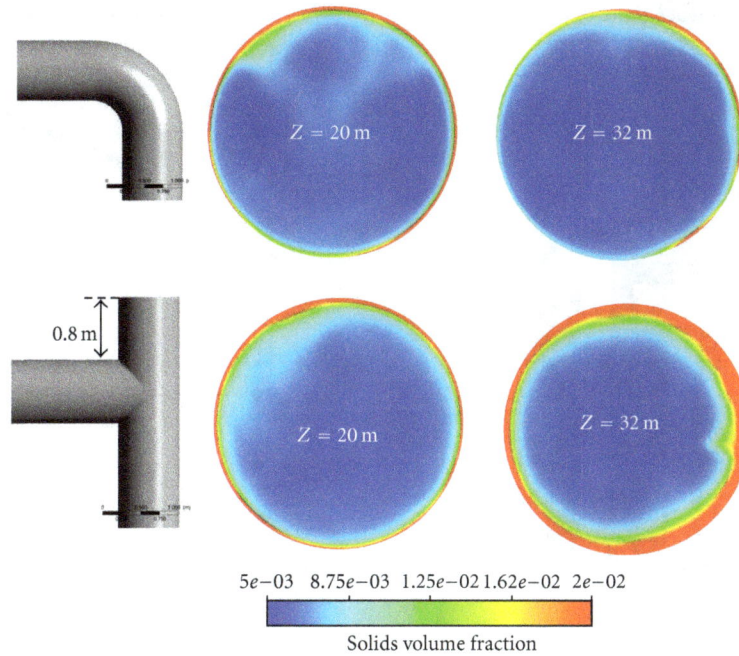

FIGURE 9: Core-annulus patterns.

TABLE 10: H_{bf} obtained for each case.

Case	H_{bf}
1	0.000
2	0.000
3	0.036
4	0.117
5	0.279

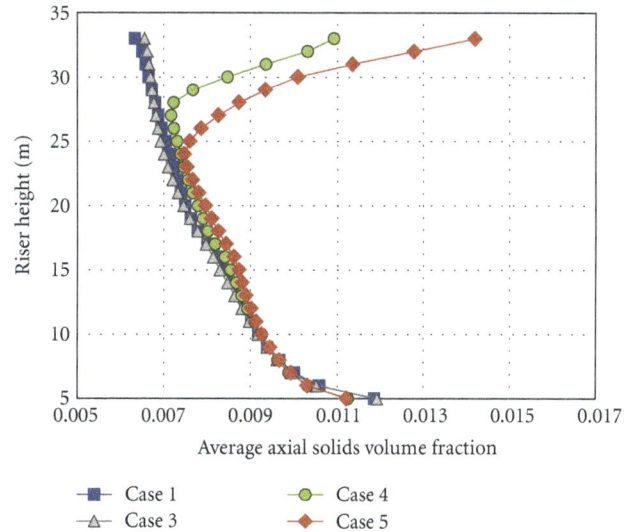

FIGURE 10: Average solids volume fraction in cross-section planes along the height.

the average volume fraction initially decreases due to the expansion of the gas phase caused by the catalytic cracking of the large hydrocarbon molecules into smaller ones during the process. The profiles continue decreasing in the cases with smooth exits, but this tendency suddenly change in risers with abrupt exits, creating a profile described as C-shape. The more restrictive the exit configuration is, the further down the riser its effect propagates. This trend is consistent with results reported in many studies on CFB risers [6, 40–42].

The recirculation of solids observed in the risers with abrupt exits results in an increase in their residence time inside the reactor. As the solid phase is treated in this study as a continuous fluid using Eulerian approach, it is not possible to calculate exactly the residence time of each particle in the system. In order to estimate an approximated time, a methodology applied by liu and Tilton [43], in which average values are estimated, was applied. Initially, a new property (Ψ), with unit of second, was defined for the solid phase. An user-defined scalar (UDS) transport equations was then introduced into the model in the form [32]

$$\frac{\partial}{\partial t}(\varepsilon_s \rho_s \Psi) + \nabla \cdot (\varepsilon_s \rho_s \mathbf{u}_s \Psi) = \nabla \cdot (\varepsilon_s \Gamma^\Psi \nabla \Psi) + S^\Psi. \quad (7)$$

If the diffusive terms were neglected, the scalar just propagates with the time and the movement of the solid phase. The source term was then implemented via user defined-function (UDF) as

$$S^\Psi = \varepsilon_s \rho_s. \quad (8)$$

As the solution advances in time, this value is added to the scalar. Thus, assuming that this property is equal to zero at the solids entrance and in the problem initialization, a cumulative function that propagates with the time and the solids movement is created. This enables an estimation

TABLE 11: Residence time.

Case	Mean value (s)
1	1.943
5	2.299

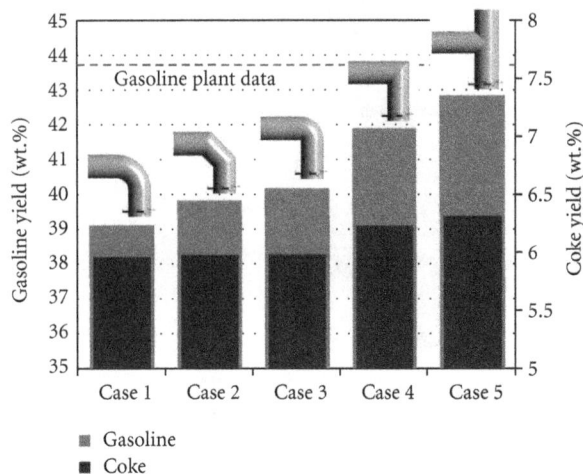

FIGURE 11: Gasoline and coke yields at the riser outlet.

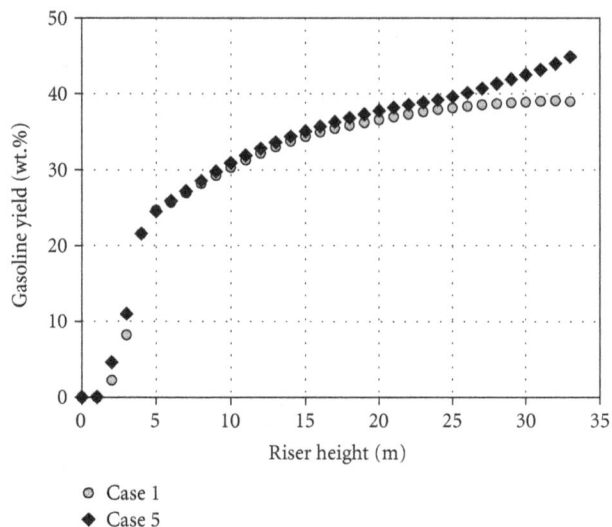

FIGURE 12: Average gasoline yield in cross-section planes along the height.

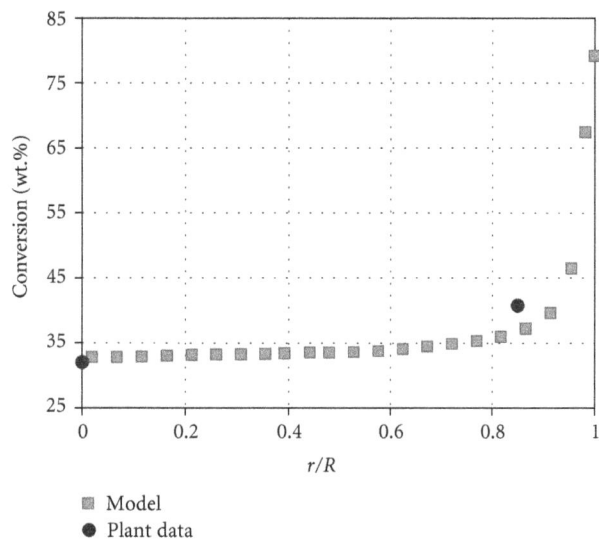

FIGURE 13: Radial profile of cracking conversion.

of an approximated residence time of the solid particles. In Table 11 the solids mean residence time obtained for Cases 1 and 5 is shown. When the abrupt T-shape exit with a projected height of 0.8 m is applied in the FCC riser simulation, the mean residence time is increased by about 18% in relation to the value obtained for the smoother exit. This difference can influence the reaction time, increasing the conversion rates.

The gasoline and coke yields at the riser outlet obtained in the simulated cases are shown in Figure 11. This value increases as resistance to the flow becomes more intense and the solids recirculation increases. Comparing Cases 1 and 5, there is an increase of about 4% in the gasoline yield.

In order to validate the simulated results with the data reported by Ali et al. [14], the operating conditions and the equipment dimensions used were the same as those of their industrial riser. Since Ali et al. [14] do not report details of the exit configuration of this riser, several configurations were tested in the simulations. However, it is known that the T-shape design, as used in the simulation of Case 5, is the most common configuration found in industry, since it reduces the erosive particle impingement on the roof of the riser. As can be seen in Figure 11, the value of the gasoline yield which most closely approximates the industrial value (43.9%) was the one obtained in the simulation of Case 5, with a deviation of only about 2%.

Average values of gasoline yield were also taken in transversal planes along the riser height in Cases 1 and 5, as shown in Figure 12. Similarly to the profiles of average solids volume fraction (Figure 10), the gasoline yield profiles coincide up to a certain height, at which point the tendency in the riser changes with the abrupt T-shape exit.

The results obtained for the cracked hydrocarbons conversion in the simulation of Case 5 were also compared with plant data sampled by Martin et al. [41] at 4 m above the upper feedstock injection point at two locations along the riser radius (Figure 13). Although they do not provide information on the geometry and operating conditions of the industrial plant for which the data were taken, the comparison of simulated and experimental local cracking conversion showed good agreement. It can be clearly seen that the conversion is higher toward the wall. This can be attributed to the larger catalyst concentration, smaller velocity and higher temperature near the wall, and to the catalyst/gas backmixing at this region. The mass fractions of coke and gasoline in the outlet region for the Cases 1 and 5 are shown in Figure 14, where noticed higher values can be

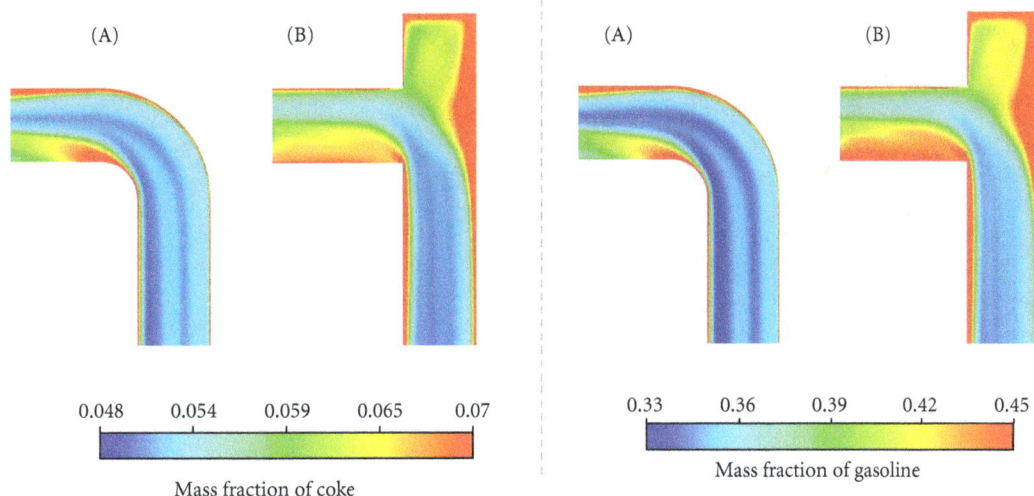

FIGURE 14: Mass fractions of coke and gasoline for Cases 1 (A) and 5 (B).

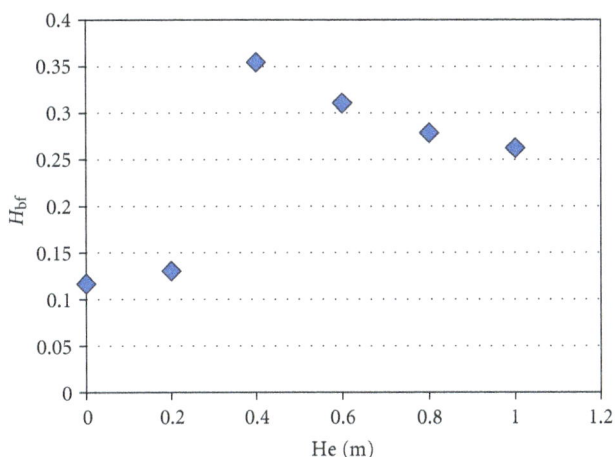

FIGURE 15: H_{bf} parameter obtained for different projected heights.

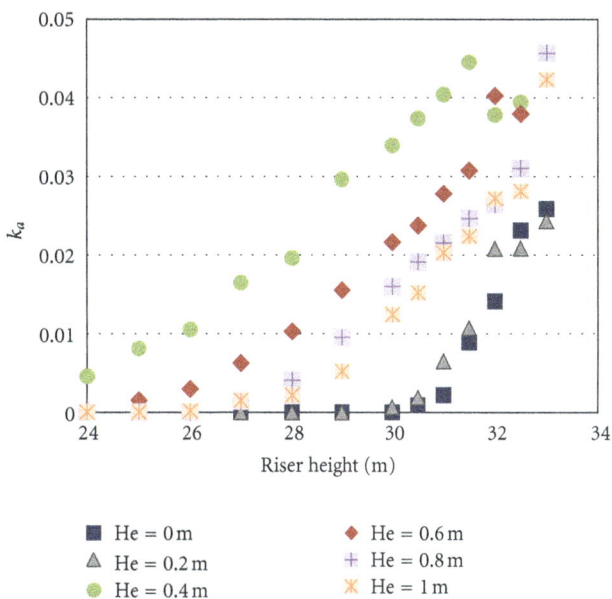

FIGURE 16: k_a parameter obtained along the height for different projected heights.

in the zones in which higher recirculation of solids was found (Figure 8).

As shown above, the geometries applied in the simulation of Cases 1, 2, and 3 (configurations A, B and C in Figure 3) have very small or null reverse flow of solids and therefore can be characterized as smooth exits for the operating conditions and feedstock used in the simulations. In the cases in which the type D configuration with different projected heights is used, there is significant solids recirculation which affects the flow even far from the outlet bend, and therefore it can be classified as abrupt exits. In order to determine the response of the flow to the variation in the projected height of configuration D, the simulation of four new cases, in which He was assumed to be 0.2 m, 0.4 m, 0.6 m, and 1.0 m, was proposed.

The relative height of backflux (represented by parameter H_{bf}) was estimated for each new case, as can be seen in Figure 15. As He increases, there is initially a sharp increase

in the value of H_{bf}, which reaches maximum value at He = 0.4 m, where the solids downward flow is detected at a distance from the outlet bend corresponding to about 35% of the length of the riser. Above He = 0.4 m, the value of H_{bf} decreases, tending to stabilize at about 0.25.

The parameters which represent the fraction of cross-sectional area of the riser in which there is a downward flow of solids (k_a) and the asymmetry of the flow (k_a^*) were also estimated for each of these cases. They are shown in Figures 16 and 17 according to riser height. There is an increase in the value of k_a along the height of the reactor. As shown in Figure 10, the solids accumulation in risers with abrupt exits leads to the enlargement of the annular

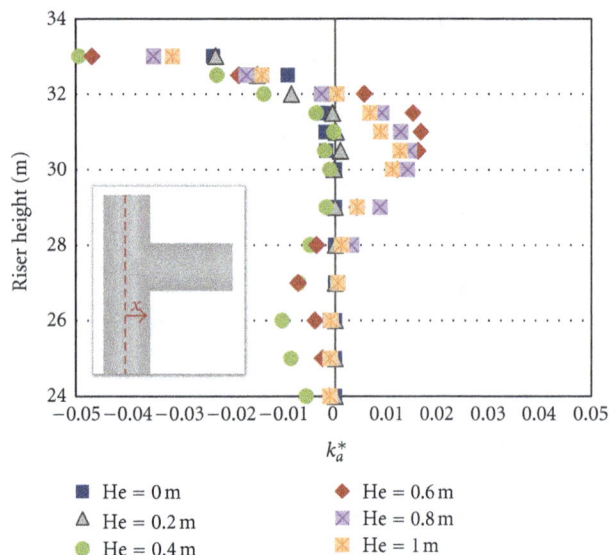

FIGURE 17: k_a^* parameter obtained along the height for different projected heights.

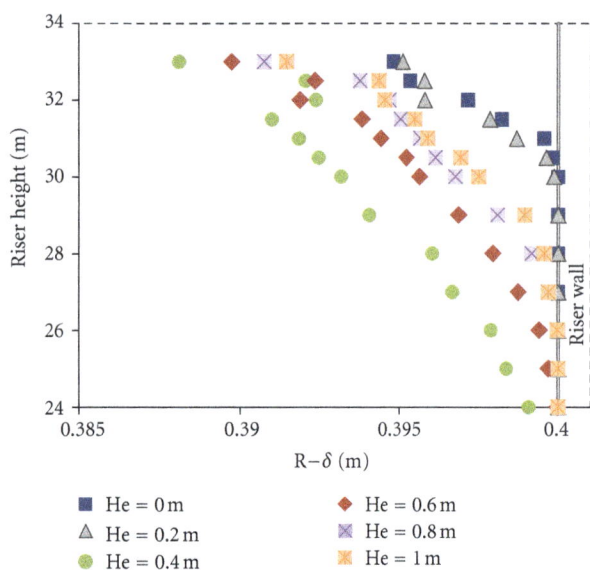

FIGURE 18: Film thickness of downflow.

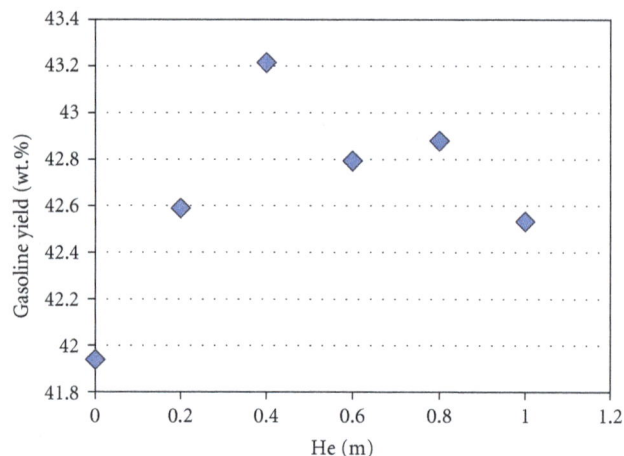

FIGURE 19: Gasoline yields at the riser outlet.

asymmetric patterns near the inner top side wall were also observed by Wang et al. [44] in an experimental study of gas-solid circulating fluidized bed using Electrical Capacitance Volume Tomography (ECVT).

The film thickness of the downflow (δ) is indicated in Figure 18, which shows the values of $R - \delta$ along the riser height. In all cases, this film is thickest immediately below the outlet bend at a height of 33 m and decreases downward from there. Because of the high solids reflux rate found in the riser with a projected height of 0.4 m, the values found for δ are also higher in this case.

In Figure 19 the average values for gasoline yield at the riser outlet obtained for the risers with different projected height are shown. These results follow almost the same tendency as that observed in Figure 15, indicating that solids recirculation increases the formation of gasoline fractions in these cases.

Under the flux conditions used in the present study, the increase in solids residence time improves the reactor efficiency, mainly in cases with a high solids reflux ratio. In addition, as the catalytic reactions depend on the presence of catalyst in the system and the rate of reaction is linearly dependent on its concentration, the increase in the solids volume fraction at the top section of the riser results in an increase in the reaction rates with higher product yields.

5. Conclusions

In the present work, different outlet bend designs for an industrial FCC riser were proposed to study their influence on the dynamic of the flow and consequently on reactor efficiency. A three-dimensional and three-phase reactive flow model was used in order to predict most of the phenomena present in this complex process.

As can be seen from the simulation results, small changes in the geometry of the reactor had a significant effect on the flow patterns and therefore on the product yields. The use of abrupt exits results in solids backmixing, enhancing the residence time of the catalyst and increasing its concentration near the top of the riser. An area of densification is then

structure, consequently increasing the fraction of the area occupied by the downward flowing solids. The configuration with He = 0.4 m has the highest values of k_a even below a height of 24 m, which indicates that the solids recirculation is more intense for this case, corroborating the results shown in Figure 15.

Analysis of parameter k_a^* (Figure 17) shows that the flow is more asymmetric in regions far from the outlet bend for the exit with He = 0.4 m. Moreover, the displacement of this parameter to the negative side of axis x above a height of 32 m in all cases confirms the patterns seen in Figure 8, in which there is a region of solids accumulation on the side opposite to the outlet bend when abrupt exits are used. These

created near the exit, which, depending on the severity of the exit restriction, extends along the length of the riser.

The increase in the amount of catalyst particles favors the heterogeneous FCC reactions, which are proportional to the solids concentration in the system. As a result, risers with abrupt exits, in addition to reduce the erosive particle impingement on the roof of the equipment, had higher gasoline yields. The T-shape exit with a projected height of 0.4 m gave the highest value for this variable (closer to the industrial data), indicating that this outlet configuration improves the reactor efficiency under the conditions applied in the present study.

It is important to emphasize that computer simulations require detailed information about the feedstock and the catalyst particles as well as the reactor design. This information is rarely found in experimental studies on industrial reactors, making necessary the adoption of some assumptions to make possible their simulation. Nevertheless, the models used and the material properties chosen gave results close to the industrial data obtained under the same operating conditions applied in the simulations, especially when configurations commonly found in industrial plants were applied.

Notations

$a_{i=1,2,3}$:	Empirical model constants [—]
A:	Superficial area [m^2]
C:	Molar concentration [kmol m^{-3}]
C_D:	Drag coefficient [—]
C_G:	Constant of elasticity modulus function [Pa]
C_p:	Specific heat [J kg^{-1} K^{-1}]
C_μ:	Model constant
d:	Diameter [m]
D_m:	Mass diffusivity [m^2 s^{-1}]
e_{ss}:	Coefficient of restitution [—]
E:	Activation energy [J mol^{-1}]
F_D:	Drag force between droplets and the gas phase [s^{-1}]
$\mathbf{g}$:	Gravitational acceleration [m^2 s^{-1}]
G:	Elasticity modulus [Pa]
$g_{0,ss}$:	Radial distribution function [—]
h:	Heat transfer coefficient [W m^{-2} K^{-1}]
H:	Static enthalpy [J mol^{-1}]
k:	Thermal conductivity [W m^{-1} K^{-1}] or turbulent kinetic energy [m^2 s^{-2}]
k_c:	Mass transfer coefficient [m s^{-1}]
k_r:	Reaction kinetic constant [m^3 kmol^{-1} s^{-1}]
k_r^0:	Preexponential factor [m^3 kmol^{-1} s^{-1}]
K_c:	Deactivation constant [kg$_{cat}$ kmol^{-1}]
$K_{i=1,2,3,4}$:	Model functions
L:	Latent heat of vaporization [kJ kg^{-1}]
m:	Mass [kg]
M_w:	Molecular mass [kg kmol^{-1}]
n:	Reaction order [—]
n_d:	Droplets density [m^{-3}]
p:	Static pressure [Pa]

p_s:	Solids pressure [Pa]
P^k:	Shear production of turbulence [Pa s^{-1}]
Pr:	Prandtl number [—]
q_1:	Specific coke concentration [kmol kg$_{cat}^{-1}$]
R:	Reaction rate [kmol m^{-3} s^{-1}] or universal gas constant [J mol^{-1} K^{-1}]
$\hat{R}$:	Net reaction rate of formation/consumption [kmol m^{-3} s^{-1}]
Re:	Reynolds number [—]
S:	Source term
T:	Static temperature [K]
$\mathbf{u}$:	Velocity vector [m s^{-1}]
$\mathbf{u}'$:	Velocity fluctuation [m s^{-1}]
x:	Position [m]
Y:	Mass fraction [—].

Greek Letters

β:	Interphase momentum transfer [kg m^{-3} s^{-1}]
δ:	Film thickness of downflow [m]
ε:	Volume fraction [—]
ϵ:	Turbulence dissipation rate [m^2 s^{-3}]
Γ:	Mass diffusion coefficient [m^2 s^{-1}]
ϕ:	Catalyst activity function [—]
Φ:	Pressure strain [kg m^{-1} s^{-3}]
$\Pi_{R,ij}$:	Interaction between the continuous and the dispersed phase turbulence [kg m^{-1} s^{-3}]
Ψ:	Residence time of the solid phase [s]
γ:	Interphase heat transfer coefficient [W m^{-2} K^{-1}]
Γ:	Diffusivity [kg m^{-1} s^{-1}]
λ:	Thermal conductivity [W m^{-1} K^{-1}]
μ:	Molecular viscosity [Pa s]
ρ:	Density [kg m^{-3}]
σ_k:	Model constant
Θ:	Granular temperature [m^2 s^{-2}].

Subscripts

d:	Droplet
g:	Gas phase
i:	Species or lump
max:	Maximum
r:	Reaction
s:	Solid phase
surf:	Surface
∞:	Bulk.

Acknowledgment

The authors gratefully acknowledge the financial support of PETROBRAS for this research.

References

[1] Y. Cheng, F. Wei, G. Yang, and J. Yong, "Inlet and outlet effects on flow patterns in gas-solid risers," *Powder Technology*, vol. 98, no. 2, pp. 151–156, 1998.

[2] S. K. Gupta and F. Berruti, "Evaluation of the gas-solid suspension density in CFB risers with exit effects," *Powder Technology*, vol. 108, no. 1, pp. 21–31, 2000.

[3] E. H. Van Der Meer, R. B. Thorpe, and J. F. Davidson, "Flow patterns in the square cross-section riser of a circulating fluidised bed and the effect of riser exit design," *Chemical Engineering Science*, vol. 55, no. 19, pp. 4079–4099, 2000.

[4] A. T. Harris, J. F. Davidson, and R. B. Thorpe, "The influence of the riser exit on the particle residence time distribution in a circulating fluidised bed riser," *Chemical Engineering Science*, vol. 58, no. 16, pp. 3669–3680, 2003.

[5] C. W. Chan, A. Brems, S. Mahmoudi et al., "PEPT study of particle motion for different riser exit geometries," *Particuology*, vol. 8, no. 6, pp. 623–630, 2010.

[6] X. Wang, L. Liao, B. Fan et al., "Experimental validation of the gas-solid flow in the CFB riser," *Fuel Processing Technology*, vol. 91, no. 8, pp. 927–933, 2010.

[7] M. T. Lim, S. Pang, and J. Nijdam, "Investigation of solids circulation in a cold model of a circulating fluidized bed," *Powder Technology*, vol. 226, pp. 57–67, 2012.

[8] J. W. Chew, R. Hays, J. G. Findlay et al., "Reverse core-annular flow of Geldart Group B particles in risers," *Powder Technology*, vol. 221, pp. 1–12, 2012.

[9] T. Pugsley, D. Lapointe, B. Hirschberg, and J. Werther, "Exit effects in circulating fluidized bed risers," *Canadian Journal of Chemical Engineering*, vol. 75, no. 6, pp. 1001–1010, 1997.

[10] G. Van engelandt, G. J. Heynderickx, J. De Wilde, and G. B. Marin, "Experimental and computational study of T- and L-outlet effects in dilute riser flow," *Chemical Engineering Science*, vol. 66, no. 21, pp. 5024–5044, 2011.

[11] J. De Wilde, G. B. Marin, and G. J. Heynderickx, "The effects of abrupt T-outlets in a riser: 3D simulation using the kinetic theory of granular flow," *Chemical Engineering Science*, vol. 58, no. 3–6, pp. 877–885, 2003.

[12] B. Chalermsinsuwan, P. Kuchonthara, and P. Piumsomboon, "Effect of circulating fluidized bed reactor riser geometries on chemical reaction rates by using CFD simulations," *Chemical Engineering and Processing*, vol. 48, no. 1, pp. 165–177, 2009.

[13] A. K. Das, J. De Wilde, G. J. Heynderickx, and G. B. Marin, "CFD simulation of dilute phase gas-solid riser reactors—part II: simultaneous adsorption of SO_2-NO_x from flue gases," *Chemical Engineering Science*, vol. 59, no. 1, pp. 187–200, 2004.

[14] H. Ali, S. Rohani, and J. P. Corriou, "Modelling and control of a riser type fluid catalytic cracking (FCC) unit," *Chemical Engineering Research and Design*, vol. 75, no. 4, pp. 401–412, 1997.

[15] I. S. Han and C. B. Chung, "Dynamic modeling and simulation of a fluidized catalytic cracking process—part I: process modeling," *Chemical Engineering Science*, vol. 56, no. 5, pp. 1951–1971, 2001.

[16] I. S. Han and C. B. Chung, "Dynamic modeling and simulation of a fluidized catalytic cracking process—part II: property estimation and simulation," *Chemical Engineering Science*, vol. 56, no. 5, pp. 1973–1990, 2001.

[17] S. V. Nayak, S. L. Joshi, and V. V. Ranade, "Modeling of vaporization and cracking of liquid oil injected in a gas-solid riser," *Chemical Engineering Science*, vol. 60, no. 22, pp. 6049–6066, 2005.

[18] J. S. Ahari, A. Farshi, and K. Forsat, "A mathematical modeling of the riser reactor in industrial FCC unit," *Petroleum and Coal*, vol. 50, no. 2, pp. 15–24, 2008.

[19] F. Van Landeghem, D. Nevicato, I. Pitault et al., "Fluid catalytic cracking: modelling of an industrial riser," *Applied Catalysis A*, vol. 138, no. 2, pp. 381–405, 1996.

[20] C. Derouin, D. Nevicato, M. Forissier, G. Wild, and J. R. Bernard, "Hydrodynamics of riser units and their impact on FCC operation," *Industrial and Engineering Chemistry Research*, vol. 36, no. 11, pp. 4504–4515, 1997.

[21] R. Deng, F. Wei, T. Liu, and Y. Jin, "Radial behavior in riser and downer during the FCC process," *Chemical Engineering and Processing*, vol. 41, no. 3, pp. 259–266, 2002.

[22] G. C. Lopes, L. M. Rosa, M. Mori, J. R. Nunhez, and W. P. Martignoni, "Three-dimensional modeling of fluid catalytic cracking industrial riser flow and reactions," *Computers and Chemical Engineering*, 2011.

[23] K. N. Theologos and N. C. Markatos, "Advanced modeling of fluid catalytic cracking riser-type reactors," *AIChE Journal*, vol. 39, no. 6, pp. 1007–1017, 1993.

[24] K. N. Theologos, A. I. Lygeros, and N. C. Markatos, "Feedstock atomization effects on FCC riser reactors selectivity," *Chemical Engineering Science*, vol. 54, no. 22, pp. 5617–5625, 1999.

[25] A. Gupta and D. S. Rao, "Model for the performance of a fluid catalytic cracking (FCC) riser reactor: effect of feed atomization," *Chemical Engineering Science*, vol. 56, no. 15, pp. 4489–4503, 2001.

[26] G. C. Lopes, L. M. Da Rosa, M. Mori, J. R. Nunhez, and W. P. Martignoni, "The importance of using three-phase 3-D model in the simulation of industrial FCC risers," *Chemical Engineering Transactions*, vol. 24, pp. 1417–1422, 2011.

[27] H. Farag, A. Blasetti, and H. De Lasa, "Catalytic cracking with FCCT loaded with tin metal traps: adsorption constants for gas oil, gasoline, and light gases," *Industrial and Engineering Chemistry Research*, vol. 33, no. 12, pp. 3131–3140, 1994.

[28] S. B. Schut, E. H. Van Der Meer, J. F. Davidson, and R. B. Thorpe, "Gas-solids flow in the diffuser of a circulating fluidised bed riser," *Powder Technology*, vol. 111, no. 1-2, pp. 94–103, 2000.

[29] L. S. Lee, Y. W. Chen, and T. N. Huang, "Four-lump kinetic model for fluid catalytic cracking process," *Canadian Journal of Chemical Engineering*, vol. 67, no. 4, pp. 615–619, 1989.

[30] J. A. Juárez, F. L. Isunza, E. A. Rodrìguez, and J. C. M. Mayorga, "A strategy for kinetic parameter estimation in the fluid catalytic cracking process," *Industrial & Engineering Chemistry Research*, vol. 36, pp. 5170–5174, 1997.

[31] S. A. Morsi and A. J. Alexander, "An investigation of particle trajectories in two-phase flow systems," *The Journal of Fluid Mechanics*, vol. 55, no. 2, pp. 193–208, 1972.

[32] Ansys Inc. (US), *ANSYS FLUENT 12. 0—Theory Guide*, Ansys, 2009.

[33] W. Zhang, Y. Tung, and F. Johnsson, "Radial voidage profiles in fast fluidized beds of different diameters," *Chemical Engineering Science*, vol. 46, no. 12, pp. 3045–3052, 1991.

[34] J. H. Pärssinen and J. X. Zhu, "Particle velocity and flow development in a long and high-flux circulating fluidized bed riser," *Chemical Engineering Science*, vol. 56, no. 18, pp. 5295–5303, 2001.

[35] J. C. S. C. Bastos, L. M. Rosa, M. Mori, F. Marini, and W. P. Martignoni, "Modelling and simulation of a gas-solids dispersion flow in a high-flux circulating fluidized bed (HFCFB) riser," *Catalysis Today*, vol. 130, no. 2–4, pp. 462–470, 2008.

[36] K. N. Theologos, I. D. Nikou, A. I. Lygeros, and N. C. Markatos, "Simulation and design of fluid-catalytic cracking riser-type reactors," *Computers and Chemical Engineering*, vol. 20, no. 1, pp. S757–S762, 1996.

[37] J. Gao, C. Xu, S. Lin, G. Yang, and Y. Guo, "Simulations of gas-liquid-solid 3-phase flow and reaction in FCC riser reactors," *AIChE Journal*, vol. 47, no. 3, pp. 677–692, 2001.

[38] W. Martignoni and H. I. De Lasa, "Heterogeneous reaction model for FCC riser units," *Chemical Engineering Science*, vol. 56, no. 2, pp. 605–612, 2001.

[39] J. M. Han and D. L. Liu, "Evaluation on commercial application of LPC type nozzle for FCC feed," *Process Engineering Resources*, vol. 22, article 49, 1992.

[40] A. T. Harris, J. F. Davidson, and R. B. Thorpe, "Influence of exit geometry in circulating fluidized-bed risers," *AIChE Journal*, vol. 49, no. 1, pp. 52–64, 2003.

[41] M. P. Martin, P. Turlier, J. R. Bernard, and G. Wild, "Gas and solid behavior in cracking circulating fluidized beds," *Powder Technology*, vol. 70, no. 3, pp. 249–258, 1992.

[42] X. Wu, F. Jiang, X. Xu, and Y. Xiao, "CFD simulation of smooth and T-abrupt exits in circulating fluidized bed risers," *Particuology*, vol. 8, no. 4, pp. 343–350, 2010.

[43] M. Liu and J. N. Tilton, "Spatial distributions of mean age and higher moments in steady continuous flows," *AIChE Journal*, vol. 56, no. 10, pp. 2561–2572, 2010.

[44] F. Wang, Q. Marashdeh, A. Wang, and L. Fan, "Electrical capacitance volume tomography imaging of three-dimensional flow structures and solids concentration distributions in a riser and a bend of a gas-solid circulating fluidized bed," *Industrial & Engineering Chemistry Research*, vol. 51, pp. 10968–10976, 2012.

11

Reactivity of Phenol Allylation Using Phase-Transfer Catalysis in Ion-Exchange Membrane Reactor

Ho Shing Wu and Yeng Shing Fu

Department of Chemical Engineering and Materials Science, Yuan Ze University, Zhongli, Taiwan

Correspondence should be addressed to Ho Shing Wu, cehswu@saturn.yzu.edu.tw

Academic Editor: Victor V. Nikonenko

This study investigates the reactivity of phenol allylation using quaternary ammonium salt as a phase-transfer catalyst in three types of membrane reactors. Optimum reactivity and turnover of phenol allylation were obtained using a respond surface methodology. The contact angle, water content, and degree of crosslinkage were measured to understand the microenvironment in the ion exchange membrane.

1. Introduction

Phase-transfer catalytic techniques have been used in manufacturing industry synthesis processes, such as insecticidal and chemical production [1–3]. However, a traditional liquid-liquid phase-transfer catalytic reaction has many disadvantages, because separating the catalyst and purifying the reaction system are difficult. Hence, the liquid-solid-liquid phase-transfer catalyst technique was developed. Although this type of catalyst is easy to use and recover from a solution, the reactant pore diffusion in the catalyst affects the reaction and decreases the reaction rate. A catalyst immobilized in an ion exchange membrane could solve these problems.

When a catalyst is immobilized in an inert membrane pore, the catalytic reactivity and separation functions are engineered in a complex system. The membrane technique offers advantages of (i) separating the catalyst from the reaction solution, (ii) maintaining phase separation to minimize the potential of emulsions forming, and (iii) a high surface area per unit volume of the reactor. Furthermore, Zaspalis et al. [4] reported that a reaction using a membrane catalyst could be 10 times more active than a pellet catalyst reaction. Yadav and Mehta [5] presented a theoretical and experimental analysis of capsule membrane phase-transfer catalysis for the alkaline hydrolysis of benzyl chloride to benzyl alcohol. Okahata and Ariga [6] examined the reaction

of sodium azide with benzyl bromide in the presence of a capsule membrane with pendant quaternary ammonium groups and polyethylene glycol groups on the outside. A capsule membrane is unsuited to mass industrial production because of the inconvenience of working with capsules.

Various methods of preparing ion-exchange membranes for different purposes have been proposed and practiced by industry. One of these methods is copolymerizing divinylbenzene and other vinyl monomers (e.g., styrene, chloromethylstyrene, and vinylpyridine) into a membranous copolymer using the paste method and then introducing ion-exchange groups into the copolymer [7–9]. The polymer solution (which contains polymers with ion-exchange groups and other polymers) is then cast on a flat plate to remove the solvent [10, 11]. Most commercial anion-exchange membranes contain benzyl trimethylammonium or N-alkyl pyridinium groups as anion-exchange groups and are cross-linked with divinylbenzene. The membrane backbone polymer is hydrophobic because of aromatic or heterocyclic groups and the active ion-exchange group is hydrophilic. A study [12, 13] that used commercially and laboratory-produced membranes as phase-transfer catalysts in the allylation of phenol showed that the reactivity of quaternary ammonium catalysts in the ion-exchange membrane was lower than that of general phase-transfer catalysts because the ion-exchange group was hydrophilic. The types

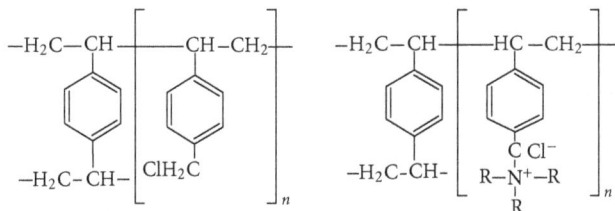

FIGURE 1: The structure of the base membrane and anion exchange membrane.

of ion-exchange groups were hydrophobic and hydrophilic to test the reactivity of phenol allylation. This study presents a discussion on the reactor design problem in a membrane reactor in a two-phase system and examines the relationship between the reactivity of phenol allylation and the membrane reactor. This study also uses membrane properties and the transfer of phenolate ions in the membrane to research the reactivity of quaternary ammonium catalysts in the membrane.

2. Experimental Section

2.1. Materials. Allyl bromide (Fluka, 99.5%), phenol (RDH, 99%), chloromethylstyrene (Aldrich, 97%), and allyl phenyl ether (PhOR, Aldrich, 99%) were provided by the indicated suppliers. Anion-exchange membrane A-172 (a polymer of 1-methyl-4-vinyl-pyridinium crosslinked with 1,4-divinylbenzene) was purchased from ASAHI CHEM Ind. Co. Ltd. (Japan). The characteristics of the A-172 membrane were thickness 0.12–0.15 mm; ion-exchange capacity 1.8–1.9 meq/g of dry membrane; water content 24%–25%; electrical resistance 1.7–20 Ω/cm^2; character: monoanion permselectivity membrane; reinforcement PP fabric.

3. Preparation of Anion Exchange Membrane

The preparation procedure was identical to that described by H. S. Wu and Y. K. Wu (2005) [13]. Four types of anion exchange membranes with different amine functional groups (trimethylamine (TMA), triethylamine (TEA), tri-*n*-propylamine (TPA), and tri-*n*-butylamine (TBA) were prepared by a reaction of a membranous copolymer composed of chloromethylstyrene (CMS type) and divinylbenzene with various tertiary amines (TMA, TEA, TPA, and TBA). Figure 1 shows the structure of the base membrane and anion exchange membrane. H. S. Wu and Y. K. Wu [13] described the synthesis of the base membrane and the process of immobilizing amine in the base membrane.

4. Water Content in the Membrane

The membrane was washed with deionized water and then immersed in deionized water for 60 min. This process was completed twice. The wet membrane was weighed after removing its surface moisture. This process was conducted at least three times to obtain accuracy within 5%. The dry membrane was weighed after drying at 60°C. Water content

was calculated using $W_C(\%) = (W_W - W)/W_W \times 100$, where W_C, W_W, and W are water content in the membrane, weight of the wet membrane after wiping, and weight of the dry membrane, respectively.

5. Kinetics of Phenol Allylation in a Membrane Reactor

Figure 2 shows the experiment apparatus of a membrane reactor. An external circulatory bath was the membrane reactor thermostat to maintain isothermal conditions. An aqueous solution (55 cm^3) of sodium hydroxide (0.00334 mol) and phenol (0.002 mol) was prepared and introduced into the membrane reactor, which was set at the desired temperature. Quantities of allyl bromide (0.03 mol), dichloroethane (55 cm^3), and diphenyl ether (internal standard) were prepared and set to the desired temperature and then introduced into the reactor. The interfacial area between the two phases was 6.0×10^{-4} m^2. The reaction temperature was 45–65°C. The reaction rate did not decrease below 5% after repeating four reaction runs.

For a kinetic run, a sample was withdrawn from the reaction solution at selected time intervals. The sample (0.1 cm^3) was immediately added to dichloroethane (0.3 cm^3) to quench the reaction. The organic phase content was then quantitatively analyzed with a high-performance liquid chromatograph using the internal standard method. The accuracy of these analytical techniques was within 2%-3% and the data were correctly reproduced within 5% of the values reported by this study. Liquid chromatography was conducted with a Shimadzu LC-SPD-10A instrument using a column packed with Phenomenex C12 (150 × 4.6 mm, SYNERGI 4u MAX-RP 80A, USA). The eluant was CH$_3$OH/H$_2$O = 3/1 with a flow rate of 1.0 cm^3 min^{-1} monitored at 254 nm (UV detector).

6. Measurement of Contact Angle

Numerous types of membranes (the A172 membrane and laboratory-produced membranes) can be used to measure contact angles. A syringe needle was used to draw solvents and then the syringe was squeezed five times until it was clean. Solvents were drawn into the syringe needle without bubbles. Membranes were placed on a table with a flat surface and the syringe needle was placed on the apparatus. One drop was dropped onto the membrane until the shape of the drop did not change. The contact angle was recorded. Measurements were taken three times.

7. Results and Discussion

Phase-transfer catalysis is a useful tool in organic synthesis and has many applications in commercial processes. However, it cannot separate a product from its catalyst. Hence, liquid-solid-liquid phase-transfer catalysis was developed to immobilize the quaternary ammonium group onto the resin and membrane [12–15]. Wu and Wang [14] proposed two

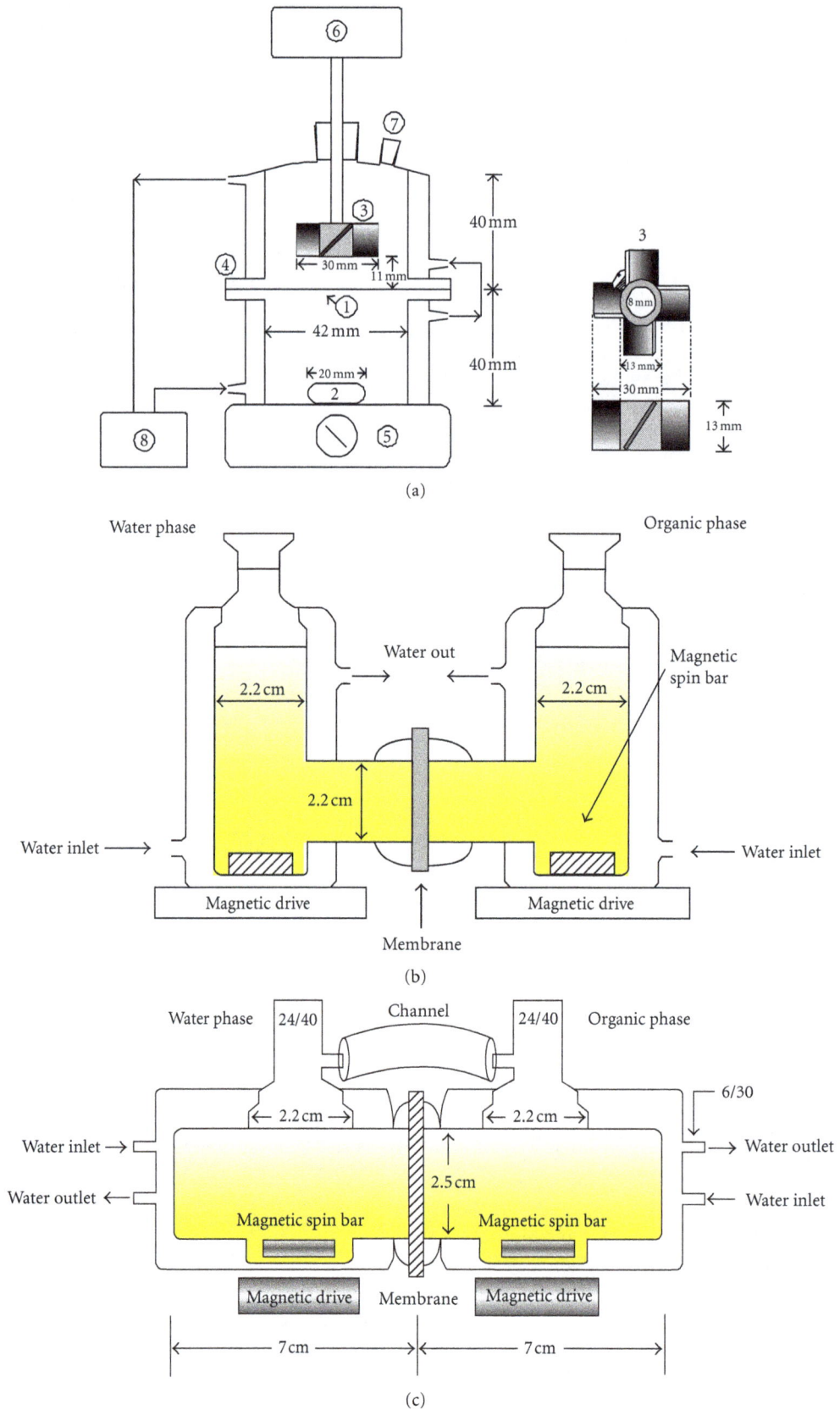

FIGURE 2: Schematic of (a) membrane reactor V1: membrane, 2; Teflon stir bar, 3; stainless stirrer, 4; membrane reactor, 5; magnetic stirrer, 6; mechanical stirrer, 7; sampling point, 8; cooling circulator bath [12]. (b) Membrane reactor H1. (c) Membrane reactor H2.

TABLE 1: Advantages and disadvantages of three reactors.

	V1 reactor	H1 reactor	H2 reactor
Diameter of membrane (cm)	6.2	2.2	2.5
Area of membrane (m^2)	$3.02 \cdot 10^{-3}$	$3.8 \cdot 10^{-4}$	$4.91 \cdot 10^{-4}$
Working volume (cm^3)	55	35	35
Activation energy (kcal/mole)	12.11	13.12	11.69
Turnover number (s^{-1})	$1.53 \cdot 10^{-3}$	$2.58 \cdot 10^{-3}$	$2.74 \cdot 10^{-3}$
Recovery rate of phenol in 90 min (ppm/(min m^2))	13794	45263	38452
Pressure drop	$\triangle$	$\triangle$	$\bigcirc$
Gravity	$\times$	$\triangle$	$\bigcirc$
Area of membrane	$\times$	$\bigcirc$	$\bigcirc$
Apparatus	$\triangle$	$\bigcirc$	$\bigcirc$
Cost	$\triangle$	$\bigcirc$	$\bigcirc$
Operating difficulty	$\triangle$	$\bigcirc$	$\bigcirc$
Reactivity	$\times$	$\bigcirc$	$\bigcirc$

$\bigcirc$ good, $\triangle$ fair, $\times$ poor.

reactors: the slurry reactor and fixed-bed reactor using quaternary ammonium poly(styrene-*co*-chloromethylstyrene) resin, to evaluate a liquid-solid-liquid triphase reaction. However, the catalyst was always suspended in the solution and could flow out of the reactor during a reaction. Hence, developing a method to secure the catalyst on the reactor bed is crucial to improve product and catalyst separation. A membrane reactor could solve these problems, but some design issues must be addressed to improve reaction rates.

8. Membrane Reactor Design Problems

Previous research [13] reported that the organic solution leaked into the aqueous solution during a reaction because gravity reduced the reactivity of the active site on the membrane in the vertical membrane reactor. Hence, the vertical membrane reactor was replaced with a horizontal membrane reactor to avoid the gravity problem, to test the reactivity of phenol allylation. The leaking phenomenon decreased in the horizontal membrane reactor. Wu and Lo [12] and H. S. Wu and Y. K. Wu [13] used the design shown in Figure 2(a). The reactor (V1) was used to recover phenol from simulated wastewater. The reaction system was prepared by pouring the aqueous solution into the bottom cell, covering the anion exchange membrane and locking the reactor. The organic solution was introduced into the top cell, starting the reaction.

Some engineering problems must be overcome in this system and the membrane structure is an important factor. Before the experiment, some bubbles may appear in the aqueous solution during the preparation step. This could influence the mass transfer of aqueous reactant from the bulk solution to the membrane. If the boiling points of both solvents (1-2 dichloroethane and water) are different, there is a large pressure drop between two cells, which forces the solution to pass through the membrane into the other phase because the membrane reactor is a closed system. If the density of the solution in the top cell is larger than in the other cells, gravity is also a problem. Gravity could

force the top solution to pass through the membrane into the bottom phase. Hence, gravity, pressure, membrane, and bubbles influence reactivity in the reactor type, as shown in Figure 2(a). The sample was also only drawn from the top cell and not from the bottom cell.

H. S. Wu and Y. K. Wu [13] proposed that the membrane reactor (H1) design shown in Figure 2(b) could solve these problems and produce better reactivity than membrane reactor V1. Moreover, the sample was simultaneously drawn from two phases in reactor H1, but from only one point in the top cell in membrane reactor V1, as shown in Figure 2(a). Reactor H1 has some disadvantages; for example, the solution is higher than the height of the membrane and the density of 1-2 dichloroethane is more than that of water. Although the gravity effect was less than in reactor V1, the organic phase still passes through the membrane during the long reaction time. Reactor H1 used a closed system for each cell; therefore, the pressure drop between two cells was large, which could be problematic.

Reactor H2 in Figure 2(c) was designed to compensate for these problems. The height of the solution was equal to that of the membrane and there was a channel tube between both phases, solving the effects of gravity and pressure. Table 1 shows the advantages and disadvantages of the three membrane reactors.

9. Optimal Reactivity in Reactors H1 and H2 Using Respond Surface Methodology

Increasing the concentration of the organic reactant (allyl-bromide) increases the reaction rate. In this case, increasing the concentration of organic reactants could also increase the mass transfer rate of the organic reactant because the mass transfer of the organic reactant from the organic phase to the membrane phase was slow. However, increasing aqueous reactant concentrations (phenol) could decrease the reaction rate. An increase in the concentration of aqueous reactants allows aqueous reactants to block the

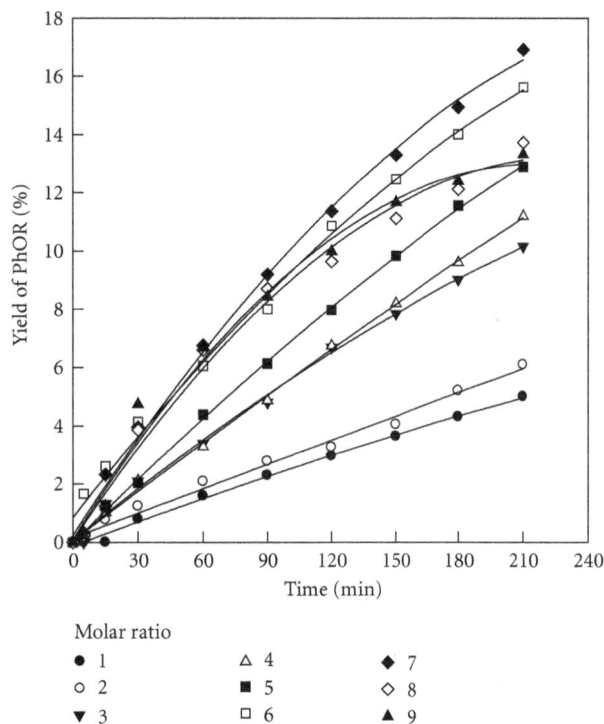

FIGURE 3: Effect of molar ratio of allylbromide to PhONa. $T = 50°C$, agitation = 400 rpm. Aqueous phase (35 cm³): deionized water, phenol = 0.011 mol, NaOH = 0.018 mole. Organic phase (35 cm³): 1,2-$C_2H_4Cl_2$, membrane reactor H1, membrane = A172.

membrane, decreasing the mass transfer rate [12]. Hence, the reaction mechanism in a liquid-liquid membrane phase-transfer catalyzed reaction is different from that in a liquid-liquid phase-transfer catalyzed reaction. This study used the Respond Surface Methodology (RSM) [16–18] to investigate the reactivity of phenol allylation in H1 and H2 reactors.

The molar ratio of organic reactant (allylbromide) to aqueous reactant (phenol) was generally an important factor. The reaction rate increased in conjunction with the molar ratio of allylbromide to phenol. Molar ratios between 1 and 9 were tested in this system. After testing, the best molar ratio (7) was used in the next step. The result is shown in Figure 3. Organic and aqueous reactant concentrations affected the reaction rate. Previously, researchers wanting to obtain optimum conditions in traditional reaction kinetics used experimental runs, which increased the number of runs. This method was time consuming and expensive. This study used RSM to test different concentrations to examine the relationship between organic and aqueous reactant concentrations. The best molar ratio of organic to aqueous reactant concentrations was found and this ratio was used in the steepest ascent path method. After the experiment design, reaction rates were calculated to obtain (9).

$$R = 2.47 \times 10^{-4} + 3.8 \times 10^{-5} \text{ [Phenol]}$$

$$+ 1.18 \times 10^{-4} \text{ [Allylbromide]}$$

$$- 8 \times 10^{-5} \text{ [Phenol]}^2$$

$$- 3.51 \times 10^{-3} \text{ [Allylbromide]}^2,$$

$$(1)$$

where R is the reaction rate (mol/(m² · s)). The respond surface of the reaction in reactor H1 is shown in Figure 4(a).

The turnover number is the number of substrate molecules converted to produce by one molecule of catalyst per unit of time when the reaction rate is maximal and the substrate is saturated. That is, the turnover number (s^{-1}) is the maximal mole of substrate consumed per catalyst per time. The turnover number is calculated using

$$\text{Turnover Number} = \frac{R}{M_c}, \qquad (2)$$

where M_c is the amount of the catalyst per area (mol/m²). Based on previous research, the yield of allyl phenyl ether increased when the excess organic concentration increased. Figure 4(a) shows that after calculation, the optimal reaction rate was 8.54×10^{-4} mol/s·m² when phenol and allylbromide were 7.41×10^{-3} mol and 0.106 mol, respectively. The average membrane weight was 0.086 g and the ion-exchange capacity was 1.6 mmol/g; therefore, the mole of the catalyst was 1.22×10^{-4} mol. The turnover number was 2.58×10^{-3} s⁻¹ at 328 K. Previous researchers [12] found a turnover number of 1.53×10^{-3} s⁻¹ in reactor V1, as shown in Figure 2(a). This value was smaller than that of reactor H1.

Table 1 shows the similar optimal reactivity of reactor H2. The optimal reaction rate and turnover number for reactor H2 are larger than those for reactor H1. Figure 4 shows that the operating concentration of allybromide or phenol in reactor H2 is also larger than those for reactor H1. Therefore, reactor H2 performs better than reactor H1.

10. Effect of Membrane Structure

Wu and Lo [12] showed that the reaction rate increased with an increasing molar ratio of allylbromide to phenol. Figure 5 shows that the allyl phenyl ether yield varies with different types of membranes. To determine what occurred in the reaction, different membrane types were tested to measure contact angles. This is shown in Figures 6 and 7.

The dry A172 membrane and laboratory-produced base membranes were polymeric membranes. Generally, the polymer membranes were hydrophobic. The contact angles of the dry A172 membrane and laboratory-produced base membranes decreased with an increasing concentration of phenol in the aqueous solution. The contact angles for the dry A172 membrane in aqueous or organic solvents were larger than those for the laboratory-produced base membranes. This could indicate that the interaction between solvents and the membrane of the laboratory-produced membranes was larger than the interaction of the A172 membrane. The reaction rate for laboratory-produced membranes could be higher than for the A172 membrane. For wet membranes, the contact angles were zero because the solvent drops were drawn into the membrane.

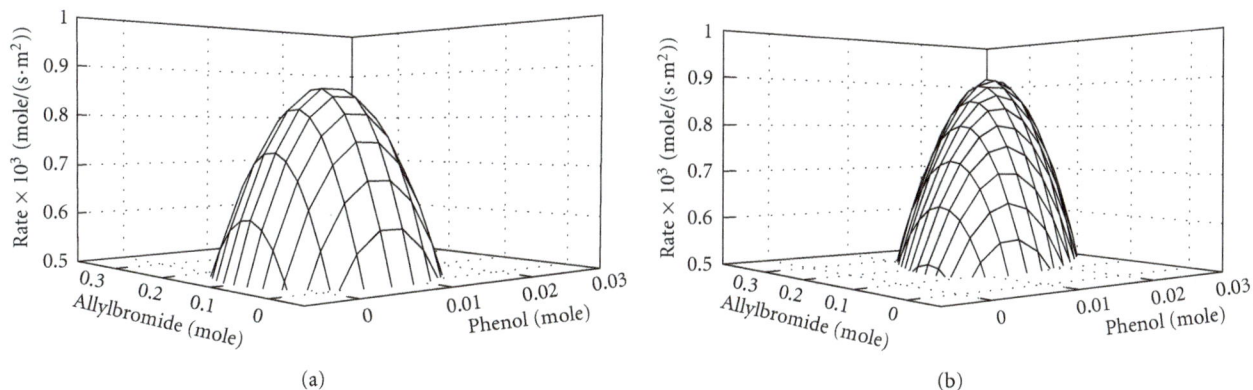

FIGURE 4: Respond surface of reaction in membrane reactors (a) H1 and (b) H2.

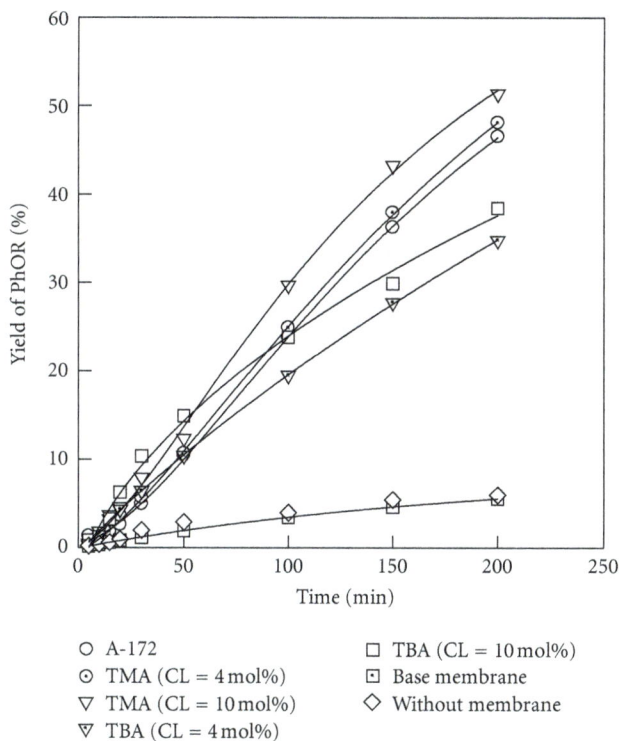

○	A-172	□	TBA (CL = 10 mol%)
⊙	TMA (CL = 4 mol%)	⊡	Base membrane
▽	TMA (CL = 10 mol%)	◇	Without membrane
▽	TBA (CL = 4 mol%)		

FIGURE 5: PhOR yield from various membranes with an excess of organic reactant. 45°C, 400 rpm. Aqueous phase (55 cm³): 0.004 mol of phenol, NaOH = 0.0064 mol. Organic phase (55 cm³): 1,2-C₂H₄Cl₂, 0.06 mol of C₃H₅Br. Data quoted from Wu and Lo [12].

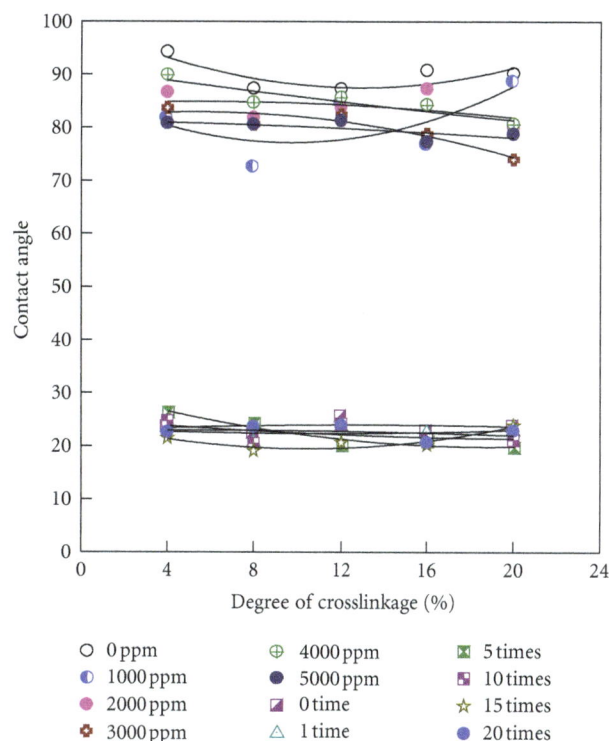

○	0 ppm	⊕	4000 ppm	⊠	5 times
◑	1000 ppm	●	5000 ppm	⊟	10 times
●	2000 ppm	◩	0 time	⬟	15 times
✴	3000 ppm	△	1 time	●	20 times

FIGURE 6: Contact angles of laboratory-produced base membranes with different degrees of crosslinkage with different concentrations of phenol and molar ratios of allylbromide to phenol.

Figure 8 shows that the water content decreased with increasing crosslinkage when the ion-exchange capacity did not change. Because the physical strength of the membrane increased with crosslinkage, it was difficult for water to exist in the membrane, and the swelling decreased.

The aqueous ion-exchange reaction of phenolate ion (PhO^-) and bromide ion (Br^-) is shown by

$$Q^+Br^- + PhO^- \xrightarrow{k_f} Q^+PhO^- + Br^-, \tag{3}$$

where Q^+ is a phase-transfer catalyst. The reaction expression is:

$$r = -\frac{d[PhO^-]}{dt} = k_f[PhO^-][Q^+Br^-]. \tag{4}$$

The k_f value was calculated from the slope in Figure 9 (upper panel) using the initial reaction rate method. The reaction rate has an optimal value. k_f decreased with increasing initial amounts of PhONa. This verifies that increasing concentrations of aqueous reactant does not increase the reaction rate, as shown in Figure 9 (lower panel). Therefore, the contact angle, water content, and ion exchange of

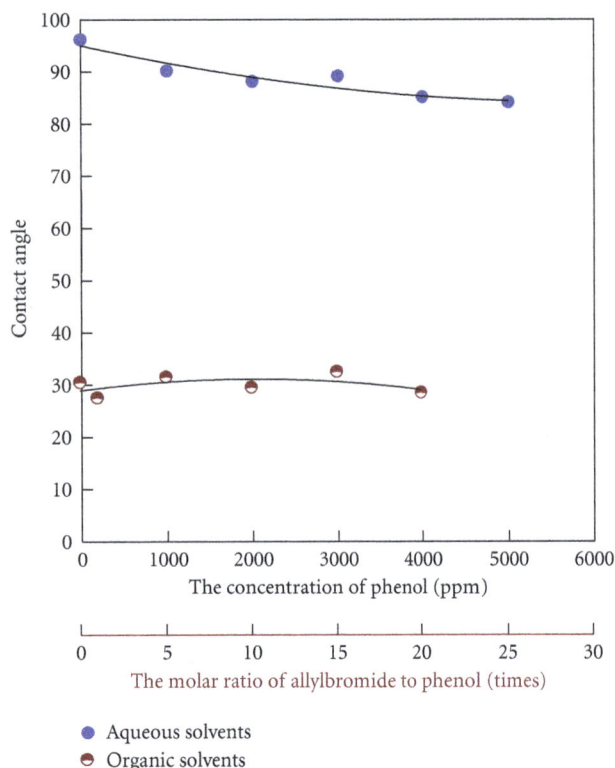

FIGURE 7: Contact angles of dry A172 membrane with different concentrations of phenol and molar ratios of allylbromide to phenol.

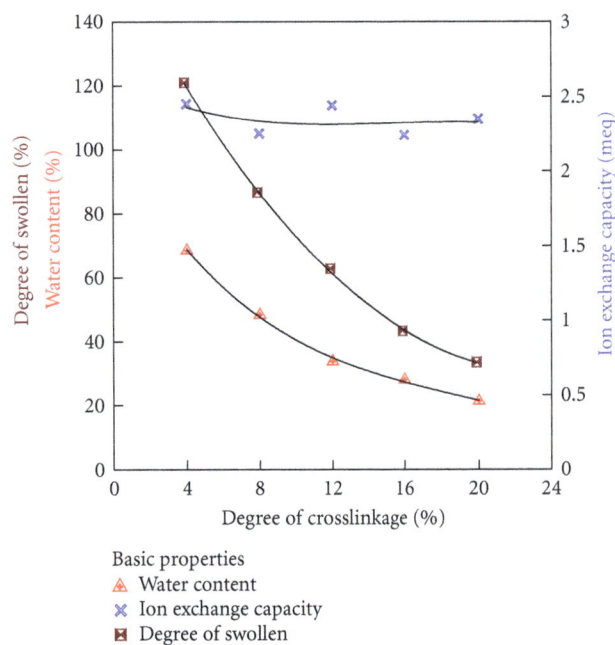

FIGURE 9: PhO⁻ in A172 versus time and the effect of PhONa on the forward ion-exchange rate. 65°C, 400 rpm, 30.2 cm² of A-172 membrane, aqueous phase 55 cm³.

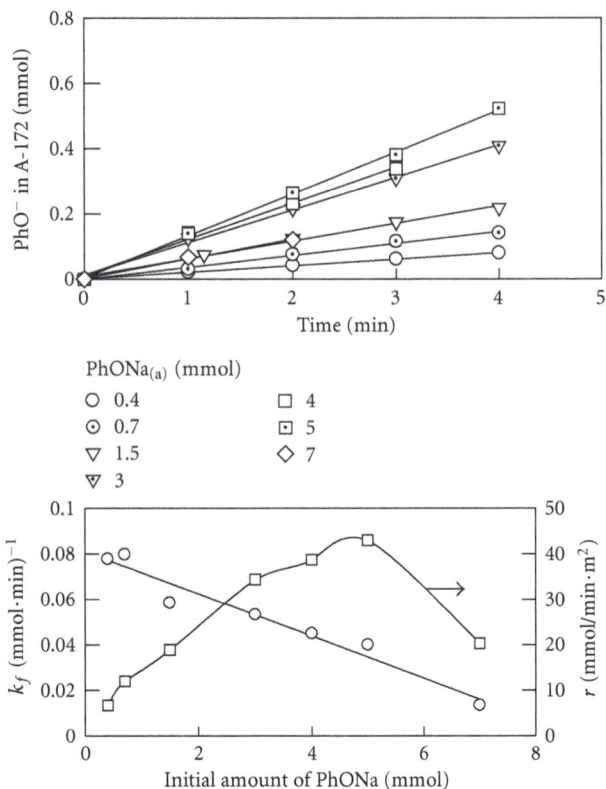

FIGURE 8: Effects of degree of crosslinkage on swelling, water content, and ion exchange capacity of a tetra-methylamine membrane.

phenolate ions with bromide in the membrane are essential for applying this membrane technique to phase-transfer catalysis.

11. Conclusion

Gravity and pressure are important factors in reactor design. The catalytic reactivity in the H2 reactor design was better than in the V1 reactor. The RSM could be used to obtain the optimal turnover number and reaction rate to verify the reactivity of a catalyst in a membrane. For phase-transfer catalytic membrane systems to perform well, the membrane structure for reactants and solvents must be studied to obtain optimal conditions.

Acknowledgment

We would like to thank the National Science Council of Taiwan for financially supporting this research under Grant nos. NSC 94-2214-E155-004 and 95-2214-E155-006.

References

[1] V. V. Dehmlow and S. S. Dehmlow, *Phase Transfer Catalysis*, Chemie, Weinheim, Germany, 1993.

[2] C. M. Starks, C. L. Liotta, and M. Halpern, *Phase-Transfer Catalysis, Fundamentals, Applications, and Industrial Perspectives*, Chapman & Hall, New York, NY, USA, 1994.

[3] H. M. Yang and H. S. Wu, "Interfacial mechanism and kinetics of phase-transfer catalysis," in *Interfacial Catalysis*, A. G. Volkov, Ed., vol. 285, chapter 11, Marcel Dekker, 2003.

[4] V. T. Zaspalis, W. Van Praag, K. Keizer, J. G. Van Ommen, J. R. H. Ross, and A. J. Burggraaf, "Reactions of methanol over catalytically active alumina membranes," *Applied Catalysis*, vol. 74, no. 2, pp. 205–222, 1991.

[5] G. D. Yadav and P. H. Mehta, "Theoretical and experimental analysis of capsule membrane phase transfer catalysis: selective alkaline hydrolysis of benzyl chloride to benzyl alcohol," *Catalysis Letters*, vol. 21, no. 3-4, pp. 391–403, 1993.

[6] Y. Okahata and K. J. Ariga, "Functional capsule membranes. 27. A new type of phase-transfer catalysts (PTC). Reaction of substrates in the inner organic phase with the outer aqueous anions catalyzed by PTC grafted on the capsule," *Journal of Organic Chemistry*, vol. 51, p. 5064, 1986.

[7] Y. Mizutani, R. Yamane, H. Ihara, and H. Motomura, "Studies of Ion Exchange Membraness. XVI. The preparation of ion exchange membranes by the "Paste Method"," *Bulletin of The Chemical Society of Japan*, pp. 361–366, 1963.

[8] Y. Mizutani, R. Yamane, and H. Motomura, "Studies of Ion Exchange Membranes. XXII. Semicontinuous preparation of ion exchange membrane by the paste method," *Bulletin of The Chemical Society of Japan*, pp. 689–694, 1964.

[9] Y. Mizutani, "Studies of Ion Exchange Membraness. The tetrahydrofuran extraction of the ion-exchange membrane and its base membrane prepared by the "Paste Method"," *Bulletin of The Chemical Society of Japan*, vol. 42, no. 1969, pp. 2459–2463, 1969.

[10] H. P. Gregor, H. Jacobson, R. C. Shair, and D. M. Wetstone, "Interpolymer ion-selective membranes. I. Preparation and characterization of polystyrenesulfonic acid-dynel membranes," *Journal of Physical Chemistry*, vol. 61, p. 141, 1957.

[11] P. Zschocke and D. Quellmatz, "Novel ion exchange membranes based on an aromatic polyethersulfone," *Journal of Membrane Science*, vol. 22, pp. 325–332, 1985.

[12] S. Wu and M. H. Lo, "Modeling and kinetics of allylation of phenol in a triphase-catalytic membrane reactor," *American Institute of Chemical Engineers*, vol. 51, pp. 960–970, 2005.

[13] H. S. Wu and Y. K. Wu, "Preliminary study on the characterization and preparation of quaternary ammonium membrane," *Industrial & Engineering Chemistry Research*, vol. 44, p. 1757, 2005.

[14] H. S. Wu and C. S. Wang, "Liquid—solid—liquid phase-transfer catalysis in sequential phosphazene reaction: kinetic investigation and reactor design," *Chemical Engineering Science*, vol. 58, no. 15, pp. 3523–3534, 2003.

[15] H. S. Wu, "Catalytic activity and kinetics of liquid—solid—liquid phase-transfer catalysts," in *New Developments in Catalysis Research*, L. P. Bevy, Ed., chapter 1, pp. 1–38, Nova Science Publishers, 2005.

[16] G. E. P. Box and J. S. Hunter, "Multi-factor experimental designs for exploring response surfaces," *Annals of Mathematical Statistics*, vol. 28, p. 195, 1957.

[17] G. E. P. Box and K. B. J. R. Wilson, "On the experimental attainment of optimum conditions," *Journal of the Royal Statistical Society B*, vol. 13, p. 1, 1951.

[18] J. J. Cilliers, R. C. Austin, and J. P. Tucker, "An evaluation of formal experimental design procedures for hydrocyclone modelling," in *Proceeding of the 4th International Conference of Hydrocyclones*, L. Svarovsky and M. T. Thew, Ed., pp. 31–49, Kluwer Academic Publishers, Southampton, UK, 1992.

12

Effects of Gravity and Inlet Location on a Two-Phase Countercurrent Imbibition in Porous Media

M. F. El-Amin,[1,2] **Amgad Salama,**[1] **and Shuyu Sun**[1]

[1] *Computational Transport Phenomena Laboratory (CTPL), Division of Physical Sciences and Engineering (PSE),*
 King Abdullah University of Science and Technology (KAUST), Thuwal 23955-6900, Saudi Arabia
[2] *Department of Mathematics, Aswan Faculty of Science, South Valley University, Aswan 81528, Egypt*

Correspondence should be addressed to M. F. El-Amin, mohamed.elamin@kaust.edu.sa

Academic Editor: Mandar Tabib

We introduce a numerical investigation of the effect of gravity on the problem of two-phase countercurrent imbibition in porous media. We consider three cases of inlet location, namely, from, side, top, and bottom. A 2D rectangular domain is considered for numerical simulation. The results indicate that gravity has a significant effect depending on open-boundary location.

1. Introduction

Oil recovery by imbibition mechanism, from fractured reservoirs, is a significant research area in multiphase flow in porous media especially for water-flooding process in fractured oil reservoirs. Fractured reservoirs are composed of the fracture network and matrix. Fractures have a higher permeability and relatively low volume compared to the matrix, whose permeability is very low but it contains the majority of the oil. Water flooding is used to increase oil recovery by increasing water pressure in fractures since water quickly surrounds oil-saturated matrices of lower permeability. The process of water flooding works well when the matrix is water-wet, and imbibition can lead to significant recoveries, while poor recoveries and early water breakthrough occur with oil-wet matrix conditions. Imbibition is defined as the displacement of the nonwetting phase (oil) by the wetting phase (water) with dominant effect of capillary forces. Imbibition can occur in both countercurrent and cocurrent flow modes, depending on the fracture network and the water injection rates. In cocurrent imbibition, water displaces oil out of the matrix; thus both water and oil flows are in the same direction. Countercurrent imbibition, on the other hand, is whereby a wetting phase imbibes into the porous matrix (rock), displacing the non-wetting phase out from one open boundary. In spite of the fact that cocurrent imbibition is faster and more efficient

than countercurrent imbibitions, the latter is often the only possible displacement mechanism for cases where a region of the matrix is exposed from one side to water filling the fracture [1–4]. Imbibition has also been investigated by several other authors either for cocurrent or countercurrent flows or both of them together (e.g., [5–8]). Reis and Cil [9] introduced one-dimensional model for oil expulsion by countercurrent water imbibition in rocks. An examination of countercurrent capillary imbibition recovery from single matrix blocks and recovery predictions by analytical matrix/fracture transfer functions was introduced by Cil et al. [10]. Lee and Kang [11] have introduced an experimental analysis of oil recovery in a fracture of variable aperture with countercurrent imbibition. Scaling of countercurrent imbibition was estimated by many authors in terms of fluid and rock properties (e.g., [12, 13]). Morrow and Mason [14] introduced a comprehensive review on recovery of oil by spontaneous imbibition. Kashchiev and Firoozabadi [15] gave analytical solutions for 1D countercurrent imbibition in water-wet media. Analytical analysis of oil recovery during countercurrent imbibition in strongly water-wet system was given by Tavassoli et al. [16]. The Barenblatt model of spontaneous countercurrent imbibition was investigated by Silin and Patzek [17]. Behbahani et al. [18] have performed a simulation of countercurrent imbibition in water-wet fractured reservoirs.

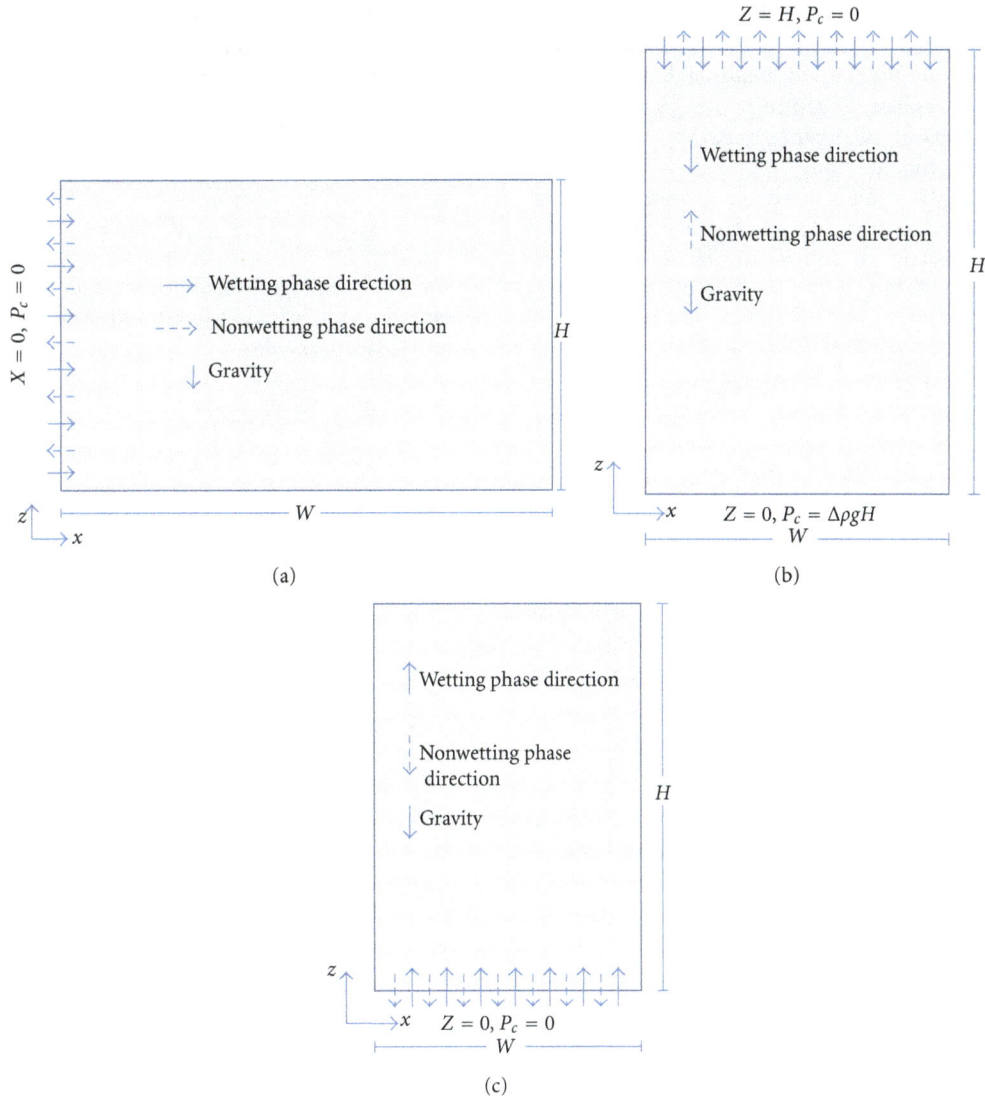

FIGURE 1: Schematic diagram of countercurrent imbibition with gravity effect for different locations of the open boundary: (a) side, (b) top, and (c) bottom.

In most of the previously mentioned imbibition studies, researchers have neglected the gravity force effect by dropping the gravity force term from the flow equations especially for the oil-water modeling. Wilkinson [19] studied the percolation model of immiscible displacement in the presence of buoyancy forces. Analytical and numerical solutions of gravity-imbibition and gravity-drainage processes were given by Bech et al. [20]. Tavassoli et al. [21] have introduced analysis of countercurrent imbibition with gravity in weakly water-wet system. A pore-scale study of gravity, capillary and viscous forces during drainage in a two-dimensional porous medium, was introduced by Løvoll et al. [22]. Effect of injection rate, initial water saturation, and gravity on water injection in slightly water-wet fractured porous media was examined experimentally by Karimaie and Torsæter [23]. Ruth et al. [24] provided an approximate analytical solution for countercurrent spontaneous imbibition. The problems of buoyancy-driven vertical migration of fluids have been treated analytically or numerically by some researchers. For example, Silin et al. [25] have introduced simple analytical solutions in a model of gas flow driven by a combination of buoyancy, viscous and capillary forces for the problem of two-phase countercurrent fluid flow.

This study is devoted to numerically investigate the influences of gravity and open-boundary location on the countercurrent imbibition of two immiscible phases in a 2D porous medium domain.

2. Formulations, Results, and Discussion

The purpose of this study is to investigate the influence of gravity for different locations of the open boundary of an incompressible, two-phase, immiscible, countercurrent imbibition in a 2D homogenous porous medium domain. In this work both buoyancy and capillarity are considered. Figures 1(a), 1(b), and 1(c) show schematic diagrams of the problem domain for different locations of the open boundary (side, top, and bottom). Wetting phase imbibes inwards in the porous medium domain of height H and width W with zero capillary pressure at the open boundary and no-flow boundary at the other boundaries.

Consider a rectangular core saturated with oil with irreducible water saturation, closed all around except at one face that is open to flow (countercurrent imbibition). The flow is governed by the combined Darcy law and the equations of mass conservation for each phase in 2D as follows:

$$\varphi \frac{dS_w}{dP_c}\left(\frac{\partial P_o}{\partial t} - \frac{\partial P_w}{\partial t}\right) = \nabla \cdot \left(\frac{Kk_{rw}}{\mu_w}(\nabla P_w - \rho_w g \nabla z)\right),$$

$$\varphi \frac{dS_w}{dP_c}\left(\frac{\partial P_o}{\partial t} - \frac{\partial P_w}{\partial t}\right) = -\nabla \cdot \left(\frac{Kk_{ro}}{\mu_o}(\nabla P_o - \rho_o g \nabla z)\right),$$

$$(1)$$

where $\nabla \equiv (\partial/\partial x, \partial/\partial z)$, subscripts w and o designate wetting phase (water) and nonwetting phase (oil), respectively. P is the phase pressure, S_w is the water saturation, k_r is relative permeability, μ is phase viscosity, ρ is phase density, and g is gravity acceleration. K is permeability and φ is porosity of the porous medium.

The capillary pressure functions are dependent on the pore geometry, fluid physical properties, and phase saturations. The two-phase capillary pressure can be expressed by the Leverett dimensionless function $J(S)$; see, for example, Chen [26], which is a function of the normalized saturation S:

$$P_c = P_o - P_w = \gamma\left(\frac{\varphi}{K}\right)^{1/2} J(S), \qquad (2)$$

where γ is the interfacial tension.

In order to consider a certain case of study, we may use a specified empirical formula of the capillary pressure in terms of normalized saturation function. The $J(S)$ function typically lies between two limiting (drainage and imbibition) curves which can be obtained experimentally. Correlation of the imbibition capillary pressure data depends on the type of application. Since our current research is concerned with the water-oil system, we use the correlation by Firoozabadi and coworkers [3, 15], in which the capillary pressure and the normalized wetting phase saturation are correlated as follows:

$$P_c = -B \ln S, \qquad (3)$$

where B is the capillary pressure parameter, which is equivalent to $\gamma(\varphi/K)^{1/2}$ in (2); thus, $B \equiv -\gamma(\varphi/K)^{1/2}$ and $J(S) \equiv \ln S$. Note that $J(S)$ is a scalar nonnegative function. Also,

$$S = \frac{S_w - S_{iw}}{1 - S_{or} - S_{iw}}, \qquad 0 \le S \le 1, \qquad (4)$$

where S_{iw} is the irreducible water saturation and S_{or} is the residual oil saturation.

For the countercurrent imbibition in which the only open end is initially in contact with oil, the ambient pressure

is considered zero. The water pressure in the core is given by the capillary pressure relationship, (2) and (3), which at $t = 0$ leads to

$$P_o = 0, \quad t = 0, \quad 0 \le x \le W, \quad 0 \le z \le H,$$

$$P_w = P_o - P_c(S_{wi}) = -P_c(S_{wi}), \qquad (5)$$

$$t = 0, \quad 0 \le x \le W, \quad 0 \le z \le H.$$

In this study we consider three different locations of the open boundary, at side, top, or bottom, namely, Case A, Case B, and Case C, as follows.

Case A. Side open-boundary

$$P_w = P_o = 0, \quad t > 0, \quad x = 0, \quad 0 \le z \le H,$$

$$q_w = q_o = 0, \quad t > 0, \quad 0 \le x \le W, \quad z = 0,$$

$$q_w = q_o = 0, \quad t > 0, \quad 0 \le x \le W, \quad z = H,$$

$$q_w = q_o = 0, \quad t > 0, \quad x = W, \quad 0 \le z \le H,$$

$$(6)$$

where q_w and q_o are the water and oil flow rate, respectively.

Case B. Top open-boundary:

$$q_w = q_o = 0, \quad t > 0, \quad x = 0, \quad 0 \le z \le H,$$

$$q_w = q_o = 0, \quad t > 0, \quad 0 \le x \le W, \quad z = 0,$$

$$P_w = P_o = 0, \quad t > 0, \quad 0 \le x \le W, \quad z = H,$$

$$q_w = q_o = 0, \quad t > 0, \quad x = W, \quad 0 \le z \le H,$$

$$(7)$$

Case C. Bottom open-boundary

$$q_w = q_o = 0, \quad t > 0, \quad x = 0, \quad 0 \le z \le H,$$

$$P_w = P_o = 0, \quad t > 0, \quad 0 \le x \le W, \quad z = 0,$$

$$q_w = q_o = 0, \quad t > 0, \quad 0 \le x \le W, \quad z = H,$$

$$q_w = q_o = 0, \quad t > 0, \quad x = W, \quad 0 \le z \le H.$$

$$(8)$$

Case A represents a domain of size $(0.2, 0.2)$ m which is meshed by 10439 nodes and 19968 triangle elements, corresponding to more than 81690 DOF (quadratic Lagrange elements), while Cases B and C are meshed by 10591 nodes and 20272 triangle elements, corresponding to more than 82906 DOF (quadratic Lagrange elements). All computations have been performed using the commercial software COMSOL version 3.5a with the direct solver UMFPACK and were running on multi(7)-core workstation using SMP mode of parallel computation. Figures 2(a) and 2(b) show mesh distributions for Case A and Cases B and C, respectively, with fine mesh on the inlet side and the opposite side.

The simulation was running for imbibition time of 40 days so that it may be compared with the study of Pooladi-Darvish and Firoozabadi [3]. We use the same values of physical properties used by Pooladi-Darvish and Firoozabadi [3] as given in Table 1. Figure 3 shows distributions of water saturation and velocity vectors of the case of incorporating

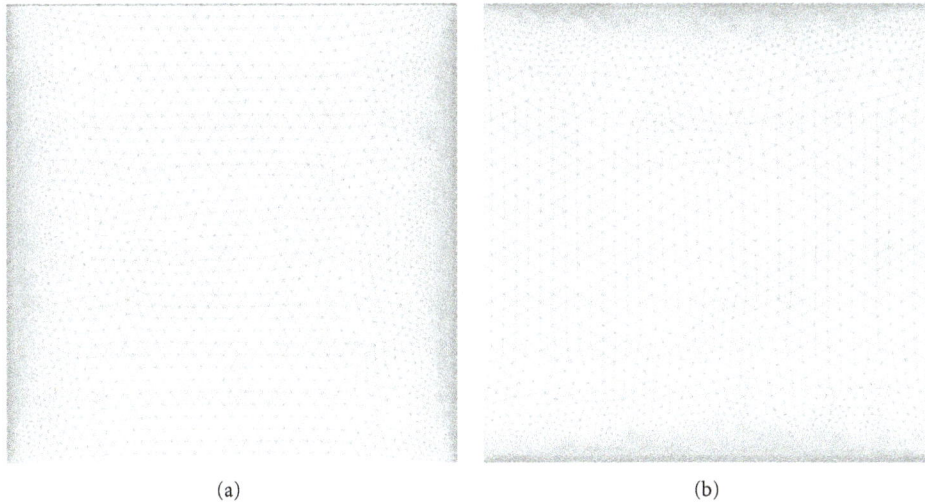

(a)   (b)

FIGURE 2: (a) Mesh distributions for the 2D dimensional Cases A. (b) Mesh distributions for the 2D dimensional Cases B and C

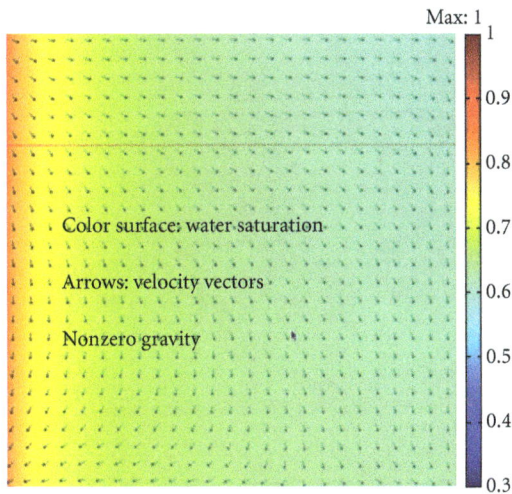

FIGURE 3: Distributions of water saturation and velocity vectors of nonzero gravity of Case A, at time imbibition of 40 days.

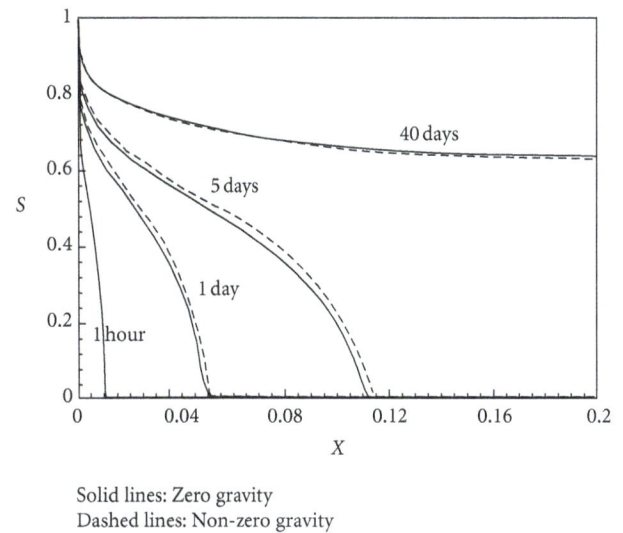

Solid lines: Zero gravity
Dashed lines: Non-zero gravity

FIGURE 4: Comparison between considering and neglecting gravity effect on water saturation against x-axis of Case A, at $z = 0.15\,\text{m}$.

gravity in the flow equations. The zero-gravity case may be matched well by 1D simulation, while the opposite is true for the case of nonzero gravity which shows a nonuniform distribution of water velocity as shown in Figure 3. Additionally, a comparison between considering and neglecting gravity effect on water saturation against x-axis of Case A is plotted in Figure 4. It can be seen from this figure that considering the gravity term in the flow equations results in a slight increase in water saturation. Also, from the same figure we may note that the saturation profiles are comparable to the 1D case as shown by Pooladi-Darvish and Firoozabadi [3]. Comparison between considering and neglecting gravity effect on water and oil pressure against distance with considering gravity effect of Case A is plotted in Figure 5. From this figure, one may note that water and oil pressure vary downstream of the saturation front with time and location. Also, it can be seen that oil pressure reaches the

TABLE 1: Primary parameters from Pooladi-Darvish and Firoozabadi [3].

a	b	B	K	k_{ro}^0	k_{rw}^0	L	S_{wi}	φ	$\mu_{w,o}$
4	4	10 kPa	0.02 μm	0.75	0.2	0.2 m	0.001	0.3	1 mPa

maximum in the two-phase region. It is interesting to note that gravity has a slight effect on water and oil pressures.

Figure 6 shows a comparison between considering and neglecting gravity effect on water x-velocity profiles against the horizontal distance for Case A. This figure indicates that at the beginning of the imbibition time the velocity is higher while after longer time of imbibition the velocity slows down as water imbibes inside the matrix. This may be interpreted based on the fact that the flow in this system is dominated by capillarity which reduces with the increase in saturation.

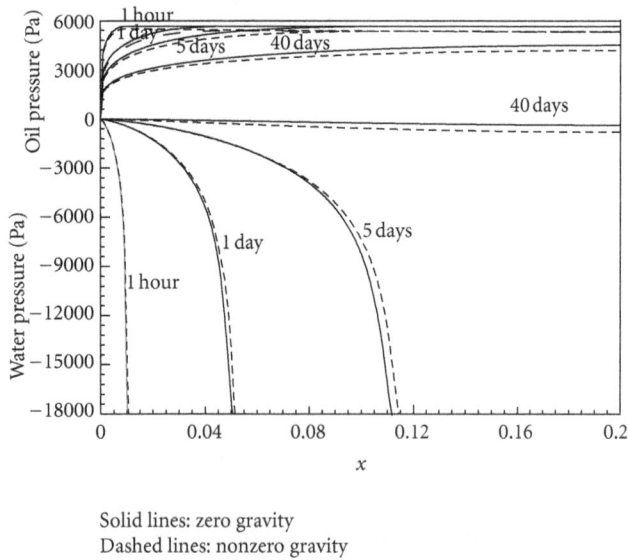

Solid lines: zero gravity
Dashed lines: nonzero gravity

FIGURE 5: Comparison between considering and neglecting gravity effect on water and oil pressure against x-axis of Case A, at $z = 0.15$ m.

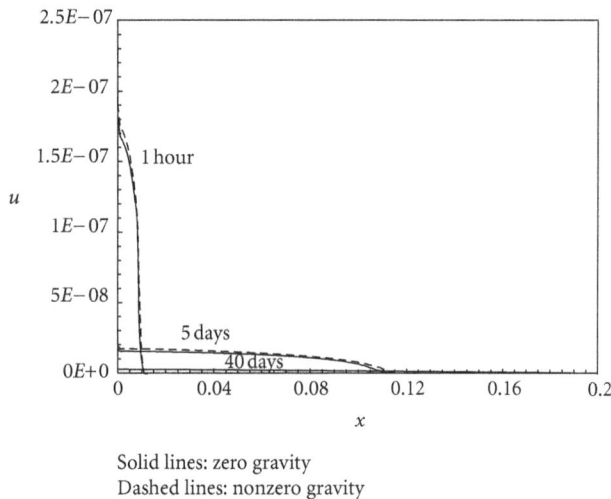

Solid lines: zero gravity
Dashed lines: nonzero gravity

FIGURE 6: Comparison between considering and neglecting gravity effect on water x-velocity against x-axis of Case A, at $z = 0.15$ m.

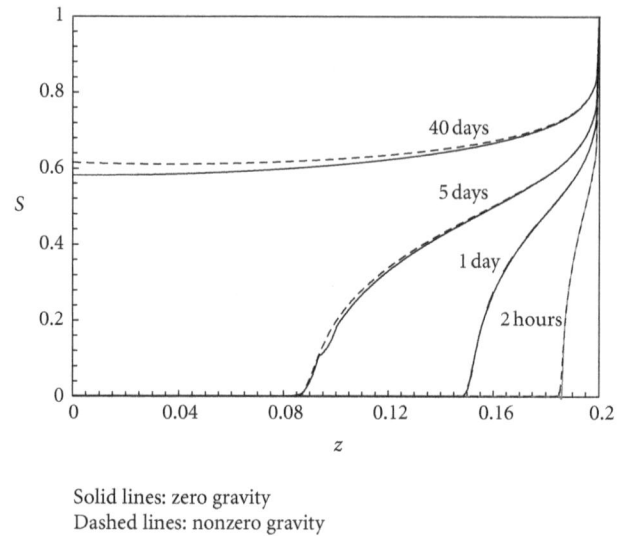

Solid lines: zero gravity
Dashed lines: nonzero gravity

FIGURE 7: Comparison between considering and neglecting gravity effect on water saturation against z-axis of Case B, at $x = 0.15$ m.

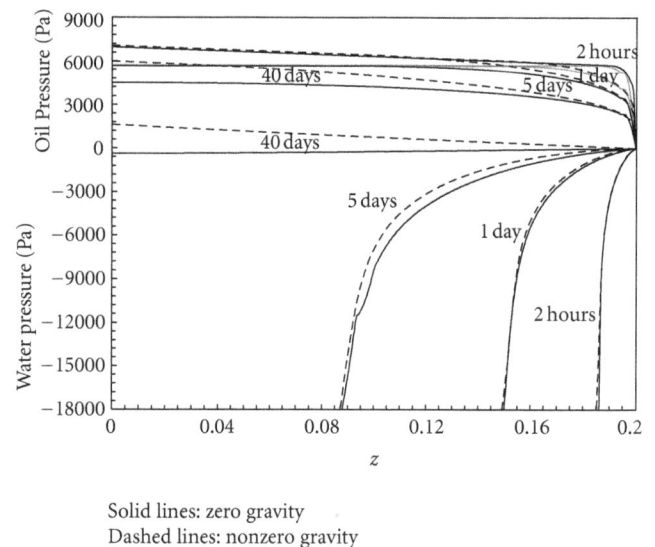

Solid lines: zero gravity
Dashed lines: nonzero gravity

FIGURE 8: Comparison between considering and neglecting gravity effect on water and oil pressure against z-axis of Case B, at $x = 0.15$ m.

In this figure, it is apparent that gravity has, generally, slight effect at early time of imbibition. However, this effect is seen to be more pronounced at later time of imbibition (e.g., after 40 days).

A comparison between considering and neglecting gravity effect on water saturation against z-axis for Case B is plotted in Figure 7. It is obvious that in the case of considering gravity the water saturation is slightly higher than that without gravity particularly after longer period of time (e.g., after 40 days of imbibition). A comparison between considering and neglecting gravity effect on water and oil pressure against z-axis of Case B is plotted in Figure 8. It is interesting to note that, for this case, both water and oil pressures are assisted by the gravity force.

Figure 9 shows a comparison between considering and neglecting gravity effect on water saturation against z-axis of Case C. It can be seen from Figure 9 that considering the gravity force in the flow equations reduces water saturation. In this case the gravity works in the opposite direction of the water flow so it resists water imbibition. Also, considering the gravity force reduces both water and oil pressures as illustrated in Figure 10.

3. Conclusions

The aim of this work is to examine the influence of gravity on countercurrent imbibition of two-phase flow in porous

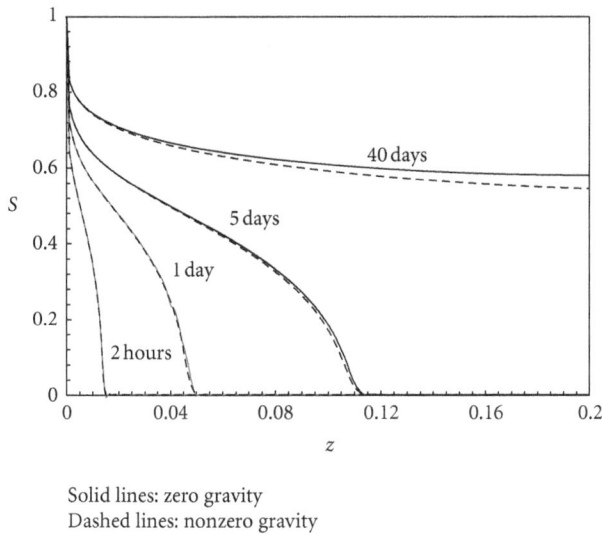

Solid lines: zero gravity
Dashed lines: nonzero gravity

FIGURE 9: Comparison between considering and neglecting gravity effect on water saturation against z-axis of Case C, at $x = 0.15$ m.

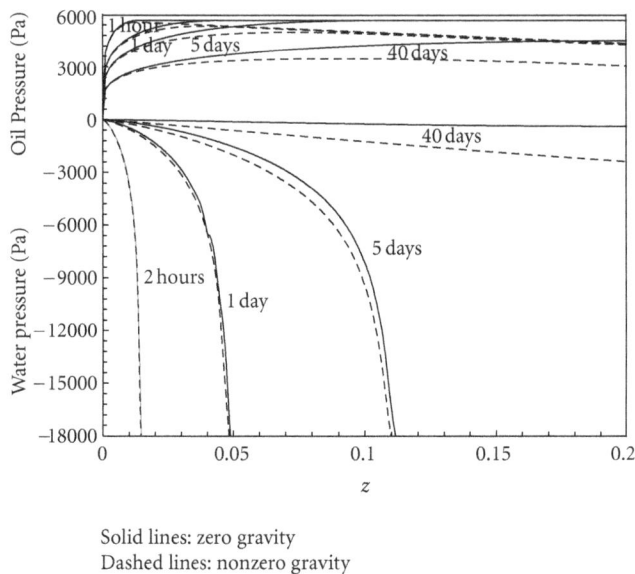

Solid lines: zero gravity
Dashed lines: nonzero gravity

FIGURE 10: Comparison between considering and neglecting gravity effect on water and oil pressure against z-axis of Case C, at $x = 0.15$ m.

media for different locations of the open boundary. A 2D simulation for three different locations of the open boundary (side, top, and bottom) is considered. A comparable study of considering and neglecting gravity in the model is done for the three different open-boundary locations. From this work one may conclude that the bottom open-boundary reduces the water imbibition in the rock matrix and therefore decreases the oil recovery, while the opposite is true for both top and side open-boundary. The results indicate that the buoyancy effects due to gravity force take place depending on the location of the open boundary.

Acknowledgment

The work was supported by the KAUST-UT Austin AEA Project (ID 7000000058).

References

[1] B. J. Bourblaux and F. J. Kalaydjian, "Experimental study of cocurrent and countercurrent flows in natural porous media," *SPE Reservoir Engineering*, vol. 5, no. 3, pp. 361–368, 1990.

[2] M. E. Chimienti, S. N. Illiano, and H. L. Najurieta, "Influence of temperature and interfacial tension on spontaneous imbibition process," in *Proceedings of the Latin American and Caribbean Petroleum Engineering Conference*, vol. 53668, SPE, Caracas, Venezuela, 1999.

[3] M. Pooladi-Darvish and A. Firoozabadi, "Cocurrent and countercurrent imbibition in a water-wet matrix block," *SPE Journal*, vol. 5, no. 1, pp. 3–11, 2000.

[4] H. L. Najurieta, N. Galacho, M. E. Chimienti, and S. N. Illiano, "Effects of temperature and interfacial tension in different production mechanisms," in *Proceedings of the Latin American and Caribbean Petroleum Engineering Conference*, vol. 69398, Buenos Aires, Argentina, 2001.

[5] R. W. Parsons and P. R. Chaney, "Imbibition model studies on water-wet carbonate rocks," *SPE Journal*, pp. 26–34, 1996.

[6] R. Iffly, D. C. Rousselet, and J. L. Vermeulen, "Fundamental study of imbibition in fissured oil fields," in *Proceedings of the Annual Technical Conference*, vol. 4102, SPE, Dallas, Tex, USA, 1972.

[7] G. Hamon and J. Vidal, "Scaling-up the capillary imbibition process from laboratory experiments on homogeneous samples," in *Proceedings of the European Petroleum Conference*, vol. 15852, SPE, London, UK, October 1986.

[8] S. Al-Lawati and S. Saleh, "Oil recovery in fractured oil reservoirs by low IFT imbibition process," in *Proceedings of the Annual Technical Conference and Exhibition*, vol. 36688, SPE, Denver, Colo, USA, 1996.

[9] J. C. Reis and M. Cil, "A model for oil expulsion by counter-current water imbibition in rocks: one-dimensional geometry," *Journal of Petroleum Science and Engineering*, vol. 10, no. 2, pp. 97–107, 1993.

[10] M. Cil, J. C. Reis, M. A. Miller, and D. Misra, "An examination of countercurrent capillary imbibition recovery from single matrix-blocks and recovery predictions by analytical matrix/fracture transfer functions," in *Proceedings of the Annual Technical Conference and Exhibition*, vol. 49005, SPE, New Orleans, La, USA, September 1998.

[11] J. Lee and J. M. Kang, "Oil recovery in a fracture of variable aperture with countercurrent imbibition: experimental Analysis," in *Proceedings of the Annual Technical Conference and Exhibition*, vol. 56416, Houston, Tex, USA, October 1999.

[12] T. Babadagli, "Scaling of cocurrent and countercurrent capillary imbibition for surfactant and polymer injection in naturally fractured reservoirs," *SPE Journal*, vol. 6, no. 4, pp. 465–478, 2001.

[13] D. Zhou, L. Jia, J. Kamath, and A. R. Kovscek, "Scaling of counter-current imbibition processes in low-permeability porous media," *Journal of Petroleum Science and Engineering*, vol. 33, no. 1–3, pp. 61–74, 2002.

[14] N. R. Morrow and G. Mason, "Recovery of oil by spontaneous imbibition," *Current Opinion in Colloid and Interface Science*, vol. 6, no. 4, pp. 321–337, 2001.

[15] D. Kashchiev and A. Firoozabadi, "Analytic solutions for 1D countercurrent imbibition in water-wet media," *SPE Journal*, vol. 8, no. 4, pp. 401–408, 2003.

[16] Z. Tavassoli, R. W. Zimmerman, and M. J. Blunt, "Analytic analysis for oil recovery during counter-current imbibition in strongly water-wet systems," *Transport in Porous Media*, vol. 58, no. 1-2, pp. 173–189, 2005.

[17] D. Silin and T. Patzek, "On Barenblatt's model of spontaneous countercurrent imbibition," *Transport in Porous Media*, vol. 54, no. 3, pp. 297–322, 2004.

[18] H. S. Behbahani, G. Di Donato, and M. J. Blunt, "Simulation of counter-current imbibition in water-wet fractured reservoirs," *Journal of Petroleum Science and Engineering*, vol. 50, no. 1, pp. 21–39, 2006.

[19] D. Wilkinson, "Percolation model of immiscible displacement in the presence of buoyancy forces," *Physical Review A*, vol. 30, no. 1, pp. 520–531, 1984.

[20] N. Bech, O. K. Jensen, and B. Nielsen, "Modeling of gravity-imbibition and gravity-drainage processes: analytic and numerical solutions," *SPE Reservoir Engineering*, vol. 6, no. 1, pp. 129–136, 1991.

[21] Z. Tavassoli, R. W. Zimmerman, and M. J. Blunt, "Analysis of counter-current imbibition with gravity in weakly water-wet systems," *Journal of Petroleum Science and Engineering*, vol. 48, no. 1-2, pp. 94–104, 2005.

[22] G. Løvoll, Y. Méheust, K. J. Måløy, E. Aker, and J. Schmittbuhl, "Competition of gravity, capillary and viscous forces during drainage in a two-dimensional porous medium, a pore scale study," *Energy*, vol. 30, no. 6, pp. 861–872, 2005.

[23] H. Karimaie and O. Torsæter, "Effect of injection rate, initial water saturation and gravity on water injection in slightly water-wet fractured porous media," *Journal of Petroleum Science and Engineering*, vol. 58, no. 1-2, pp. 293–308, 2007.

[24] D. W. Ruth, Y. Li, G. Mason, and N. R. Morrow, "An approximate analytical solution for counter-current spontaneous imbibition," *Transport in Porous Media*, vol. 66, no. 3, pp. 373–390, 2007.

[25] D. Silin, T. Patzek, and S. M. Benson, "A model of buoyancy-driven two-phase countercurrent fluid flow," *Transport in Porous Media*, vol. 76, no. 3, pp. 449–469, 2009.

[26] Z. Chen, *Reservoir Simulation: Mathematical Techniques in Oil Recovery*, SIAM, Philadelphia, Pa, USA, 2007.

13

Catalytic Transformation of Tall Oil into Biocomponent of Diesel Fuel

Jozef Mikulec,[1] **Andrea Kleinová,**[2] **Ján Cvengroš,**[2] **Ľudmila Joríková,**[1] **and Marek Banič**[1]

[1] *VÚRUP a.s., Vlčie hrdlo, 820 03 Bratislava, Slovakia*
[2] *Faculty of Chemical and Food Technology, Slovak University of Technology, Radlinského 9, 812 37 Bratislava, Slovakia*

Correspondence should be addressed to Jozef Mikulec, jozef.mikulec@vurup.sk

Academic Editor: David Kubička

One of the conventional kraft pulp mills produce crude tall oil which is a mixture of free fatty acids, resin acids, sterols, terpenoid compounds, and many others. This study is devoted to the issue of direct transformation of crude tall oil in a mixture with straight-run atmospheric gas oil to liquid fuels using three different commercial hydrotreating catalysts. Diesel fuel production is an alternative to incineration of these materials. High catalytic activity was achieved for all tested catalysts in temperature range 360–380°C, under 5.5 MPa hydrogen pressure and ratio H_2/feedstock 500–1000 l/l. Crude tall oil can be converted to diesel oil component via simultaneous refining with straight-run atmospheric gas oil on $NiMo/Al_2O_3$ and NiW/Al_2O_3-zeolite catalysts. All tested catalysts had very good hydrodenitrogenation activity and high liquid yield were at tested conditions.

1. Introduction

One of the modes of reducing the share of green house gases (GHG) emissions at energy production lies in the utilization of biomass and wastes. In the initial step, the known technological processes and feedstock commonly applied in foodstuff processing industry were utilized to produce first generation biocomponents. The use of fatty acid esters, bioalcohol, and ETBE is common at present. First-generation biocomponents are usually more expensive when compared to petroleum-based fuels. Competition with foodstuff production is questionable as well. The actual system based on indicative targets of reaching the total energy content or volume of biocomponents does not support priorities of biocomponents utilization with low-cost GHG emission decrease.

The regulatory mechanisms should be stipulated in a way that allow finds a possible reduction in the GHG emission for various biofuels and foodstuffs. Biofuels should be supported through an efficiency increase of current biocomponents and development of new improved procedures. The regulatory mechanism should not act as a barrier for new biofuel types.

One of the possible solutions is represented by the introduction of second generation biocomponents originated

from wastes. Crude tall oil (CTO) [1] is a byproduct of paper production from coniferous wood by the Kraft pulping process. As an average, 20–30 kg tall oil/ton wood is produced. It contains 30–50% wt. of free fatty (mainly oleic and linolic) acids, 40–60% of rosin acids (abietic and pimaric acids), and 10–15% of unsaponifiables containing 2–4% of sterols, fatty alcohols, phenols and hydrocarbons. Free fatty acids (FFAs) and rosin acids (RAs) can be separated by vacuum rectification. Tall pitch as a distillation residue is used as a source of phytosterols or simply for energy production purposes. The FFA fraction may be converted through esterification by methanol to fatty acids methyl esters (FAME) usable in engine fuels.

Crude tall oil can be directly used in the field of biofuels by the following procedure as well [2]. Performing FFA esterification with methanol, FFAs are converted to FAME which can be distilled off from the mixture, preferably using a film-type vacuum evaporator. Distillate (48–52% yield relating to CTO input) complies with the standard EN 14 214 except for one parameter—sulphur content reaching about 1200 mg/kg, while according to the standard, not more than 10 mg/kg is acceptable.

Within our up-to-now performed study on CTO esterification with methanol (MeOH) in a batch reactor and isolation of the formed tall oil fatty acid methyl esters (TOFAME), next significant facts have been observed.

CTO can undergo esterification without any pretreatment. Esterification carried out in a classic batch arrangement without formed water withdrawal proceeds relatively smoothly. The reaction time is 4–5 hours, the optimum MeOH : FFA molar ratio is about 7. Preferably, FFAs undergo esterification while RAs are not subjected to esterification. In a GL chromatogram, the peaks corresponding to FAME are present, those of RAME (rosin acids methyl esters) are absent, that is, all present FFA were completely transformed to FAME. The course of reaction can be cosily monitored as a decrease of acid value (AV). The final AV corresponding to conversion of all present FFAs to their methyl esters can be calculated in advance. H_2SO_4 acts as an efficient catalyst of the reaction; its 0.5% addition relative to CTO is adequate. After esterification and elimination of the formed water and unreacted MeOH, the reaction mixture can be rather efficiently separated by vacuum distillation in a film-type evaporator. TOFAME distillate is obtained in 48–52% yield relative to the CTO feed. The AV of the distilled TOFAME varies in the range of about 8 to 12 mg KOH/g and results from free RA. Esterification performed by an unconventional process arrangement is described in detail in [3]. Into a bottom part of the catalyst-containing CTO charge with the temperature of 110–120°C the liquid MeOH is continuously delivered, it evaporates and penetrates through CTO layer reacting intensively with FFA, simultaneously stirring the reaction mixture and, at the same time, carrying away the formed water. Due to the high temperature, the high local excess of MeOH to FFA, and due to the efficient elimination of water, the rate of the reaction is high and the desired conversion is reached within a short time [2].

Another option lies in technological processes of catalytic hydrogenation. Coll et al. [4] investigated transformation of rosin acids in the fraction of crude tall oil using chosen commercial catalysts of Ni-Mo or Co-Mo types. Obtained yields of the liquid were high (68–82%), in particular at temperatures over 350°C and higher hydrogen partial pressure. Myllyoja et al. [5] tested free fatty acids fraction from tall oil using NiMo catalyst at hydrogen pressure 5,0 MPa, LHSV = 1.5 h^{-1}, reaction temperatures 340–360°C, and a ratio of 900 NL H_2/L feedstock. Under these conditions, the catalyst underwent fast deactivation and high-molecular substances and aromatic compounds were formed. Bromine index increased during the test. In a patent application [6] of Stigsson and Naydenov the decarboxylation/oxygen elimination from modified crude tall oil is described. Impurities were eliminated from crude tall oil; light and heavy components were distilled off and the mixture of the 170–400°C fraction with vegetable oil and/or mineral oil was subjected to catalytic decarboxylation at 320–240°C using a hydrorefining catalyst.

Catalytic deoxygenation of triacylglycerides and other oxygen-containing bio-oils is an alternative to biogas oil production [7]. There are few ways to eliminate oxygen from the feed: (i) deoxygenation over supported noble metal catalysts; (ii) hydrodeoxygenation over metal sulphide catalysts. Metal catalyzed deoxygenation in the absence of hydrogen is an effective method for the conversion of triacylglycerides to liquid hydrocarbons. The distribution of products is influenced by the degree of unsaturation of feed and by the nature of the catalyst [8–12]. Renewable gas oil component can be produced by catalytic hydrotreating of triacylglycerides (TAGs) over supported sulphide catalysts in presence of hydrogen. Two main routes were observed simultaneously—hydrodeoxygenation (HDO) an hydrodecarboxylation [12–15]. The advantage of this catalysts is possibility to process waste TAG to clean hydrocarbons component [16–18]. Many authors verified a possibility of simultaneous hydrodesulphurization of straight run atmospheric gas oil and hydrodeoxygenation of vegetable oils and animal fats using hydrorefining catalysts—coprocessing [19–21]. One of the disadvantage is deactivation of metal sulphide catalysts. Hydrocracking of vacuum gas oil with TAG is another way to produce clean biogas oil component [22, 23]. The technology of coprocessing allows the decrease of both investment and operational costs of biocomponent production and use wide variety of feed particularly nonfood origin. Some catalyst producer offers catalyst for stand-alone process and/or coprocessing. Stand-alone processing of TAG was realized by Neste Oil.

In this contribution, possibilities to process the mixtures of unseparated crude tall oil with straight-run atmospheric gas oil atmospheric gas oil using hydrocracking Ni-W catalyst and Ni-Mo hydrotreating catalysts were examined.

2. Experimental Part

The tests were performed in a tubular downflow isothermal fixed-bed reactor (the total volume 250 mL) with catalytic bed of 100 mL, feed range 100–1000 mL/h, operation temperature 360–380°C, and operation pressure 5,5 MPa. Feed stock container and product lines were heated to prevent feed and product solidification. Catalyst was diluted with glass pellet to avoid channelling and minimize backmixing. The dimensions of the reactor were 500 mm (length) and 25 mm (inner diameter). The reactor had three electrically heated and controlled sections.

Reaction feed stock was fed into the reactor by a high-pressure piston pump and mixed with hydrogen (Messer, 99,9%) on the head of the reactor. Formed mixture, depending on the amount of the catalyst, passes through a bed of a catalyst. The temperature was measured by means of thermocouples and controlled by a PID regulators. The pressure was controlled by a backpressure regulator. The formed product passes subsequently through a cooler to a separator, where reaction gas is separated from the product. The liquid products were withdrawn after stabilization of reaction conditions (6 h) in two two-hour intervals and analyzed by offline gas chromatography after separation of the water phase. Reaction gas, after being discharged from the separator, passes through a gas flow meter allowing both to regulate and measure its amount. H_2S and NH_3 were stripped by counter current nitrogen flow.

During the test, a representative sample of the gas was withdrawn and complex material balance was elaborated. In the liquid sample, a portion of aqueous and organic

TABLE 1: Physical properties of straight-run atmospheric gas oil and crude tall oil.

Parameter	Unit	AGO	CTO
Density at 20°C	kg/m^3	843	952
Sulphur content	mg/kg	9 535	1 610
Nitrogen content	mg/kg	168.4	n.a.
Bromine value	g Br$_2$/100 g	n.a.	81.1
Iodine value	g I$_2$/100 g	n.a.	113
Acid value	mg KOH/g	0.02	138
Sodium content	mg/kg	n.a.	27.7
Potassium content	mg/kg	n.a.	6.5
Calcium content	mg/kg	n.a.	1.1
Magnesium content	mg/kg	n.a.	0.1
Phosphorus content	mg/kg	n.a.	22.7

AGO: straight-run atmospheric gas oil, CTO: crude tall oil.

phases was determined. In some experiments the feedstock was diluted with isooctane facilitating thus to follow the mechanism of transformation of tall oil components.

2.1. Feedstock. Crude depitched tall oil (Smurfit Kappa, Piteå, Sweden) was used in the tests. Straight-run atmospheric gas oil (atmospheric gas oil AGO) from the production unit AVD 6 of the refinery Slovnaft Bratislava, Slovak Republic was used within testing the simultaneous desulphurization of gas oil and crude tall oil. Some physical properties of straight-run atmospheric gas oil and crude tall oil are in Table 1. The portions of FFA present in TO are summarized in Table 2. Resin acids and unsaponifiable compounds were not analyzed in detail. According to CTO supplier the content of FFA was 45%, of RA 30%, and of the unsaponifiables 25%.

2.2. Catalysts. Three commercial hydrocracking catalysts were used by the tests: NiW/γ-Al$_2$O$_3$ + zeolit catalyst in sulphidized form, Ni-Mo hydrotreating catalyst, and Ni-W hydrotreating catalyst. Their approximate composition is in Table 3. Sulphidization occurred directly in a pressure vessel in hydrogen atmosphere at the pressure of 3 MPa using 5% solution of dimethyldisulphide in gas oil. The catalyst was dried at 120°C in a stream of nitrogen. The temperature gradually increased with the gradient 100°C/h up to 350°C, kept isothermal at 250°C for 1 h and at 350°C for 4 h.

2.3. Analytics. Reaction gas was withdrawn into glass sample flasks and analyzed by gas chromatography following the standard UOP 539-87 on an instrument Shimadzu GC 17 A. The chromatograph was equipped with an injection loop, flame-ionization, and thermal conductivity detectors, software for pressure and temperature control and data collection in the PC. After reaching steady-state operational conditions and performing sufficient flushing of the injection loop with the sample, the analysed sample was injected into the instrument using a 10 μL injection loop. Temperature and pressure program/mode was put into operation. In a three columns and two switch valves system, oxygen,

nitrogen, carbon dioxide, carbon oxide, hydrogen, and light hydrocarbons (from methane to n-pentane) were separated. Hydrocarbons higher than n-pentane were eluated in one peak. Individual components were identified based on the elution times of the components present in a standard mixture and determined comparing responses of the components present in the standard mixture and in the sample. Column types: molecular sieve, precolumn: 5 m × 0.53 mm × 3 μm SE 54, and analytical column: 60 m × 0.53 mm CP-SilicaPLOT.

Boiling point-based distribution (simulated distillation) was determined according to ASTM D 2887 with a Network GC System 6890 N (Agilent Technologies) equipped with an autosampler, on-column injector, FID detector, and separation column: RMX1 15 m × 0.53 mm × 2.65 μm.

The content of fatty acids in tall oil (TOFAME) was determined following ISO 5509 and EN 14103. Tall oil sample was saponified under a reflux condenser with methanolic solution of sodium hydroxide; the formed soaps were converted to fatty acids methyl esters by the reaction with a methanolic solution of boron trifluoride. Subsequent to addition of internal standard in isooctane, the formed fatty acids esters were extracted to isooctane layer by agitation. The content and abundance of individual TOFAME were determined using GC analysis of the dried isooctane layer. A gas chromatograph HEWLET PACKARD G 1800A GCD System with a FID detector and split injector was used for the analysis. A capillary column ZB-WAX (30 m × 0.32 mm, df 0.5 μm) was used for the separation. The content of TOFAME was obtained by the internal standard method.

The content of aromatic hydrocarbons in the final product and the used AGO was determined according to EN 12916 by HPLC method with RID detection on an instrument HPLC System, Agilent Technologies, 1200 Series. As a mobile phase, n-heptane was used; aromatic hydrocarbons were separated in groups by a number of aromatic rings using a medium polar column ZORBAX NH$_2$ 5 analytical column (4.6 × 250) mm. The content of individual aromatic groups was determined by the external standard method using o-xylene, fluorene, and phenathrene as reference compounds.

The sulphur content was determined by ultraviolet fluorescence technique according to STN EN ISO 20846 with an instrument ANTEK 9000S Pyro-Fluorescent Sulfur.

Nitrogen content was determined on an instrument TN/TSuv3000. The used operation procedure is described in ASTM D 4629 Standard Test Method.

Cetane number was evaluated using a portable diesel fuel analyser providing fast analysis of fuel composition. The Turbine/Diesel Fuel Analyzer (PetroSpec) uses the technique of infrared spectroscopic analysis for resolution and determination of individual components in the fuel sample. After obtaining the infrared absorption spectra, the instrument converts the spectral data to the parameter values.

Density of the modified samples was determined between individual steps of the experiment by oscilometric method according to the standard STN EN ISO 12185 with a commercially available instrument DMA 4500 M (Anton Paar) equipped with an oscillating U-tube.

TABLE 2: Crude tall oil—content of free fatty acids.

Acid	C14:0	C14:1	C16:0	C16:1	C18:0	C18:1	C18:2	C18:2 conj.
% wt.	0.07	0.04	1.99	0.13	0.83	11.53	24.52	5.47
Acid	C18:3	C20:0	C20:1	C22:0	C22:1	C24:0	C24:1	total
% wt.	0.90	0.54	0.49	0.70	0.10	0.05	0.01	47.36

TABLE 3: Approximate composition of catalysts.

Component	Catalyst A content, % wt.	Catalyst B content, % wt.	Catalyst C content, % wt.
NiO	5–10	3.0	1–5
MoO3	20–24	—	—
WO3	—	19	10–15
Al2O3	40–50	Balance 78	84–70
SiO2	25–30*	—	—
Aluminum phosphate	—	—	5–10

* SiO2 is in form of aluminosilicate.

FIGURE 1: GC chromatogram of liquid product—hydrotreating CTO in isooctane on catalyst A.

Acid values were determined by potentiometry according to the method described in standards STN 65 6214 or ASTM D 664. When determining the acid number, a commercially available instrument, automatic titrator 716 DMS TITRINO (Metrohm), was used.

At assessing or evaluating the samples obtained after individual steps of the whole experiment in the fuel testing laboratory, IR spectroscopy was used, too. To scan IR spectra in the region from 4000 cm^{-1} to 400 cm^{-1}, Fourier transform infrared spectrometer AVATAR 330 (Thermo Nicolet) and optical KBr cell of 0.040 mm were used.

To determine the share of RA and FA in CTO it is possible to benefit from ^{13}C NMR spectroscopy [1] exploiting a different chemical shift of the carboxylic carbon atom in RA and FA. The RA carboxyl is bonded to quarternary carbon while that of FA to secondary carbon. Chemical shifts of RA and FA carboxyls are in ^{13}C NMR spectra in the regions of 184–186 ppm and 180-181 ppm, respectively. The share of RA and FA can be easily determined based on the integrated intensities of the corresponding peaks. ^{13}C NMR spectral studies were measured with the apparatus VARIAN VXR 600 in the presence of deuterated chloroform as a solvent.

3. Results and Discussion

3.1. Hydrotreating of CTO in Solvent. Crude tall oil is dark brown viscous liquid of characteristic odour. CTO was dissolved in isooctane (5 and 20% vol.) and was kept by heating at 50°C preventing thus the formation and deposition of solid rosin acids. CTO was in isooctane fully soluble.

Catalyst A was used to test 5% vol. solution of CTO in isooctane in the temperature range 320–380°C, hydrogen partial pressure of 5.5 MPa, LHSV = 1 h^{-1}, and the ratio of hydrogen to the tall oil 1000 l/l and 500 l/l. With increasing temperature, the increased concentration of CO and CO2 as a result of decarboxylation reactions of acids present is registered. In the same ratio, the concentration of isobutane

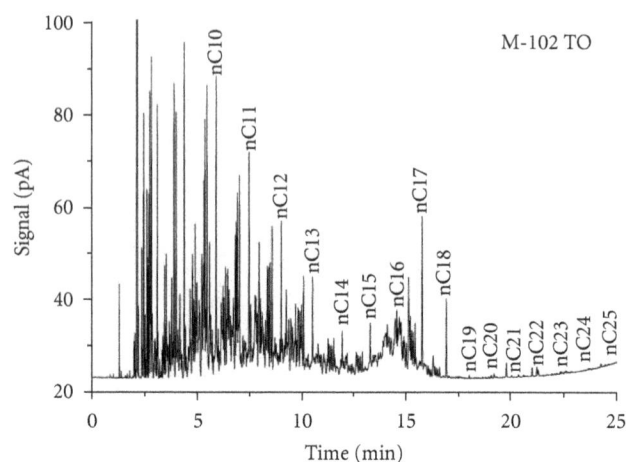

in the gas was increased. The reaction was carried out mainly at 380°C. The change of the hydrogen/tall oil ratio from the value of 1000 to 500 l/l had not a significant impact on the reactions.

The formation of CO can be explained by secondary reactions of CO2 and high excess hydrogen in the presence of a catalytic system. In the gaseous reaction products, methane, propane and water as a result hydrodeoxygenation free fatty acids were observed, respectively. The progress of reaction was checked with a higher concentration of CTO in isooctane (20% vol.).

At a reaction temperature of 380°C, hydrogen partial pressure of 5.5 MPa, LHSV = 1 h^{-1} and the ratio of hydrogen to the CTO 1096 l/l, the light brown solution with characteristic odour changed to a clear solution, which had a strong diesel smell. Reaction water content was 0.8 wt%. The gas share was 3.1 wt%. and they had the highest content of isobutane, CO, CO2, and propane. GC analysis indicates that on the catalyst A takes place the hydrodeoxygenation and hydrodecarboxylation of fatty acids to n-alkanes. The chemical structure of CTO resin acids is similar to that of abietic or dehydroabietic acids [4]. The resin acids were hydrogenated to the cyclanes hydrocarbons and water. Second mechanism was hydrodecarboxylation to cyclic hydrocarbons and CO2

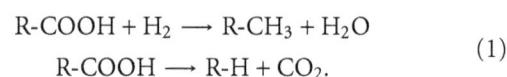

$$\begin{aligned} \text{R-COOH} + \text{H}_2 &\longrightarrow \text{R-CH}_3 + \text{H}_2\text{O} \\ \text{R-COOH} &\longrightarrow \text{R-H} + \text{CO}_2. \end{aligned} \tag{1}$$

Figure 1 shows GC chromatogram of product from hydrotreatment mixtures CTO in isooctane on catalyst A. As a side reaction cracking reaction to lighter hydrocarbons and isomerization takes place also on acidic centre of

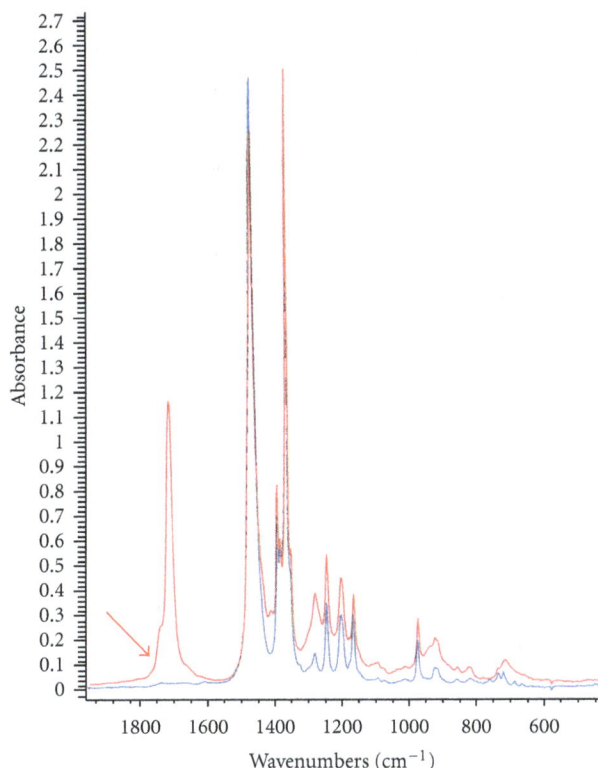

FIGURE 2: IR spectrum of 20% solution of CTO before (red line) and after reaction (blue line).

the catalyst (presence of the zeolite). The isomerization reaction was indicated by very low cloud point ($-20°$C).

In Figure 2, IR spectrum of 20% solution of tall oil and reaction product is displayed indicating a total elimination of C=O group at about 1680–1720 cm^{-1}.

The degree of oxygen elimination was confirmed also by ^{13}C NMR spectra (Figure 3). Figure 2 shows a selected spectral part of CTO used for hydrogenation in isooctane. The lower spectra correspond to the introduced feedstock while the upper ones document the situation after hydrogenation. Comparison of the spectra leads to a conclusion that the peaks of functional groups of both acid types (FA + RA) fully disappeared.

To verify a possible mechanism of the conversion of resin acids to hydrocarbons, test was made with a solution of rosin in a mixture of isooctane and acetone. The largest part of the gas was isobutane, which suggests a predominant hydrocracking reaction of alkyl groups associated with the saturation of double bonds and ring opening of cycloaliphatic rings. In the liquid products cycloalkanes were dominant with carbon numbers of C_5 through C_{12}. Moreover, in liquid product were present saturated n- and iso-alkanes but not olefinic compounds. Oxygen compounds were not detected.

Analysis of the total sulphur content manifested that the elimination of sulphur compounds from CTO by means of hydrogenation represents a more serious issue. In the feedstock was 322 mg/kg of total sulphur while in the product

after reaction was 122 mg/kg. The results indicate a content of refractory sulphur compounds.

3.2. Hydrotreating of Mixture CTO with AGO. The tests of hydrorefining and hydrodeoxygenation of the mixture of AGO with CTO were performed at 360 and 380°C, hydrogen pressure 5.5 MPa, and two value of ration hydrogen to the feedstock ratios with three different hydrotreating catalysts A, B, and C. For testing, concentrations of CTO 20 vol. were used and 30% vol. were used also for catalyst A. CTO was fully soluble in AGO. The main target of the experiments was to compare the performance of the three different hydrotreating catalysts with respect to some selected performance criteria as liquid yield, cetane number, HDS, HDO and HDN activity, aromatics content, and density. The test results are listed in Tables 4, 5, and 6.

3.3. Boiling Point Distribution. The method of simulated distillation was used for comparison of the catalysts performance. Shows Figure 4 the comparison of simulated distillation curve for catalysts A, B, and C with feed containing 20% vol. CTO in straight-run AGO at 380°C, partial pressure hydrogen 5,5 MPa, LHSV = 1 h^{-1}, and ratio hydrogen to feed about 1000 NL/L. The distillation curves for the studied catalysts were inherently different. Catalyst A is suitable for hydrocracking of vacuum distillates and even at relatively low temperature and pressure the reaction of cracking took place. It is obvious from Figure 4 that the used hydrocracking catalyst A influences the formation of lighter components and lowers the end of distillation. The product from a catalyst A contains 17% vol. gasoline fraction to 150°C. The initial boiling point and end of distillation were, however, lowered which indicate the substantial contribution of hydrocracking reactions. Products from the catalysts B and C were in the distillation range of gas oil. It is evident that the acidity of the catalyst determines the cracking reaction byproducts and must be balanced. The best catalyst from boiling point distribution point of view was the catalyst B.

3.4. HDS Activity of the Catalysts. Investigated mixture CTO in AGO contains different types of sulphur compounds. HDS activity was decreased from catalyst C (99.26%) to B (98,38) and A (96,99). In neither of the experiments, desulphurization down to 10 mg/kg was not reached which documents the presence of resistant sulphur compounds and necessity to exploit a combination of the higher active catalysts.

3.5. HDN Activity. Nitrogen content was reduced below 2 mg/kg. All tested catalyst had very good HDN activity but different one. The best HDN activity had catalyst C (97,49%), catalyst A has 95.67% HDN efficiency and HDN efficiency, of catalyst B was 89,62%.

3.6. HDO Activity. The oxygen removal was very high (99,9%), highest HDO activity had catalyst C.

3.7. HDA Activity. The influence of tungsten on hydrogenation of aromatic compounds and aromatic rings opening

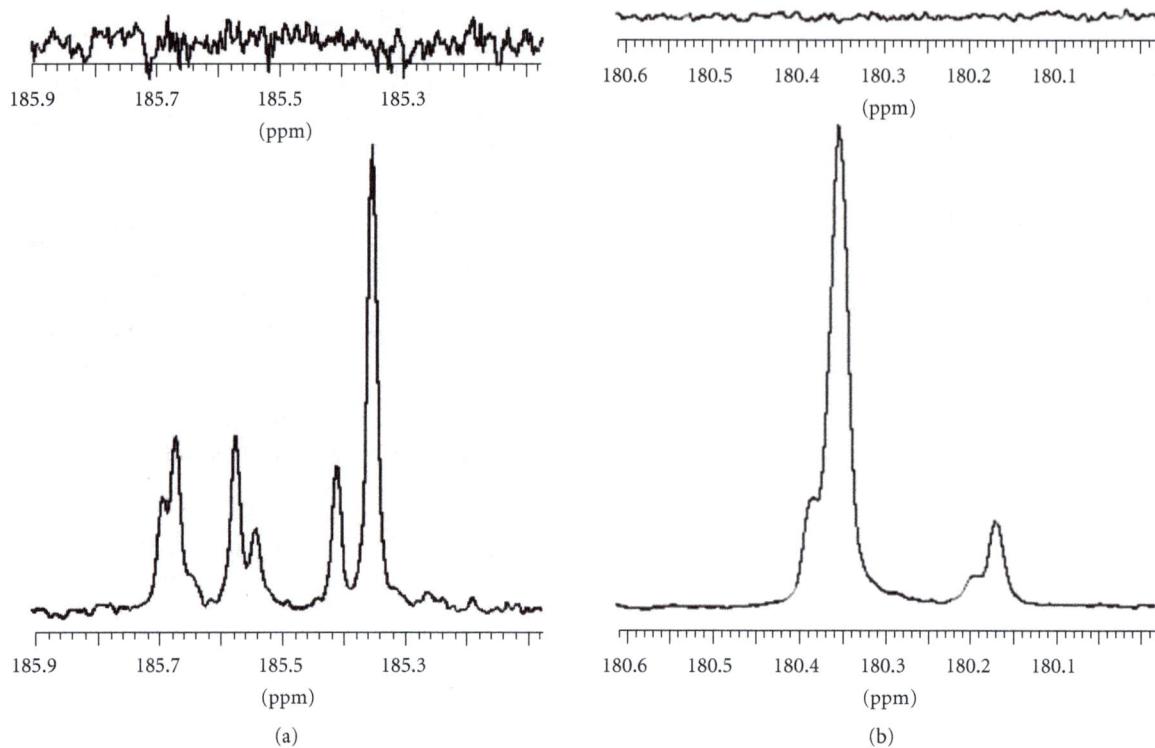

FIGURE 3: ^{13}C NMR spectrum of CTO after and before hydrogenation. The left spectra belong to RA (184–186 ppm). The right spectra to FFA (180-181 ppm).

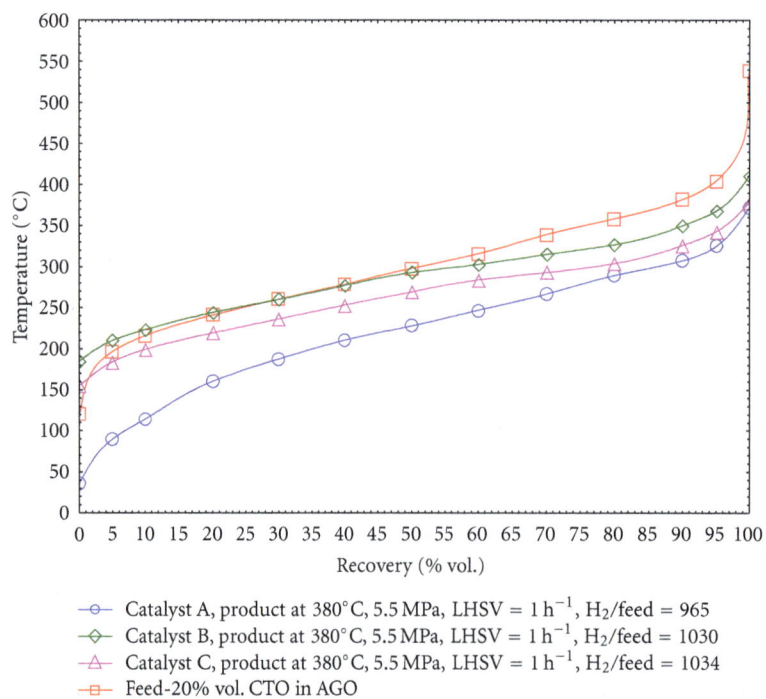

FIGURE 4: Simulated distillation curves of the feedstock and selected products of hydrorefining.

TABLE 4: Results of hydrorefining of the mixtures of gas oil and crude tall oil; catalyst A.

Parameter	Products					
Temperature, °C	380		360		380	
Pressure, MPa			5.5			
LHSV, h^{-1}			1			
H_2/feed, L/L·h	965	504	970	525	985	544
CTO content in feed, % vol.			20		30	
Hydrocarbons C_1–C_5, % wt.	1,42	0,93	0,57	1,19	1,19	0,45
Propane, % wt.	0,60	0,40	0,24	0,51	0,51	0,32
CO content, % wt.	0,84	0,55	0,34	0,71	0,70	1,27
CO_2 content, % wt.	0,28	0,18	0,11	0,23	0,23	0,96
Liquid product, % wt.	96.00	97.21	98.05	97.36	96.64	95.85
Reaction water, % wt.	0.86	0.73	0.69	1.07	0.73	1.15
Sulphur content, mg/kg	239	120	176	291	84	97
Acid value, mg KOH/g	0.01	0.01	0.04	0.02	0.02	0.04
Nitrogen content, mg/kg	1	1.2	1.5	1.6	0.8	0.9
Density at 15°C. kg/m³	801	813	814	814	826	830
Monoaromatics, % wt.	16.5	19.6	20	21.2	19.8	20.4
Σ Di- and Tri-aromatics, % wt.	1.66	3.32	2.68	2.39	3.23	3.99
Cetane number	56,5	55	53,5	54	57	56,8

TABLE 5: Results of hydrorefining of the mixtures of gas oil and crude tall oil; catalyst B.

Parameter	Products			
Temperature, °C	380		360	
Pressure, MPa	5.5		5,5	
LHSV, h^{-1}	1		1	
H_2/feed, L/L·h	1030	541	1034	531
CTO content in feed, % vol.		20		
Hydrocarbons C_1–C_5, % wt.	0,98	0,48	0,55	0,37
Propane, % wt.	0,38	0,33	0,69	0,46
CO content, % wt.	0,65	0,32	0,68	0,48
CO_2 content, % wt.	0,38	1,25	2,07	1,40
Liquid product, % wt.	96.74	97.73	96.00	95.67
Reaction water, % wt.	0.86	1.02	1.93	2.93
Sulphur content, mg/kg	128.8	308.3	114.2	38.2
Acid value, mg KOH/g	0.15	0.03	0.11	0.04
Nitrogen content, mg/kg	2.4	2.8	1.5	1.8
Density at 15°C. kg/m³	839	835	832	833
Monoaromatics, % wt.	25	23.2	20.7	21.4
Σ Di- and Tri-aromatics, % wt.	2.9	2.6	2	2
Cetane number	55	56.4	55.8	56.7

was significant. The content of compounds containing three and two aromatic rings decreased, while that with one ring increased. HDA activity of the catalysts B and C was not significant.

3.8. Catalyst Activity. Within a short-term test lasting 72 h no deactivation of the catalyst expressed by means of the acid value of the product was observed in spite of the fact that no precatalyst intended to eliminate catalytic poisons from tall oil was used.

3.9. Cetane Number. Cetane number is very important value for evaluation of combustion of motor fuel used in compression-ignition engines. Cetane number of the products was high in all catalyst tests. This result only confirms changes in composition of products due to the rising of n-alkanes concentration.

3.10. Density. The value of the density is an important parameter for the assessment of the possibilities of mixing of the fuel. Products of the joint coprocessing AGO and

TABLE 6: Results of hydrorefining of the mixtures of gas oil and crude tall oil; catalyst C.

Parameter	Products			
Temperature, °C	380		360	
Pressure, MPa	5.5		5,5	
LHSV, h^{-1}	1		1	
H_2/feed, L/L·h	1034	511	1054	503
CTO content in feed, % vol.	20			
Hydrocarbons C_1–C_5, % wt.	1,27	1,14	0,68	0,44
Propane, % wt.	0,49	0,44	0,53	0,27
CO content, % wt.	0,28	0,99	0,22	0,60
CO_2 content, % wt.	2,57	2,93	1,66	0,67
Liquid product, % wt.	95,12	94,52	95,72	96,64
Reaction water, % wt.	2,31	2,55	2,62	1,38
Sulphur content, mg/kg	58.8	100.5	218.4	473.4
Acid value, mg KOH/g	0.14	0.17	0.05	0.07
Nitrogen content, mg/kg	0.58	0.8	1	1.2
Density at 15°C. kg/m^3	831	837	830	829
Monoaromatics, % wt.	21.2	24.5	16.8	16.8
Σ Di- and Tri-aromatics, % wt.	2.3	2.9	1.4	1.5
Cetane number	56.6	56.3	59	59

CTO comply with the standard for diesel fuel in the case of catalysts B and C. The straight chain alkanes can undergo isomerization and partly cracking to produce isomerised alkanes with lighter alkanes in case of catalyst A. In this case light fraction must be eliminated by stripping before use in diesel fuel.

4. Conclusions

NiW and NiMo hydrocracking/hydrotreating catalysts were found to be active for hydrotreating of crude depitched tall oil and in coprocessing with atmospheric gas oil. The reaction pathway of crude tall oil involves hydrogenation of double bonds, decarboxylation, hydrodeoxygenation, isomerization, and hydrocracking of alkane and cyclic structures. Crude tall oil can be converted to a component of diesel fuel via simultaneous refining of atmospheric gas oil and tall oil using the commercial hydrotreating catalysts at 360–380°C and hydrogen pressure of 5.5 MPa. Small amounts of remaining impurities in feedstock (metals) must be eliminated through guard bed catalyst. The difference between the catalysts was in HDS activity of refractory sulphur species. The selectivity for the maximum diesel range hydrocarbons must be controlled by appropriate acidity of catalyst support.

Acknowledgment

This work was supported by Slovak Research and Development Agency, Project no. APVV 10-0665-2010

References

[1] J. Cvengroš, Ľ Malík, M. Košík, and I. Šurina, "Fractionation of tall oil," *Chemicky Prumysl*, vol. 35, pp. 542–545, 1985 (Slovak).

[2] R. Mikulášik, I. Šurina, S. Katuščák, J. Cvengroš, and M. Polovka, "Preparation of biodiesel from tall oil," *Chemical Papers*, vol. 99, pp. 1234–2345, 2008.

[3] T. Kocsisová, J. Cvengroš, and J. Lutišan, "High-temperature esterification of fatty acids with methanol at ambient pressure," *European Journal of Lipid Science and Technology*, vol. 107, no. 2, pp. 87–92, 2005.

[4] R. Coll, S. Udas, and W. A. Jacoby, "Conversion of the rosin acid fraction of crude tall oil into fuels and chemicals," *Energy and Fuels*, vol. 15, no. 5, pp. 1166–1172, 2001.

[5] J. Myllyoja, J. Aalto, and E. Harlin, "Process for the manufacture of diesel range hydrocarbons," WO 2007003708 (A1), 2007.

[6] L. Stigsson and V. Naydenov, "Conversion of crude tall oil to renewable feedstock for diesel range fuel compositions," WO 2009131510 (A1), 2009.

[7] I. Kubičková and D. Kubička, "Utilization of triglycerides and related feedstocks for production of clean hydrocarbon fuels and petrochemicals: a review," *Waste and Biomass Valorization*, vol. 1, no. 3, pp. 293–308, 2010.

[8] T. Morgan, D. Grubb, E. Santillan-Jimenez, and M. Crocker, "Conversion of triglycerides to hydrocarbons over supported metal catalysts," *Topics in Catalysis*, vol. 53, no. 11-12, pp. 820–829, 2010.

[9] P. T. Do, M. Chiappero, L. L. Lobban, and D. E. Resasco, "Catalytic deoxygenation of methyl-octanoate and methyl-stearate on Pt/Al_2O_3," *Catalysis Letters*, vol. 130, no. 1-2, pp. 9–18, 2009.

[10] J. Wildschut, F. H. Mahfud, R. H. Venderbosch, and H. J. Heeres, "Hydrotreatment of fast pyrolysis oil using heterogeneous noble-metal catalysts," *Industrial and Engineering Chemistry Research*, vol. 48, no. 23, pp. 10324–10334, 2009.

[11] J. G. Na, B. E. Yi, J. N. Kim et al., "Hydrocarbon production from decarboxylation of fatty acid without hydrogen," *Catalysis Today*, vol. 156, no. 1-2, pp. 44–48, 2010.

[12] S. Lestari, P. Mäki-Arvela, I. Simakova, J. Beltramini, G. Q. M. Lu, and D. Y. Murzin, "Catalytic deoxygenation of stearic acid

and palmitic acid in semibatch mode," *Catalysis Letters*, vol. 130, no. 1-2, pp. 48–51, 2009.

[13] B. Donnis, R. G. Egeberg, P. Blom, and K. G. Knudsen, "Hydroprocessing of bio-oils and oxygenates to hydrocarbons. Understanding the reaction routes," *Topics in Catalysis*, vol. 52, no. 3, pp. 229–240, 2009.

[14] P. Šimáček, D. Kubička, G. Šebor, and M. Pospíšil, "Hydroprocessed rapeseed oil as a source of hydrocarbon-based biodiesel," *Fuel*, vol. 88, no. 3, pp. 456–460, 2009.

[15] D. Kubička and J. Horáček, "Deactivation of HDS catalysts in deoxygenation of vegetable oils," *Applied Catalysis A*, vol. 394, no. 1-2, pp. 9–17, 2011.

[16] S. Bezergianni, A. Kalogianni, and A. Dimitriadis, "Catalyst evaluation for waste cooking oil hydroprocessing," *Fuel*, vol. 93, pp. 638–641, 2012.

[17] S. Bezergianni, A. Dimitriadis, A. Kalogianni, and P. A. Pilavachi, "Hydrotreating of waste cooking oil for biodiesel production. Part I: effect of temperature on product yields and heteroatom removal," *Bioresource Technology*, vol. 101, no. 17, pp. 6651–6656, 2010.

[18] S. Bezergianni, A. Dimitriadis, T. Sfetsas, and A. Kalogianni, "Hydrotreating of waste cooking oil for biodiesel production. Part II: effect of temperature on hydrocarbon composition," *Bioresource Technology*, vol. 101, no. 19, pp. 7658–7660, 2010.

[19] J. Mikulec, J. Cvengroš, Ľ. Joríková, M. Banič, and A. Kleinová, "Second generation diesel fuel from renewable sources," *Journal of Cleaner Production*, vol. 18, no. 9, pp. 917–926, 2010.

[20] Ch. Templis, A. Vonortas, I. Sebos, and N. Papayannakos, "Vegetable oil effect on gasoil HDS in their catalytic co-hydroprocessing," *Applied Catalysis B*, vol. 104, no. 3-4, pp. 324–329, 2011.

[21] J. Walendziewski, M. Stolarski, R. Łużny, and B. Klimek, "Hydroprocesssing of light gas oil—rape oil mixtures," *Fuel Processing Technology*, vol. 90, no. 5, pp. 686–691, 2009.

[22] R. Tiwari, B. S. Rana, R. Kumar et al., "Hydrotreating and hydrocracking catalysts for processing of waste soya-oil and refinery-oil mixtures," *Catalysis Communications*, vol. 12, no. 6, pp. 559–562, 2011.

[23] S. Bezergianni, A. Kalogianni, and I. A. Vasalos, "Hydrocracking of vacuum gas oil-vegetable oil mixtures for biofuels production," *Bioresource Technology*, vol. 100, no. 12, pp. 3036–3042, 2009.

14

Effect of Additives on Characterization and Photocatalytic Activity of TiO$_2$/ZnO Nanocomposite Prepared via Sol-Gel Process

Shahram Moradi,[1] Parviz Aberoomand Azar,[2] Sanaz Raeis Farshid,[1] Saeed Abedini Khorrami,[1] and Mohammad Hadi Givianrad[2]

[1] *Department of Chemistry, Tehran North Branch, Islamic Azad University, Tehran 1913674711, Iran*
[2] *Department of Chemistry, Science and Research Branch, Islamic Azad University, Tehran, Iran*

Correspondence should be addressed to Shahram Moradi, shm⁻moradi@yahoo.com

Academic Editor: Donald L. Feke

TiO$_2$/ZnO nanocomposites were prepared by the sol-gel method with and without addatives such as carboxy methyl cellulose (CMC), poly(ethylene glycol) (PEG), polyvinylpyrrolidon, (PVP), and hydroxylpropylcellulose (HPC). The characteristics of the prepared TiO$_2$/ZnO nanocomposites were identified by IR spectra, X-ray diffraction (XRD), scanning electron microscopy (SEM), and energy dispersive X-ray spectroscopy (EDS) methods. The additives have a significant effect on the particle size distribution and photocatalytic activity of TiO$_2$/ZnO nanocomposites. The photocatalytic activity of the synthesized nanocomposites was investigated for decolorization of methyl orange (MO) in water under UV-irradiation in a batch reactor and the results showed that the photocatalytic activity of the nanocomposites have been increased by CMC, PEG, PVP, and HPC, respectively. SEM has shown that the particle size distribution of TiO$_2$/ZnO nanocomposite in the presence of HPC was better than the other samples.

1. Introduction

Azo dyes constitute the largest group of coloring materials in the textile industry [1–3]. Release of these substances in nature is the largest source of pollution for natural ecosystems. In recent decades, semiconductor photocatalyst TiO$_2$ has been investigated as one of the most promising candidate for a photocatalyst and it has been attracted due to its potential application in removing of all types of organic pollutants in water [4–6]. In order to improve the photocatalytic activity and the response into visible part of the spectrum, TiO$_2$ doping with metal ions or metal oxides has been applied [1, 2].

Several methods on the preparation of TiO$_2$/ZnO were reported, such as solid-state method, impregnation method, and chemical coprecipitation method [7–9]. Another method is the sol-gel method that has significant advantages such as, high purity, good uniformity of the powder microstructure, low-temperature synthesis, and easily controlled reaction condition and therefore it has been used for preparing of TiO$_2$/ZnO nanocomposite [4, 7, 10–12].

In this work, TiO$_2$/ZnO nanocomposites were prepared by the sol-gel method in presence and absence of CMC, PEG, PVP, and HPC as additives. The synthesized nanocomposites were characterized by means of XRD, SEM, EDS, and IR spectroscopy. Since the particle size is an important parameter in photocatalytic activity, the effects of mentioned additives have been surveyed on the particle size distribution. The photocatalytic activity of TiO$_2$/ZnO nanocomposites were assessed for decolorization of methyl orange (MO) in water under UV irradiation in a batch reactor.

2. Materials and Methods

2.1. Materials. The chemicals used in this study were Tetra isopropyl orthotitanate (TTIP) (for synthesis), zinc nitrate tetrahydrate, diethanolamine (DEA) (for synthesis), Glacial acetic acid, methyl orange (MO), absolute ethanol, deionized water from Merck Chemical Company, and carboxymethyl cellulose (CMC), poly(ethylene glycol) (PEG), poly vinyl pyrrolidone (PVP), and hydroxylpropyl cellulose (HPC) from Sigma-Aldrich Company.

TABLE 1: Different samples of TiO_2/ZnO.

Sample	a	b	c	d	e
Additive type	—	CMC	PEG	PVP	HPC

2.2. Preparation of Nanocomposites.

In this study, TiO_2/ZnO nanocomposite powders were prepared by the sol-gel process. The TiO_2 sol was made at room temperature, and TTIP was used as a precursor as follows. In the first stage, additive (CMC or PEG or PVP or HPC = 30 g/L) was dissolved in ethanol under fast stirring for 5 minutes. Then TTIP was added into ethanol with a 1 : 9 molar ratio of TTIP to ethanol and was stirred for 15 minutes, to obtain a precursor solution. After that, a mixture of absolute ethanol, acetic acid, and deionized water with the molar ratio of 10 : 6 : 1 was added slowly into the precursor by a fast stirring and it was continuously stirred for 15 minutes to achieve a yellow transparent sol. Here, acetic acid was used as an inhibitor to reduce quick hydrolysis of TTIP, and so the pH value was adjusted on 5.

In the second stage, ZnO sol was prepared as follows. Firstly, zinc nitrate tetrahydrate was dissolved in absolute ethanol with the molar ratio of 0.1 : 110. After that, stirred for 5 minutes, then a mixture of absolute ethanol, diethanolamine, and deionized water with the molar ratio of 10 : 2 : 1 was added slowly into the precursor by a fast stirring and it was continuously stirred for 15 minutes to achieve transparent sol.

The prepared ZnO sol was directly added into the TiO_2 acidic sol with the molar ratio of 1 : 50 to get TiO_2/ZnO sol. This sol aged for 24 hours. After that, the prepared sol was dried in the air, then heat treated at 350°C for 10 minutes and then at 500°C for 5 hours. During this process, the temperature was raised at speed of 5°C/Sec. The samples were naturally cooled after the heat treatment (Table 1) [2, 5, 11, 13, 14].

2.3. Characterization of Nanocomposites.

The characteristics of TiO_2/ZnO nanocomposites were investigated as follows. FT-IR spectra were obtained as KBr pellets in the wave number range of 400 to 4000 cm^{-1} using Thermo Nicolet Nexus 870 FT-IR spectroscopy. Phase identification of the nanocomposites was done by XRD from STADI P, STOE with CuK_α radiation from 0 to 100 (2θ) at room temperature. The morphology and microanalysis of the nanocomposites were studied by SEM (SEM-XL30, Philips) and EDS. Additionally; UV-Vis absorption spectrum was obtained by means of Varian UV-Vis spectrophotometer.

2.4. Photocatalytic Activity Measurement.

Photocatalytic activity of the nanocomposites was investigated for the decolorization of MO. All of the experiments were accomplished in a rectangular cube glass reactor with 1 liter capacity. A 15 W UV lamp (Osram) was applied as a light source and it was placed in a quartz tube, which was installed inside the reactor (Figure 1). Initially, 1 g of photocatalyst was added into a 1 liter solution of MO with initial concentration of 5 ppm. Before irradiation, the suspension was stirred for 24

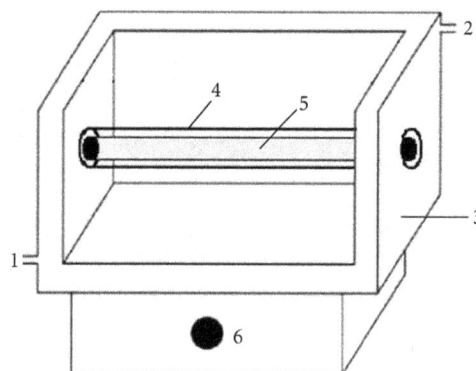

FIGURE 1: Schematic diagram of the photoreactor system. 1.Water entrance, 2.Water exit, 3.Glass jacket, 4.quartz cover, 5.UV lamp, 6.Stirrer.

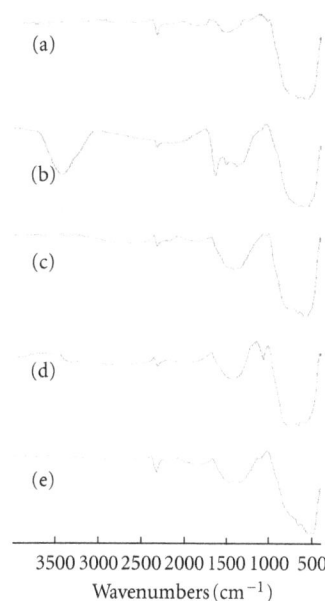

FIGURE 2: FTIR spectra of powder samples.

hours in darkness, due to elimination of absorption effect of the solution in the catalyst. After that, the lamp was switched on for starting the reaction. During irradiation, the suspensions were sampled at regular intervals and immediately centrifuged to remove catalyst particles [3, 4, 15–17].

3. Results and Discussion

3.1. FT-IR Spectra.

FT-IR spectra of samples have been presented in Figure 2 in the wave number range from 400 to 4000 cm^{-1}. Four significant peaks were observed around 650, 800, 1450, and 3450 cm^{-1}. The peak around 650 and 800 cm^{-1} can be devoted to symmetric stretching vibration of the Ti-O-Ti and vibration mode of Zn-O-Ti groups [8–10]. The peak around 1450 cm^{-1} was ascribed to the vibration mode of Ti-O and Ti-O-C that the Ti-O-C may result from the interaction between the Ti–O network and

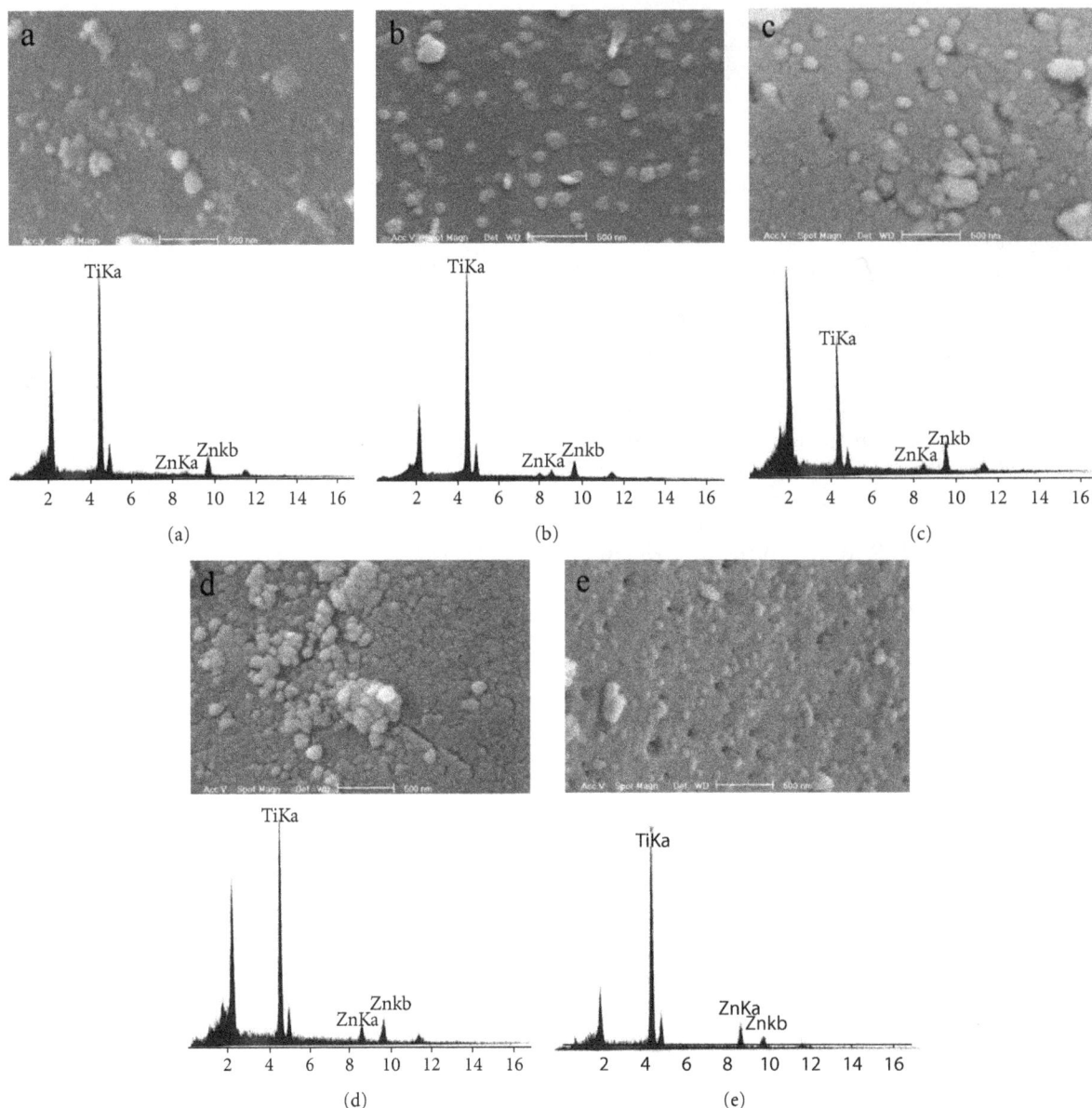

FIGURE 3: SEM images and EDS analysis of powder samples.

the organic polymers (CMC, PEG, PVP, or HPC). The wide peak around 3450 cm^{-1} which observed in sample b has been assigned to the OH stretching vibration of surface hydroxyl group. During the hydrolysis of TTIP, a large amount of ethanol lead to the appearance of hydroxyl bond [2, 4, 18, 19].

3.2. SEM. The SEM images and EDS analysis on the prepared nanocomposites were carried out and the results were shown in Figure 3.

Sample **a** without any additives, contains scattered particles which have different sizes. Sample **b** with CMC, is almost similar to sample **a**. In sample **c,** in presence of PEG, particle size becomes small and its distribution is more monotonous than previous samples [5, 9, 11, 20]. In sample **d,** with

PVP the number and density of particles are increased but the aggregation and the sticking together of the particles are observed. The last sample, **e,** which contains HPC has the most uniform particle distribution and it shows no agglomeration in comparison with other samples [7, 8, 21].

Figure 3. also demonstrates EDS analysis of TiO$_2$/ZnO nanocomposites in absence and presence of additives. In all of samples, the nanoparticles were composed of Ti and Zn. This proves that Zn was incorporated into the TiO$_2$ nanoparticles to form nanocomposite [9, 14].

3.3. XRD. XRD patterns of all samples have been shown in Figure 4. The peaks were observed at 2θ = (25), (27), (31, 48, 57, 63) and (35, 39, 43) which were related to anatase, rutile, zincite, and zinc phases, respectively. As can be seen, anatase

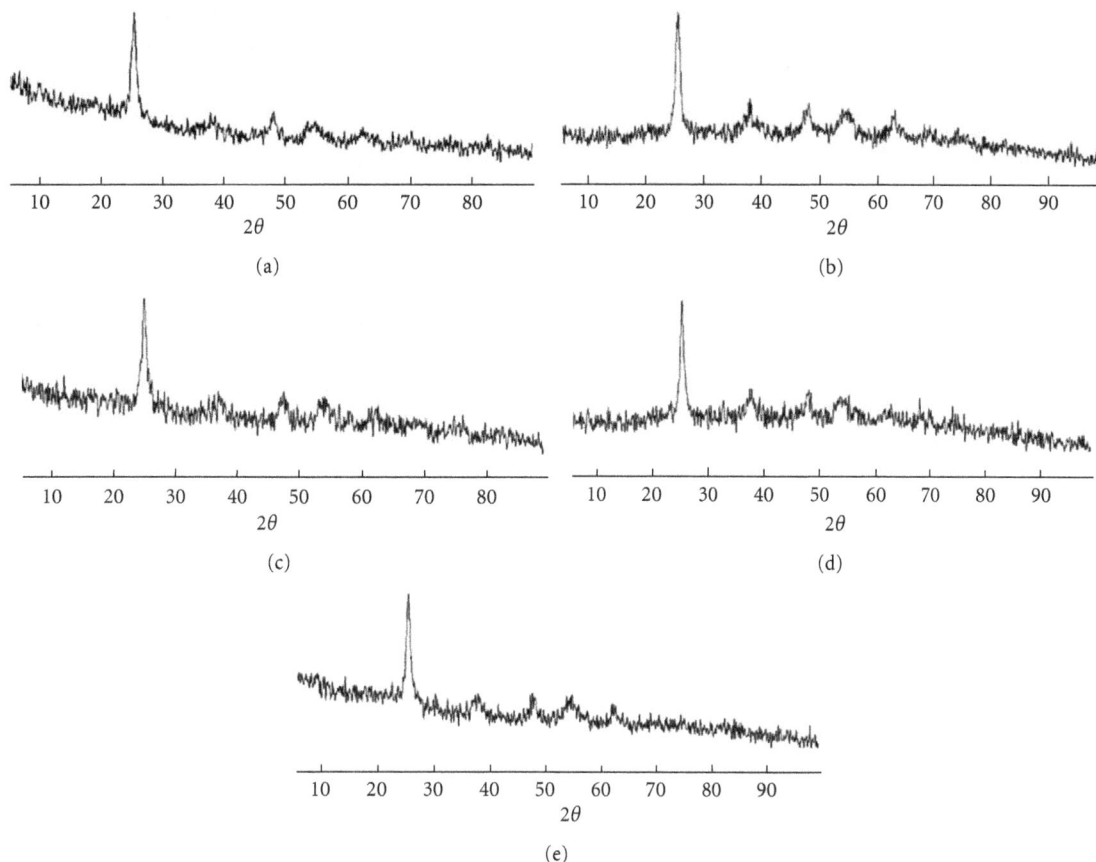

(a)

(b)

(c)

(d)

(e)

FIGURE 4: XRD patterns of powder samples.

phase as the dominant phase has observed in all samples. In all samples; TiO$_2$ and ZnO were not doped, however, separate crystallization of them was observed [5, 18, 20].

The calculated anatase phase percentages of the samples with different additives are between 80–90%. Sample **e** with 85% anatase phase and 15% rutile phase has showed a good catalytic activity [21–23].

3.4. Photocatalytic Activity. Figure 5. displays photocatalytic activity of the nanocomposites for decolorization of MO as a function of time at $\lambda = 465$ nm. The photocatalytic activity of the synthesized nanocomposites was investigated for decolorization of MO (5 mgL^{-1}) in water under UV irradiation in a batch reactor. According to Figure 5, the photocatalytic activity is enhanced by using additives [6–8]. Sample **e** with HPC has the best photocatalytic activity in comparison of other samples and about 3.5 h after starting the reaction, the absorbance of MO solution has been reached to 0. It indicates that organic polymers have been used as a dispersed factor to avoid accumulation of nanocomposite particles [23–25].

4. Conclusions

Five samples of TiO$_2$/ZnO nanocomposites have been prepared with and without CMC, PEG, PVP, and HPC

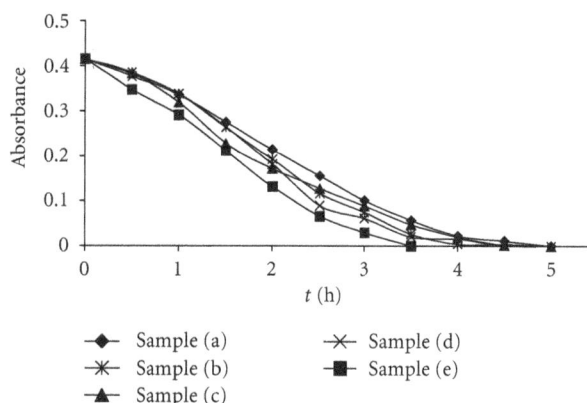

FIGURE 5: Decolorization of MO solution under UV radiation.

by the sol-gel method. The XRD results exhibit that in all samples anatase phase has observed as the dominant phase and sample **e** with 85% anatase phase and 15% rutile phase has showed the best photocatalytic activity. The SEM images indicated that in presence of HPC, density of particles and their size distribution have been improved. It is important for photocatalytic activity that the particle size of the photocatalyst be homogeneous. Finally, all studies show that photocatalytic activity of the nanocomposites has been enhanced in presence of additives and HPC was more effective than others.

References

[1] G. Colón, M. Maicu, M. C. Hidalgo, and J. A. Navío, "Cu-doped TiO$_2$ systems with improved photocatalytic activity," *Applied Catalysis B*, vol. 67, no. 1-2, pp. 41–51, 2006.

[2] H. Yu, X. J. Li, S. J. Zheng, and W. Xu, "Photocatalytic activity of TiO$_2$ thin film non-uniformly doped by Ni," *Materials Chemistry and Physics*, vol. 97, no. 1, pp. 59–63, 2006.

[3] P. Aberoomand Azar, S. Moradi Dehaghi, S. Samadi, S. Kamyar, and M. Saber Tehrani, "Effect of Nd^{3+}, pectin and poly(ethylene glycol) on the photocatalytic activity of TiO$_2$/SiO$_2$ film," *Asian Journal of Chemistry*, vol. 22, pp. 1619–1627, 2010.

[4] J. Tian, L. Chen, J. Dai, X. Wang, Y. Yin, and P. Wu, "Preparation and characterization of TiO$_2$, ZnO, and TiO$_2$/ZnO nanofilms via sol-gel process," *Ceramics International*, vol. 35, no. 6, pp. 2261–2270, 2009.

[5] X. T. Wang, S. H. Zhong, and X. F. Xiao, "Photo-catalysis of ethane and carbon dioxide to produce hydrocarbon oxygenates over ZnO-TiO$_2$/SiO$_2$ catalyst," *Journal of Molecular Catalysis A*, vol. 229, no. 1-2, pp. 87–93, 2005.

[6] M. H. Liao, C. H. Hsu, and D. H. Chen, "Preparation and properties of amorphous titania-coated zinc oxide nanoparticles," *Journal of Solid State Chemistry*, vol. 179, no. 7, pp. 2020–2026, 2006.

[7] D. L. Liao, C. A. Badour, and B. Q. Liao, "Preparation of nanosized TiO$_2$/ZnO composite catalyst and its photocatalytic activity for degradation of methyl orange," *Journal of Photochemistry and Photobiology A*, vol. 194, no. 1, pp. 11–19, 2008.

[8] J. Tian, L. Chen, Y. Yin et al., "Photocatalyst of TiO$_2$/ZnO nano composite film: preparation, characterization, and photodegradation activity of methyl orange," *Surface and Coatings Technology*, vol. 204, no. 1-2, pp. 205–214, 2009.

[9] K. Karthik, S. K. Pandian, and N. V. Jaya, "Effect of nickel doping on structural, optical and electrical properties of TiO$_2$ nanoparticles by sol-gel method," *Applied Surface Science*, vol. 256, no. 22, pp. 6829–6833, 2010.

[10] J. Wang, J. Li, Y. Xie et al., "Investigation on solar photocatalytic degradation of various dyes in the presence of Er^{3+} : YAlO$_3$/ZnO-TiO$_2$ composite," *Journal of Environmental Management*, vol. 91, no. 3, pp. 677–684, 2010.

[11] C. Chen, Z. Wang, S. Ruan, B. Zou, M. Zhao, and F. Wu, "Photocatalytic degradation of C.I. Acid Orange 52 in the presence of Zn-doped TiO$_2$ prepared by a stearic acid gel method," *Dyes and Pigments*, vol. 77, no. 1, pp. 204–209, 2008.

[12] Z. Zhang, Y. Yuan, Y. Fang, L. Liang, H. Ding, and L. Jin, "Preparation of photocatalytic nano-ZnO/TiO$_2$ film and application for determination of chemical oxygen demand," *Talanta*, vol. 73, no. 3, pp. 523–528, 2007.

[13] Z. Liu, Z. Jin, W. Li, and J. Qiu, "Preparation of ZnO porous thin films by sol-gel method using PEG template," *Materials Letters*, vol. 59, no. 28, pp. 3620–3625, 2005.

[14] C. Hariharan, "Photocatalytic degradation of organic contaminants in water by ZnO nanoparticles: revisited," *Applied Catalysis A*, vol. 304, no. 1-2, pp. 55–61, 2006.

[15] Y. Jiang, M. Wu, X. Wu, Y. Sun, and H. Yin, "Low-temperature hydrothermal synthesis of flower-like ZnO microstructure and nanorod array on nanoporous TiO$_2$ film," *Materials Letters*, vol. 63, no. 2, pp. 275–278, 2009.

[16] S. Janitabar-Darzi and A. R. Mahjoub, "Investigation of phase transformations and photocatalytic properties of sol-gel prepared nanostructured ZnO/TiO$_2$ composites," *Journal of Alloys and Compounds*, vol. 486, no. 1-2, pp. 805–808, 2009.

[17] S. Rengaraj, S. Venkataraj, J. W. Yeon, Y. Kim, X. Z. Li, and G. K. H. Pang, "Preparation, characterization and application of Nd-TiO$_2$ photocatalyst for the reduction of Cr(VI) under UV light illumination," *Applied Catalysis B*, vol. 77, no. 1-2, pp. 157–165, 2007.

[18] Z. Zhang, Y. Yuan, L. Liang, Y. Cheng, G. Shi, and L. Jin, "Preparation and photoelectrocatalytic activity of ZnO nanorods embedded in highly ordered TiO$_2$ nanotube arrays electrode for azo dye degradation," *Journal of Hazardous Materials*, vol. 158, no. 2-3, pp. 517–522, 2008.

[19] A. Abdel Aal, M. A. Barakat, and R. M. Mohamed, "Electrophoreted Zn-TiO$_2$-ZnO nanocomposite coating films for photocatalytic degradation of 2-chlorophenol," *Applied Surface Science*, vol. 254, no. 15, pp. 4577–4583, 2008.

[20] C. C. Chan, C. C. Chang, W. C. Hsu, S. K. Wang, and J. Lin, "Photocatalytic activities of Pd-loaded mesoporous TiO$_2$ thin films," *Chemical Engineering Journal*, vol. 152, no. 2-3, pp. 492–497, 2009.

[21] J. Chen, N. Yao, R. Wang, and J. Zhang, "Hydrogenation of chloronitrobenzene to chloroaniline over Ni/TiO$_2$ catalysts prepared by sol-gel method," *Chemical Engineering Journal*, vol. 148, no. 1, pp. 164–172, 2009.

[22] B. A. Sava, A. Diaconu, M. Elisa, C. E. A. Grigorescu, I. C. Vasiliu, and A. Manea, "Structural characterization of the sol-gel oxide powders from the ZnO-TiO$_2$-SiO$_2$ system," *Superlattices and Microstructures*, vol. 42, no. 1–6, pp. 314–321, 2007.

[23] S. Liao, H. Donggen, D. Yu, Y. Su, and G. Yuan, "Preparation and characterization of ZnO/TiO$_2$, SO$_4$$^{2-}$/ZnO/TiO$_2$ photocatalyst and their photocatalysis," *Journal of Photochemistry and Photobiology A*, vol. 168, no. 1-2, pp. 7–13, 2004.

[24] M. D. Snel, F. Snijkers, J. Luyten, A. Kodentsov, and G. de With, "Tape casting and reaction sintering of titanium-titanium oxide-nickel oxide mixtures," *Journal of the European Ceramic Society*, vol. 28, no. 6, pp. 1185–1190, 2008.

[25] J. Qiu, Z. Jin, Z. Liu et al., "Fabrication of TiO$_2$ nanotube film by well-aligned ZnO nanorod array film and sol-gel process," *Thin Solid Films*, vol. 515, no. 5, pp. 2897–2902, 2007.

15

Computational Fluid Dynamics of Two-Opposed-Jet Microextractor

Pritam V. Hule, B. N. Murthy, and Channamallikarjun S. Mathpati

Department of Chemical Engineering, Institute of Chemical Technology, Matunga, Mumbai 400 019, India

Correspondence should be addressed to Channamallikarjun S. Mathpati, cs.mathpati@ictmumbai.edu.in

Academic Editor: Sreepriya Vedantam

Liquid-liquid extraction is an important unit operation in chemical engineering. The conventional designs such as mixer settler have lower-energy efficiency as the input energy is dissipated everywhere. Experimental studies have proved that the novel designs such as two-opposed-jet contacting device (TOJCD) microextractor allow energy to be dissipated close to the interface, and major part of energy is used for drop breakup and enhancement of surface renewal rates. It is very difficult to estimate the local variation of energy dissipation (ε) using experiments. Computational fluid dynamics (CFD) has been used to obtain ε at different rotating speed of the top disc and nozzle velocity. In this work, performance analysis of TOJCD microextractor has been carried out using Reynolds stress model. The overall ε value was found in the range of 50 to 400 W/kg and shear rate in the range of 100000 1/s. A semiempirical correlation for $k_L a$ is proposed, and parity plot with experimental data has been plotted.

1. Introduction

Liquid-liquid extraction (LLE) is an important unit operation in chemical engineering. Typical applications of the LLE are in metal extraction, aromatics nitration and sulfonation, polymer processing, waste water treatment as well as food and petroleum industries. LLE is a mass transfer operation in which a liquid solution (the feed) is contacted with an immiscible or nearly immiscible liquid (solvent) that exhibits preferential affinity or selectivity towards one or more of the components in the feed. Two streams result from this contact: the extract, which is the solvent-rich solution containing the desired extracted solute, and the raffinate, the residual feed solution containing little solute. The conventional designs such as mixer settler often have lower energy efficiency as the input energy is dissipated everywhere in the extractor.

The transfer of solute from one phase to another is controlled by diffusion across the interface and often rate limiting. The process can be intensified and energy efficiency can be improved by novel designs such as two-opposed-jet contact device [1, 2], annular centrifugal extractors [3], impinging jet contactors [4], pulsed sieve plate extraction columns [5], and so forth. Experimental studies have proved

that these novel designs offer higher energy efficiency. High energy dissipation and shear rates are used for breakup and enhancement of surface renewal rates. However, it is very difficult to estimate the local variation of energy dissipation using experiments. Computational fluid dynamics (CFD) can be used in such cases to estimate the local variation of energy dissipation and optimize the hardware configuration and selection of operating conditions [6].

The present work deals with Reynolds stress modelling of two-opposed-jet contact device designed by Dehkordi [1, 2]. This equipment can also be classified as microextractor and hence termed as "two-opposed-jet microextractor" in the present study. The principle of opposed jets is to bring the two jets flowing along the same axis in the opposite direction into collision. As the result of such a collision, a relatively narrow zone, called the impingement zone of high turbulence intensity, is created which offers excellent conditions for intensifying heat- and mass-transfer rates. The extensive details of impinging jet technique and applications for various processes can be found in Tamir [7] and Saien et al. [4].

Impinging jet system offers lot of complexities in modelling due to stagnation point, significant variations, and redistribution of Reynolds stress and highly anisotropic flow.

The standard k-ε model will not work in simulating these types of flow accurately due to inherent assumption of isotropy. As the stagnation point is approached, there is a significant redistribution of energy between various stress components and hence Reynolds stress modelling (RSM) with appropriate pressure strain model is required. It is well known that for single-phase cases CFD tool is proven to be reliable; for instance, flow pattern in stirred tank [8], jet reactor [9], and centrifugal extractor [10] was well predicted. In view of this, in the present work a very fine and structured mesh with good quality has been employed to ensure better predictions.

The main objectives of present work were to obtain energy dissipation rate and shear rate distributions in two-opposed-jet microextractor for various geometric and operating conditions. Further, the energy dissipation from CFD data has been used to develop a semiempirical model for overall mass transfer coefficient which takes into account fundamental basis of turbulence phenomena. The energy supplied in the extractor is utilized for (i) creating liquid motion, (ii) creation of new surface, that is, drop breakup and increasing interfacial area, and (iii) turbulent fluctuations at liquid-liquid interface which improve true mass transfer coefficient (k_L). Point (ii) contributes to major utilization of energy input in case of microextractor. To quantify this, two-phase simulations with interphase forces need to be carried out; however, CFD tool is still not reliable for realistic predictions of immiscible mixing behaviour. Therefore, in the present work, an attempt has been made to understand this system using single phase simulations. The experimental mass transfer coefficient is correlated with total energy dissipated in the microextractor. The focus is not on getting individual components of energy utilization (i), (ii), and (iii), but to get overall dissipation rate. This can be obtained using single-phase simulations. In the present work, single-phase CFD simulation using Reynolds stress model has been carried out for preliminary analysis of microextractor.

2. Previous Work

In this section, numerical and experimental efforts to understand hydrodynamics and turbulent statistics of impinging jet have been discussed. Wang and Mujumdar [11–13] studied mixing characteristics of multiple and multiset turbulent opposing jets using standard k-ε model. The effects of turbulence models, model constants, operating conditions, geometric parameters, flow conditions at the nozzle exit, turbulent Schmidt number as well as unequal opposing jets on mixing in the three-dimensional confined turbulent opposing jet flow were examined systematically. They observed that multiple opposing jets achieve better mixing than single opposing jets in the mixer studied. Compared to the single opposing jets, the multiple opposing jets yield mixing which is poorer in the dome, the impingement zone, and its vicinity, but better in the downstream zones after a critical value of axial distance. The total pressure drop for $n = 2$ is roughly 3.5 times higher than that for $n = 4$ due largely to the higher inlet velocity in the former at a given total mass flow rate. This demonstrates a significant power reduction without

detriment to mixing performance for multiple opposing jets ($n \geq 3$) compared to single opposing jets ($n = 2$) and thus the economic benefits of such mixers.

Abdel-Fattah [14] studied two-dimensional impinging circular twin-jet flow numerically (standard k-ε model) and experimentally. The parameters studied were jet Reynolds number, nozzle to plate spacing, nozzle to nozzle centerline spacing, and jet angle. It was concluded that the stagnation primary point moves away in the radial main flow direction by increasing the jet angle. This shift becomes stronger by increasing the nozzle to nozzle centerline spacing. A secondary stagnation point was set up between two jets. The value of pressure at this point decreases by decreasing Reynolds number and/or increasing the jet angle. Turbulent kinetic energy increases within each vortex region; this increment decreases by increasing of jet angle and/or the nozzle to plate spacing.

Gavi et al. [15] carried out CFD study and scale-up of confined impinging jet reactors for precipitation of nanoparticles. Mixing at the molecular level is modelled with a presumed probability density function approach: the direct quadrature method of moments coupled with the interaction by exchange with the mean (DQMOM-IEM) model. Comparison between experimental data and simulations in a wide range of operating conditions showed excellent agreement. Best agreement with experimental data was found when the RSM and the Standard k-ε model coupled with enhanced wall treatment were used.

Kleingeld et al. [16] carried out numerical modelling of impinging jet reactor. The model for the prediction of the interfacial area production in IJ reactor was developed and implemented in the form of a Monte Carlo simulation, based on the fact that bubble breakup in a turbulent environment is governed by the interactions of bubbles with turbulent eddies. Due to this intimate contact between phases, mass transfer coefficients (k_L) of up to 1.5×10^{-3} m/s have been realised, which, coupled with values of the specific interfacial area (a) of 8–18000 m^2/m^3, have yielded volumetric mass transfer coefficients ($k_L a$) of up to 22 s^{-1} which are orders of magnitude higher than typical values obtained by conventional systems.

Marchisio [17] carried out LES of mixing and reaction in confined IJ reactor. Subgrid-scale mixing is described with a presumed PDF approach, namely, DQMOM-IEM. Model predictions compare well with experimental data (and Reynolds averaged Navier Stokes equation predictions from our previous work); comparison was carried out for two sets of initial concentration and two different reactor geometry. The analysis shows that CIJRs are indeed interesting devices, because of the high mixing efficiency and because of the absence of stagnant and recirculation zones. Niamnuy and Devahastin [18] have studied the effects of geometry and operating conditions on the mixing behavior of IJ mixer. For a fixed value of d/D, an increase in the value of the jet Reynolds number led to a better mixing in the impingement zone and its vicinity. This mixing behavior persisted until a critical value of dimensionless axial distance (x/D) was reached beyond which the mixing quality changed. For a mixer with two sets of inlet jets, it was found that a larger spacing between the two sets of inlet jets (higher S/D) resulted in a better mixing in the region between the sets of inlet jets but yielded no significant difference in the required main flow channel length to obtain a well-mixed condition.

Li et al. [19] studied the stagnation point offsets of turbulent opposed jets at various exit velocity ratios, and nozzle separations were experimentally studied by a hot-wire anemometer, smoke-wire technique, and numerically simulated by Reynolds stress model (RSM). Results show that, for $2D \leq L \leq 4D$ (where L is nozzle separation and D is nozzle diameter), the position of the impingement plane is unstable and oscillates within a region between two relative stable positions when the exit velocities are equal. The instability and sensitivity of the stagnation point offset to the small difference of the exit velocities of opposed jets may ascribe to the instability of the large-scale vortices in the boundary layers of opposed jets. The study of the stagnation point offset of impinging streams is crucial for the effective use of such flow in industrial applications because imbalance of the exit flux of impinging streams is inevitable practically. The most important find of our study is that there exists a region of $2D \leq L \leq 8D$, in which the stagnation point of opposed jets is very sensitive to the exit velocity ratio, and small difference (3% or less) of exit velocity can cause the stagnation point to deviate obviously.

Unger and Muzzio [20] have measured concentration profiles and mixing performance in IJ reactors using laser-induced fluorescence (LIF) technique. Flow structures were visualized by imaging concentration distributions at five vertical planes throughout the mixers. Mixing is quantified for each Reynolds number examined by calculating the overall intensity of segregation. Mixing performance varies substantially as a function of Reynolds number. The results demonstrated the ability of laser-induced fluorescence to quantitatively capture small- and large-scale flow structures and accurately and reproducibly quantify mixing performance in real time for industrially relevant mixing devices. For unsteady-laminar flows $80 < Re_j < 300$, the jet oscillations which occur in the standard geometry result in better mixing than the swirling motion of the asymmetric jets; flow for the asymmetric jets is mostly steady and similar to flow for $Re_j < 80$. For more highly unsteady flows $Re_j > 300$, efficient mixing occurs in both geometries although the asymmetric eliminates the dead region in the bottom of the mixer and results in slightly greater overall homogeneity.

Sun et al. [21] studied velocity distribution of two opposed jets using hot film anemometry and CFD. The radial velocities of opposed jets with various exit velocities, nozzle diameters, and nozzle separations were measured experimentally. The normalized radial velocities are self-similar across various radial sections at $r \geq 1.5D$, and the radial velocity profiles can be described by a Gaussian distribution function. The half-width increases linearly with increasing radial distance at $r \geq 1.5D$, and spreading rates of radial jet are about 0.121. The normalized radial velocity at impingement plane increases firstly and then decreases with the increasing normalized radial distance. The normalized radial velocity is independent on nozzle diameter, nozzle separation, and exit velocity. The maximum radial velocity

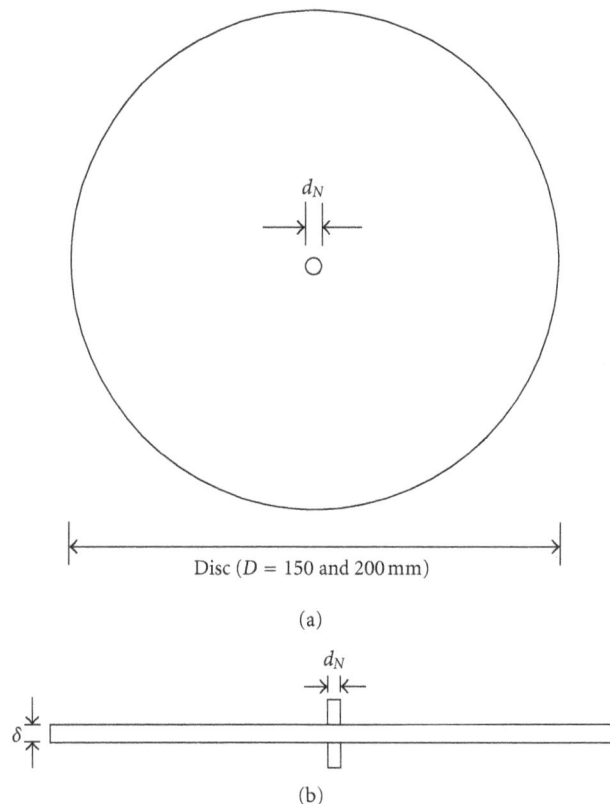

FIGURE 1: Geometric details. (a) Top view and (b) side view.

at impingement plane is proportional to the exit velocity, and it is inversely proportional to the 0.551th power of the normalized nozzle separation. The position of the maximum radial velocity increases with the nozzle separation at $L/D < 1$ and keeps invariant at $L/D \geq 1$.

It can be pointed out from the above-mentioned literature survey that impinging jet technique has been studied widely, and knowledge about the transport phenomena under turbulent conditions has been improved to satisfactory level using experimental and computational fluid dynamics. This understanding can be effectively utilized for design of various equipments as well as performance optimization. In this work, two-opposed-jet microextractor has been studied using Reynolds stress model. The Reynolds stress model is very effective in solving stagnation flows with very strong anisotropy.

3. Systems under Consideration

The microextractor geometry is shown in Figures 1(a) and 1(b). It consists of the following parts: (1) two circular disks with dimension of D (m) $\times \delta$ (m) $= 0.15 \times 0.0015$ and 0.2×0.00015, where D and δ are disk diameter and the distance between disks, respectively. The upper disk was rotated at various speeds (600, 1000, and 1400 rpm), and the lower disk was fixed. The aqueous and organic phases were fed into the lower and upper disks, respectively, through nozzle of 0.001 m diameter. Thus, the contact between the

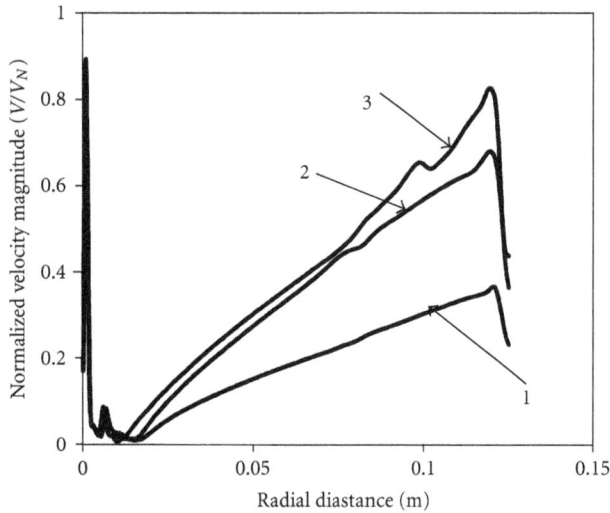

FIGURE 2: Variation of normalized velocity magnitude (V/V_N) along radial coordinate [$z = 0.00075$ m] disc diameter: 0.15 m, flow rate: 300 mL/min, line 1: 600 rpm, line 2: 1000 rpm, and line 3: 1400 rpm.

TABLE 1: Simulation details and results of two-opposed-jet microextractor using RSM.

Sr. no.	RPM	Aq flow rate Q_A (mL per min.)	Disc diameter (m)	Nozzle velocity m/s
1	600	300	0.15	6.36
2	1000	300	0.15	6.36
3	1400	300	0.15	6.36
4	600	190	0.15	4.03
5	1000	190	0.15	4.03
6	1400	190	0.15	4.03
7	600	110	0.15	2.33
8	1000	110	0.15	2.33
9	1400	110	0.15	2.33
10	600	300	0.20	6.36
11	1000	300	0.20	6.36
12	1400	300	0.20	6.36
13	600	190	0.20	4.03
14	1000	190	0.20	4.03
15	1400	190	0.20	4.03
16	600	110	0.20	2.33
17	1000	110	0.20	2.33
18	1400	110	0.20	2.33

two phases took place within the extraction compartment only. The present study is restricted to single phase using water as working fluid. The reason for using single phase simulations has already been explained in introduction. The ratio of flow rates through the top and bottom nozzles was kept unity for all the simulations. Flow rates considered in the present work were (i) 5×10^{-5} m³/s (300 mL/min), (ii) 3.2×10^{-5} m³/s (190 mL/min), and (iii) 1.8×10^{-5} m³/s (110 mL/min). The geometric and operating parameters

are taken from the experimental work of Dehkordi [1]. Dehkordi [1] has considered a system where iso-butyric acid in aqueous solution was getting transferred to organic phase (cumene). The objective of present work was to study the hydrodynamics in such microextractors which predict the mass transfer rates. The mass transfer rate is strongly dependent on the energy dissipation profile. The performance of extractor in dependent on (i) jet velocity, (ii) disc rotation speed, (iii) density, and (iv) viscosity of the phase. In the work, effects of disc rotation speed and jet velocity are analyzed.

4. Governing Equations and Boundary Conditions

The standard k-ε model inherently fails to predict properly the anisotropic flow situations (Reynolds [22], Launder [23], and Hanjalić [24]). Reynolds stress model, in theory, can circumvent most of the deficiencies of standard k-ε model and also it has an ability to predict more accurately each individual stress. A Reynolds stress model solves continuity equation (1), momentum equation (2) six equations for the Reynolds stress (3), and another equation for the dissipation rate (4). The pressure strain term (Π_{ij}) in (3) is the most uncertain term in the RSM. This term is responsible for making turbulence isotropic and redistribution of energy between components $\langle u_1'^2 \rangle$, $\langle u_2'^2 \rangle$, and $\langle u_3'^2 \rangle$. This improves the accuracy of prediction of turbulence production rate as well as local turbulent kinetic energy dissipation rate (ε). One has the following:

$$\frac{\partial \rho}{\partial t} + \frac{\partial \langle u_i \rangle}{\partial x_i} = 0, \tag{1}$$

$$\rho \frac{\partial \langle u_i \rangle}{\partial t} + \rho \langle u_j \rangle \frac{\partial \langle u_i \rangle}{\partial x_j} = -\frac{\partial \langle p \rangle}{\partial x_i} + \frac{\partial}{\partial x_j}\left(\mu \frac{\partial \langle u_i \rangle}{\partial x_j} - \rho \langle u_i' u_j' \rangle \right), \tag{2}$$

$$\left\{ \rho \frac{\partial \tau_{ij}}{\partial t} \right\} + \left\{ \rho \langle u_k \rangle \frac{\partial \tau_{ij}}{\partial x_k} \right\}$$
$$= \left\{ -\rho \left(\tau_{ik} \frac{\partial \langle u_j \rangle}{\partial x_k} + \tau_{jk} \frac{\partial \langle u_i \rangle}{\partial x_k} \right) \right\} + \left\{ \frac{\partial}{\partial x_k} \left(\frac{\mu_t}{\sigma_k} \frac{\partial \tau_{ij}}{\partial x_k} \right) \right\}$$
$$+ \left\{ \frac{\partial}{\partial x_k} \left(\mu \frac{\partial \tau_{ij}}{\partial x_k} \right) \right\} + \left\{ -\frac{2}{3} \varepsilon \delta_{ij} \right\} + \left\{ \Pi_{ij} \right\}, \tag{3}$$

$$\left\{ \rho \frac{\partial \varepsilon}{\partial t} \right\} + \left\{ \rho \langle u_j \rangle \frac{\partial \varepsilon}{\partial x_j} \right\}$$
$$= \left\{ \rho C_{\varepsilon 1} \frac{\varepsilon}{k} \tau_{ij} \frac{\partial \langle u_i \rangle}{\partial x_j} \right\} + \left\{ \frac{\partial}{\partial x_j} \left(\frac{\mu_t}{\sigma_\varepsilon} \frac{\partial \varepsilon}{\partial x_j} \right) \right\}$$
$$+ \left\{ \frac{\partial}{\partial x_j} \left(\mu \frac{\partial \varepsilon}{\partial x_j} \right) \right\} + \left\{ -C_{\varepsilon 2} \rho \frac{\varepsilon^2}{k} \right\}. \tag{4}$$

4.1. Boundary Conditions. The top disc has been given rotational boundary condition whereas bottom disc is stationary.

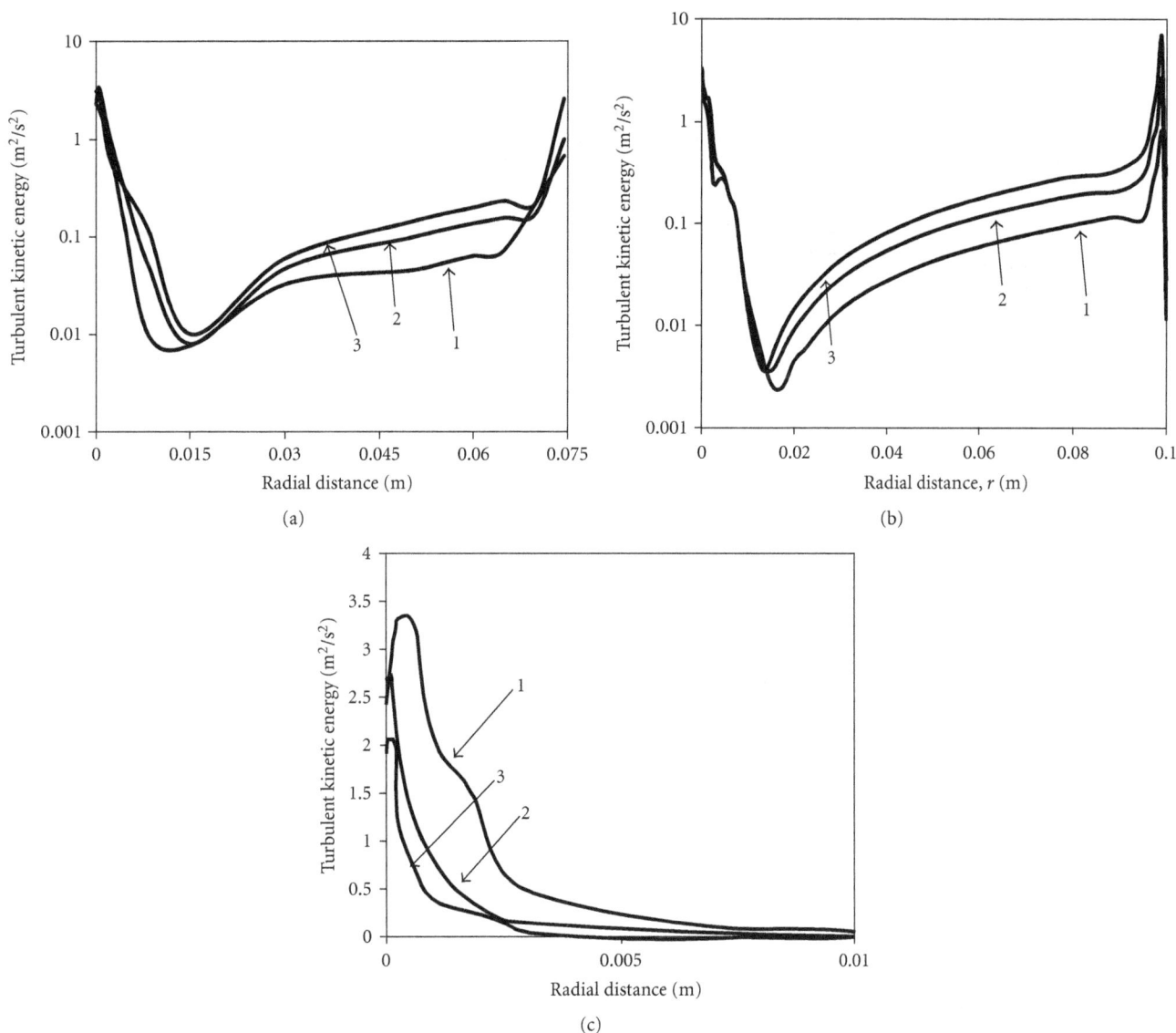

FIGURE 3: Variation of turbulent kinetic energy (k) along radial coordinate [$z = 0.00075$ m]. (a) Disc diameter: 0.15 m, flow rate: 300 mL/min, line 1: 600 rpm, line 2: 1000 rpm, line 3: 1400 rpm. (b) Disc diameter: 0.20 m, flow rate: 300 mL/min, line 1: 600 rpm, line 2: 1000 rpm, line 3: 1400 rpm. (c) Disc diameter: 0.15 m, rotation speed: 1000 rpm, line 1: 300 mL/min, line 2: 190 mL/min, line 3: 110 mL/min.

No slip condition is imposed on both the discs. Velocity inlet boundary condition is used at top and bottom nozzles. The opening at the end of the discs is given pressure outlet condition.

5. Simulation Details

Table 1 summarizes all the cases considered in this study. Hexahedral elements were used for meshing the geometry, and a good quality of mesh was ensured throughout the computational domain using GAMBIT mesh generation tool. In this work, all the computational work has been carried out with finite volume approach using the commercially available software FLUENT 6.2. Further, the second-order upwind scheme was used for continuity, momentum, and turbulence equations. All the discretised equations were solved in a segregated manner with the Semi-Implicit Method for Pressure Linked Equations (SIMPLEs) algorithm. In the present work, all the solutions were considered to be fully converged when repeated iterations do not decrease the sum of residuals below 1×10^{-4}. Default model constants have been used for all the RSM parameters. For each case, grid independency study has been carried out using 0.8, 1.2, and 1.5 million cells for disc diameter of 0.15 m and 1.2, 1.4, and 1.6 million cells for another disc diameter of 0.20 m. The predicted mean shear and turbulent kinetic energy dissipation rate profiles were found to be same for all the three cases. Therefore, the present study employed 1.2 to 1.4 million cells. Further, the clearance between the discs; that is, 0.0015 m was resolved using 30 elements.

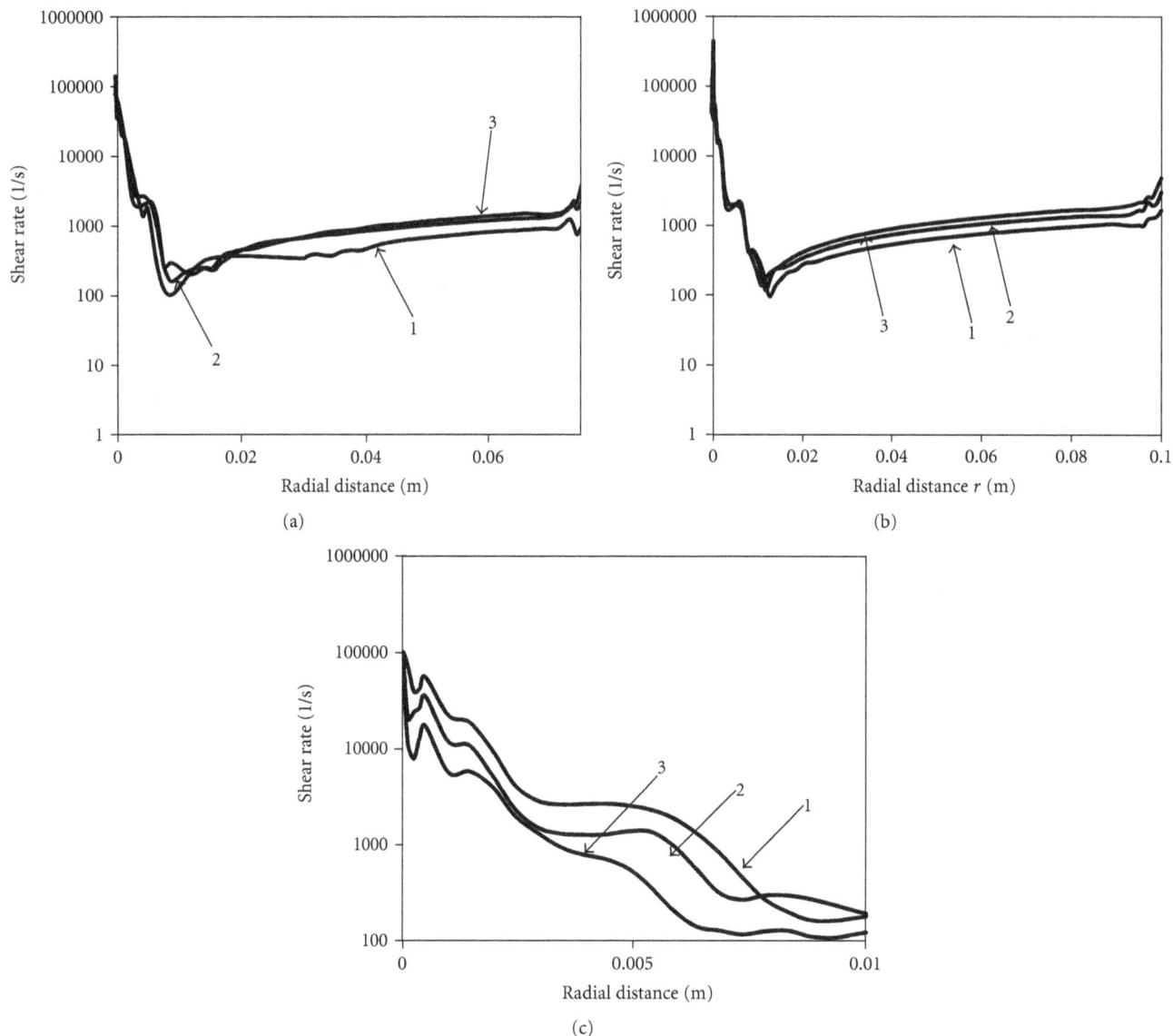

(a)

(b)

(c)

FIGURE 4: Variation of shear rate along radial coordinate [z = 0.00075 m]. (a) Disc diameter: 0.15 m, flow rate: 300 mL/min, line 1: 600 rpm, line 2: 1000 rpm, line 3: 1400 rpm. (b) Disc diameter: 0.20 m, flow rate: 300 mL/min, line 1: 600 rpm, line 2: 1000 rpm, line 3: 1400 rpm. (c) Disc diameter: 0.15 m, rotation speed: 1000 rpm, line 1: 300 mL/min, line 2: 190 mL/min, line 3: 110 mL/min.

The simulations were performed on desktop machines with i3 processor and 4 GB RAM. Each simulation took 20 hours on single processor.

6. Results and Discussion

In the present study, as stated earlier water flow rates were same through both the top and bottom nozzles which are of the same diameter. Therefore, the jets collide each other at exact mid plane. Hence, all the results are provided at the midplane (z = 0.00075 m). CFD results are presented in the form of velocity, shear rate, turbulence kinetic energy, and its dissipation rate with respect to various geometric (disc diameters) and operating parameters (flow rate = 110 mL/min to 300 mL/min and rotational speed = 600 to 1400 rpm).

Figure 2 shows zero velocity at the point of collision. The velocity shows a maxima close to the point of collision and suddenly drops down, then it gradually increased till r = $D/2$. This strong variation of velocity generates high shearing action which is in turn responsible for rate of surface renewal as well as drop breakup. It can be seen that flow rate significantly affects the velocity and shear in the central region, and rotation of top disc plays important role away from the centre. For D = 0.15 m and flow rate of 300 mL/min, Figure 3(a) shows the turbulent kinetic energy profiles at various disc rotational speeds (600, 1000, and 1400 rpm). The turbulent kinetic energy shows maxima at the centre (collision point) and periphery of the disc (due to highest centrifugal force). At the periphery it is a strong function of rotation speed and disc diameter (Figures 3(a) and 3(b)).

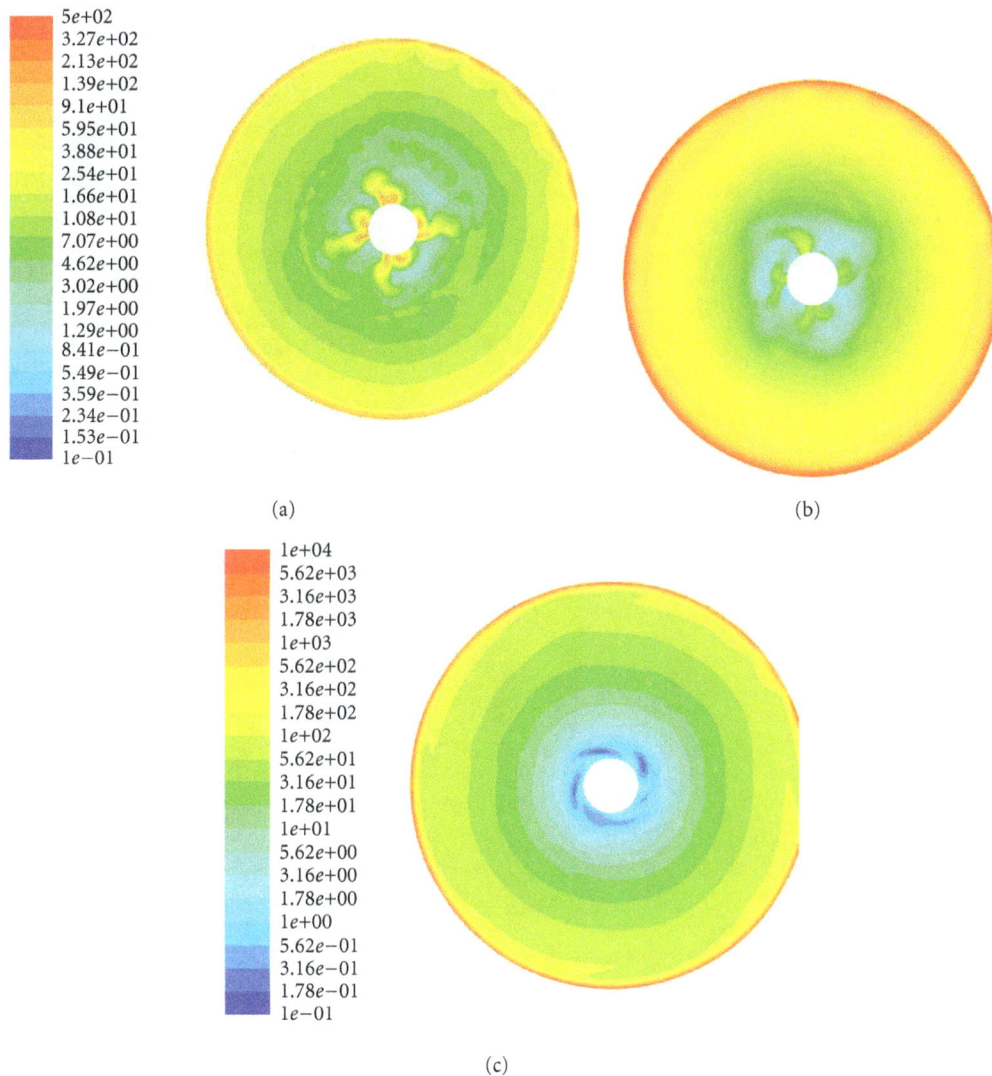

FIGURE 5: Contour plot of turbulent dissipation rate (ε) outer region ($r = 0.01$ m to $r = 0.075$ m) [$D = 0.2$ m] [$z = 0.00075$ m]. Flow rate: 300 mL/min, (a) 600 rpm, (b) 1000 rpm, and (c) 1400 rpm.

Figure 3(c) shows the variation of turbulent kinetic energy as a function of flow rate, and it can be observed that the maximum kinetic energy value decreases with decrease in flow rate. Combination of both the high shear and turbulent fluctuations in the central region is responsible for highest turbulence production (turbulence stress × shear rate) which in turn provides very high dissipation rate. Figures 4(a)–4(c) shows variation of shear rate with flow rate and rotational speed. Change in flow rate from 110 mL/min ($\sim 10,000$ s^{-1}) to 300 mL/min ($\sim 100,000$ s^{-1}) has marked effect on shear rate. Figures 5 and 6 show the contour plot of turbulent dissipation rate. Figures 5(a)–5(c) show the effect of rotational speed at constant flow rate of 300 mL/min for $D = 0.2$ m. A very high dissipation is obtained at the periphery of the disc. At the flow rate of 300 mL/min, the average turbulent kinetic energy dissipation rate (TKED) of 10000 W/kg is obtained for 1400 rpm and about 500 W/kg for 600 rpm. In case of 110 mL/min, the average TKED rates

for 600, 1000, and 1400 rpm are 58, 150, and 377 W/kg, respectively. Figures 6(a)–6(c) show contour plots of the TKE, and it is observed that, with increasing flow rate, the region experiencing high dissipation rate is found to increase.

Figures 7(a) and 7(b) show the effect of rotational speed for $D = 0.15$ and 0.2 m, respectively. For a given flow rate, dissipation in the central region does not have any effect of rotational speed. At $r > 0.02$ m, effect of rotational speed is prominent. Figure 7(c) shows the effect of flow rate on dissipation rate in central region. The central zone ($\sim 10^5$ W/kg) has almost order of magnitude higher turbulent dissipation rate compared to periphery. The increase of flow rate from 110 mL/min to 300 mL/min increased the dissipation rate from 13,000 to 37,000 W/kg. This value is comparable to those obtained in high shear mixers and emulsifiers used in chemical and allied industries to get microemulsions. Typical correlation for droplet diameter in

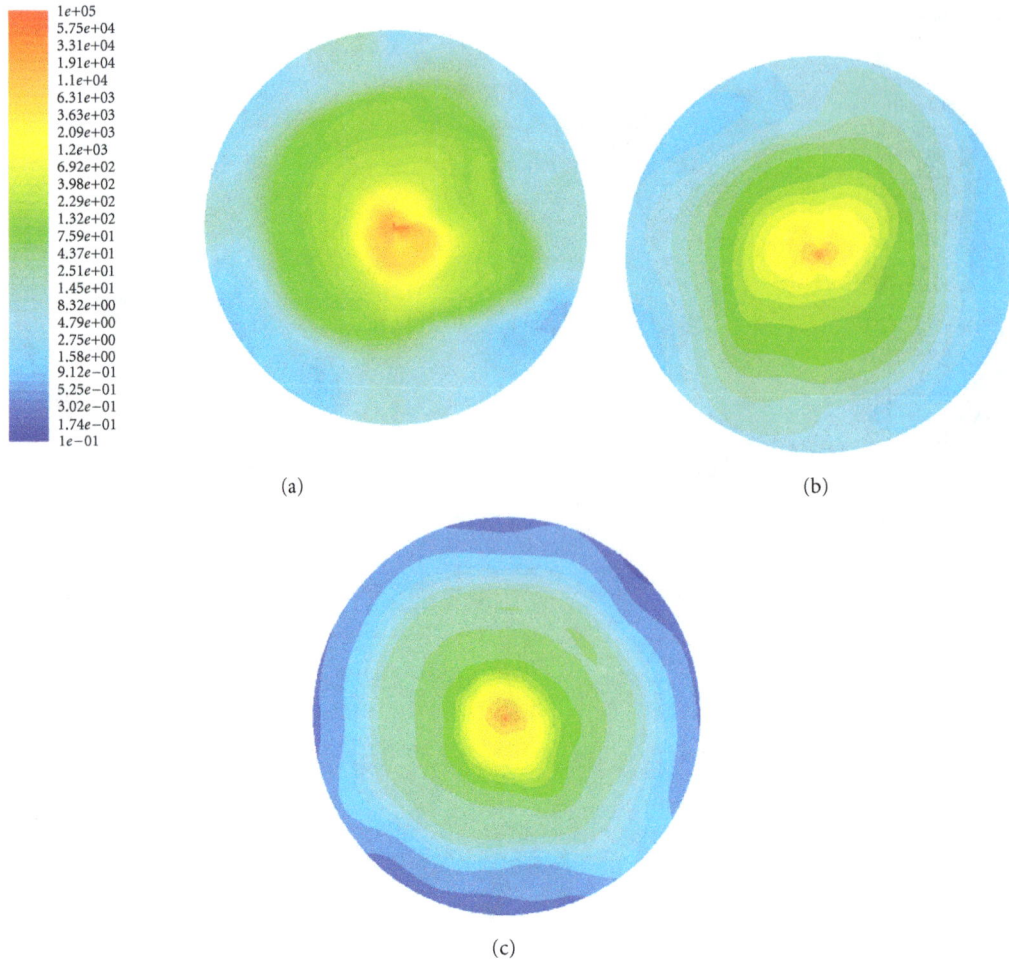

FIGURE 6: contour plot of turbulent dissipation rate (ε) inner region ($r = 0$ to $r = 0.01$ m) [$z = 0.00075$ m] [$D = 0.2$ m]. Upper disk rotation speed = 1000 rpm, (a) 300 mL/min, (b) 190 mL/min, (c) 110 mL/min.

high shear is provided by Davies [25]. Many of the qualitative observations made by Dehkordi [1] can be explained in quantitative manner by present CFD simulations. Some of their observations are reported below

> "An increase in extraction efficiency may be noticed by increasing the upper disk speed. Such a behavior is the consequence of increasing the mixing and turbulence, which control the present extraction process."

> "In addition, an increase in the upper disk speed, N, increases the overall volumetric mass transfer coefficient, $k_L\underline{a}$. This behavior may be explained by increasing the shear forces exerted on the phases and the turbulence that leads to an increase in the surface renewal mechanism and, hence, an increase in the interfacial mass-transfer area, $\underline{a}$."

The volume average dissipation rate as a function of rotation speed shows 550% increase in dissipation rate from 600 to 1400 rpm. In case of very high shear flows, the transfer coefficients are observed to be proportional to square root of local turbulent dissipation rate. The interfacial area (i.e., drop size) is governed by the flow rate compared to rotation speed as maximum shearing takes place in the central region and drop breakup as well. The upper disc rotation stabilizes the drops to certain extent and overall dissipation may not have a control on the interfacial area. Hence, overall $k_L a$ varies as 0.5 power of turbulent dissipation rate. Hence the expected improvement in extraction efficiency is $(377/58)^{0.5}$ = 2.55. The ratio of experimental $k_L a$ at 1400 and 600 rpm is $(13.5/8)$ =1.7.

Dehkordi [1] fitted a correlation based on experimental data for overall mass transfer coefficient is given by (5). V_d is microextractor volume in liter, Q_A is the flow rate in liter/min, D is disc diameter in m, and N is rotational speed in rpm

$$k_L\underline{a} \ (\text{s}^{-1}) = 0.003D^{-0.1}V_d^{-0.6}Q_A^{0.59}N^{0.41}. \tag{5}$$

It can be seen from this correlation that aqueous and organic phase flow rate has significant effect on the overall mass transfer coefficient (as it governs droplet breakup rate and hence interfacial area) and followed by rotation speed (affects true mass transfer coefficient in the domain and

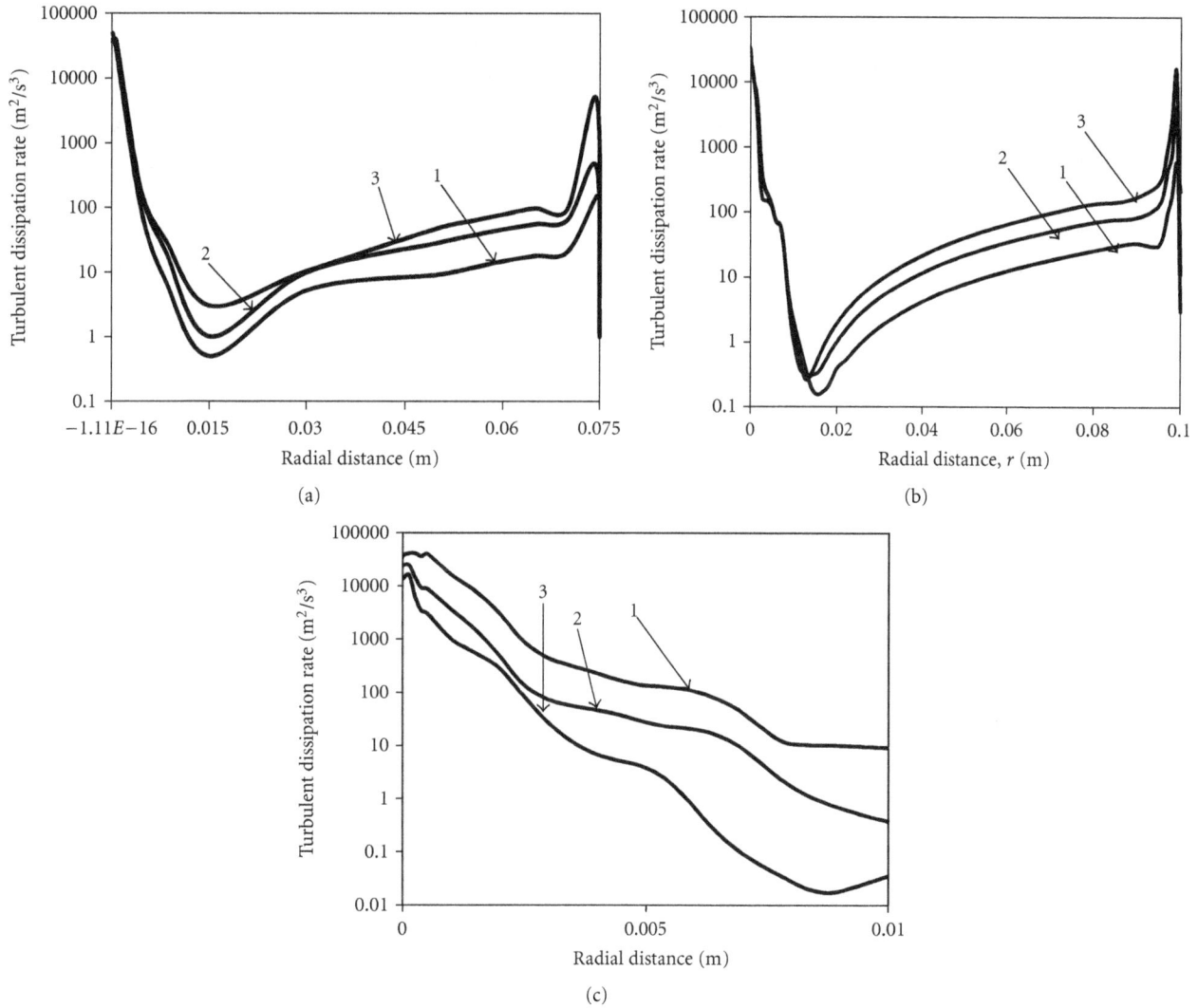

FIGURE 7: Variation of turbulent dissipation rate (ε) along radial coordinate [z = 0.00075 m]. (a) Disc diameter: 0.15 m, flow rate: 300 mL/min, line 1: 600 rpm, line 2: 1000 rpm, line 3: 1400 rpm. (b) Disc diameter: 0.20 m, flow rate: 300 mL/min, line 1: 600 rpm, line 2: 1000 rpm, line 3: 1400 rpm. (c) Disc diameter: 0.15 m, rotation speed: 1000 rpm,: line 1: 300 mL/min, line 2: 190 mL/min, line 3: 110 mL/min.

stabilizing the drops). Experimental observations in various conventional contactors such as bubble columns, stirred vessels, jet mixers, high shear mixers, annular centrifugal extractors, and so forth suggest that the overall mass transfer coefficient is a function of power input per unit volume under turbulent conditions. However, in case of TOJCD, the power input per unit volume affects the true mass transfer coefficient (k_L), and the interfacial area is governed by the impinging jets which have energy proportional to nozzle velocity. Equation (6) is modified correlation to incorporate this theory of turbulence for microextractor. Table 2 shows the power consumed (P) per unit mass (M) for different cases along with the experimental mass transfer coefficient obtained by Dehkordi [1, 2]. We have

$$k_L\underline{a} \ (\text{s}^{-1}) = A\left(\frac{P}{M}\right)^B V_N^C. \tag{6}$$

MATLAB software is used to estimate the best fit values of A, B, and C based on the data provided in Table 2. The best fit values obtained are A = 0.018, B = 0.22, and C = 0.78. The revised correlation is provided as (7). These values clearly indicate that the nozzle velocity plays much more important role compared to the rotation speed of upper disc. We have

$$k_L\underline{a} \ (\text{s}^{-1}) = 0.018\left(\frac{P}{M}\right)^{0.22} V_N^{0.78}. \tag{7}$$

The parity plot of experimental and predicted (from (7)) mass transfer coefficient is shown in Figure 8. It can be seen that the predictions are in very good agreement with the experimental mass transfer coefficient. Figure 9 shows the variation of mass transfer coefficient with rotation speed and jet velocity. Jet velocity has significant impact on mass transfer than rotation speed. The CFD simulation has provided valuable information about the power dissipation

TABLE 2: Power consumption per unit mass from CFD and experimental $k_L a$ [1, 2].

Sr. no.	RPM	Aq flow rate Q_A (mL per min)	D (m)	V_N (m/s)	P/M W/kg (present CFD)	$k_L a$ (s^{-1}) $\times$ 10^4 (Dehkordi [1, 2])
1	600	300	0.20	6.36	58	1837
2	1000	300	0.20	6.36	149	2275
3	1400	300	0.20	6.36	376	2819
4	600	190	0.20	4.03	50	1354
5	1000	190	0.20	4.03	180	1520
6	1400	190	0.20	4.03	370	2005
7	600	110	0.20	2.33	50	837
8	1000	110	0.20	2.33	177	1033
9	1400	110	0.20	2.33	397	1328

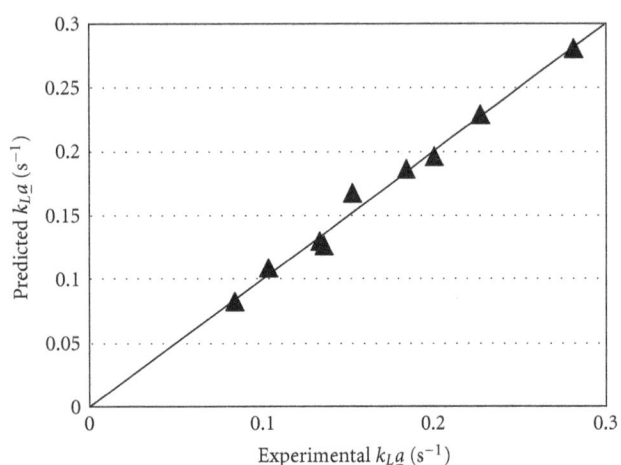

FIGURE 8: Parity plot of overall mass transfer coefficient prediction from proposed correlation with experimental data of Dehkordi [1, 2].

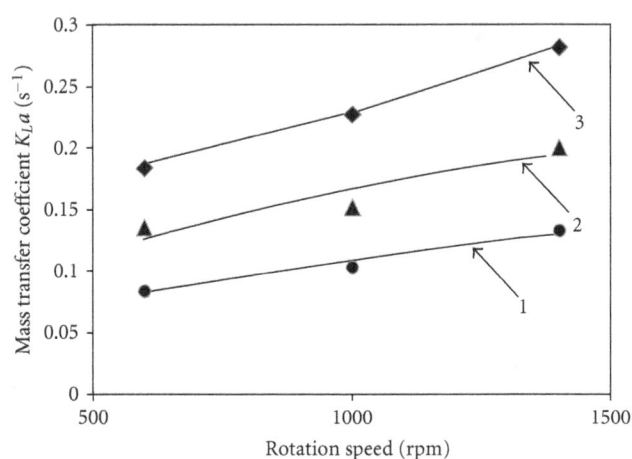

FIGURE 9: Variation of mass transfer coefficient with disk rotation speed; Prediction (correlation using CFD data); Line 1: 110 mL/min; Line 2: 190 mL/min; Line 3: 300 mL/min, • EXPT 110 mL/min, ▲ EXPT 190 mL/min, ♦ EXPT 300 mL/min.

per unit mass, which can be used to obtain semiempirical models with deeper insight into transport phenomena.

7. Conclusion

Microreactor technology is becoming popular in chemical process industries due to ease of scale-up and better energy efficiency. The microextractor design considered in the present work gives almost one theoretical stage with excellent extraction efficiency. The major turbulence generation (subsequent breakup of drops and mass transfer) is dominated by the rotation of top disc in peripheral region and jet velocity in the central region. The inertial force of the jet and the centrifugal force are going to be few orders of magnitude higher than the viscous forces. In this work, two-opposed-jet microextractor has been studied using Reynolds stress model. The Reynolds stress model is very effective in solving stagnation flows with very strong anisotropy. The average turbulent dissipation rate for 600, 1000, and 1400 rpm was 58, 150, and 377 W/kg, respectively. The CFD simulation with small eddy model is capable of predicting transport phenomena in this microextractor. The transport

phenomena are governed by small-scale eddies, and drop size is controlled by the shear in extractor volume. The understanding gained from the single phase CFD simulation can give very useful insights into two-phase extraction processes. CFD simulations clearly bring out the fact that the extractor efficiency is strongly dependent on the jet velocity and rotation speed of the upper disc. However, the effect of change of the disc diameter is not significant.

Notations

a: Interfacial area (m^2/m^3)
d_N: Nozzle diameter (m)
D: Disc diameter (m)
D_t: Turbulent diffusion coefficient (m^2/s)
k: Turbulent kinetic energy (m^2/s^2)
k_L: True mass transfer coefficient (m/s)
M: Mass of fluid in extractor, kg
N: Rotation speed of upper disc, rpm
p_i: Pressure (N/m^2)
$\langle p_i \rangle$: Time-averaged pressure (N/m^2)
P: Power dissipated (W)

Q_A: Volumetric flow rate (m^3/s)

S_{ki}:   Shear rate (s^{-1}) =
    $1/2(\partial\langle u_k\rangle/\partial x_i + \partial\langle u_i\rangle/\partial x_k)$

$\langle u_i\rangle$:   Time average of velocity (m/s)

V_d: Extractor volume, m^3

V_N: Nozzle velocity, m/s

x_i: Coordinate distances in i direction where
    $i = 1, 2, 3$ (m).

Greek Letters

δ_{ij}:   Kronekar delta = 1 if $i = j$; = 0 if $i \neq j$

δ: Gap width (m)

ε: Energy dissipation rate (m^2/s^3)

μ: Dynamic viscosity (kg/ms)

μ_t: Turbulent viscosity (kg/ms)

v: Kinematic viscosity (m^2/s)

v_t: Eddy viscosity (m^2/s)

Π_{ij}:   Pressure strain tensor, (N/m^2 s)

ρ: Density of liquid (kg/m^3)

σ_ε: Parameter in ε equation in RSM

τ_{ij}:   Reynolds Stress (m^2/s^2) = $-\langle u_i' u_j'\rangle$.

Abbreviations

CFD: Computational fluid dynamics

RSM: Reynolds stress modelling

TOJCD: Twin-opposed-jet contacting device.

References

[1] A. M. Dehkordi, "Application of a novel-opposed-jets contacting device in liquid-liquid extraction," *Chemical Engineering and Processing*, vol. 41, no. 3, pp. 251–258, 2002.

[2] A. M. Dehkordi, "Liquid-liquid extraction with chemical reaction in a novel impinging-jets reactor," *AIChE Journal*, vol. 48, no. 10, pp. 2230–2239, 2002.

[3] B. D. Kadam, J. B. Joshi, S. B. Koganti, and R. N. Patil, "Hydrodynamic and mass transfer characteristics of annular centrifugal extractors," *Chemical Engineering Research and Design*, vol. 86, no. 3, pp. 233–244, 2008.

[4] J. Saien, S. A. E. Zonouzian, and A. M. Dehkordi, "Investigation of a two impinging-jets contacting device for liquid-liquid extraction processes," *Chemical Engineering Science*, vol. 61, no. 12, pp. 3942–3950, 2006.

[5] R. L. Yadav and A. W. Patwardhan, "Design aspects of pulsed sieve plate columns," *Chemical Engineering Journal*, vol. 138, no. 1–3, pp. 389–415, 2008.

[6] C. S. Mathpatii, M. V. Tabib, S. S. Deshpande, and J. B. Joshi, "Dynamics of flow structures and transport phenomena, 2. Relationship with design objectives and design optimization," *Industrial and Engineering Chemistry Research*, vol. 48, no. 17, pp. 8285–8311, 2009.

[7] A. Tamir, *Impinging Streams Reactors: Fundamentals and Applications*, Elsevier, Amsterdam, The Netherlands, 1994.

[8] B. N. Murthy and J. B. Joshi, "Assessment of standard k-ε, RSM and LES turbulence models in a baffled stirred vessel agitated by various impeller designs," *Chemical Engineering Science*, vol. 63, no. 22, pp. 5468–5495, 2008.

[9] C. S. Mathpati, S. S. Deshpande, and J. B. Joshi, "Computational and experimental fluid dynamics of jet loop reactor," *AIChE Journal*, vol. 55, no. 10, pp. 2526–2544, 2009.

[10] S. S. Deshmukh, S. Vedantam, J. B. Joshi, and S. B. Koganti, "Computational flow modeling and visualization in the annular region of annular centrifugal extractor," *Industrial and Engineering Chemistry Research*, vol. 46, no. 25, pp. 8343–8354, 2007.

[11] S. J. Wang and A. S. Mujumdar, "Mixing characteristics of 3D confined turbulent opposing jets," in *International Workshop and Symposium on Industrial Drying*, pp. 309–316, Mumbai, India, December 2004.

[12] S. J. Wang, S. Devahastin, and A. S. Mujumdar, "Effect of temperature difference on flow and mixing characteristics of laminar confined opposing jets," *Applied Thermal Engineering*, vol. 26, no. 5-6, pp. 519–529, 2006.

[13] S. J. Wang and A. S. Mujumdar, "Flow and mixing characteristics of multiple and multi-set opposing jets," *Chemical Engineering and Processing*, vol. 46, no. 8, pp. 703–712, 2007.

[14] A. Abdel-Fattah, "Numerical and experimental study of turbulent impinging twin-jet flow," *Experimental Thermal and Fluid Science*, vol. 31, no. 8, pp. 1061–1072, 2007.

[15] E. Gavi, D. L. Marchisio, and A. A. Barresi, "CFD modelling and scale-up of Confined Impinging Jet Reactors," *Chemical Engineering Science*, vol. 62, no. 8, pp. 2228–2241, 2007.

[16] A. W. Kleingeld, L. Lorenzen, and F. G. Botes, "The development and modelling of high-intensity impinging stream jet reactors for effective mass transfer in heterogeneous systems," *Chemical Engineering Science*, vol. 54, no. 21, pp. 4991–4995, 1999.

[17] D. L. Marchisio, "Large Eddy Simulation of mixing and reaction in a Confined Impinging Jets Reactor," *Computers and Chemical Engineering*, vol. 33, no. 2, pp. 408–420, 2009.

[18] C. Niamnuy and S. Devahastin, "Effects of geometry and operating conditions on the mixing behavior of an in-line impinging stream mixer," *Chemical Engineering Science*, vol. 60, no. 6, pp. 1701–1708, 2005.

[19] W. Li, Z. Sun, H. Liu, F. Wang, and Z. Yu, "Experimental and numerical study on stagnation point offset of turbulent opposed jets," *Chemical Engineering Journal*, vol. 138, no. 1-3, pp. 283–294, 2008.

[20] D. R. Unger and F. J. Muzzio, "Laser-induced fluorescence technique for the quantification of mixing in impinging jets," *AIChE Journal*, vol. 45, no. 12, pp. 2477–2486, 1999.

[21] Z. G. Sun, W. F. Li, and H. F. Liu, "Study on the radial jet velocity distribution of two closely spaced opposed jets," *International Journal of Heat and Fluid Flow*, vol. 30, no. 6, pp. 1106–1113, 2009.

[22] W. C. Reynolds, "Fundamentals of Turbulence for Turbulence Modeling and Simulation," Report No. 755, Lecture Notes for Von Karman Institute, Agard., 1987.

[23] B. E. Launder, "Second-moment closure and its use in modelling turbulent industrial flows," *International Journal for Numerical Methods in Fluids*, vol. 9, no. 8, pp. 963–985, 1989.

[24] K. Hanjalić, "Advanced turbulence closure models: a view of current status and future prospects," *International Journal of Heat and Fluid Flow*, vol. 15, no. 3, pp. 178–203, 1994.

[25] J. T. Davies, "Drop sizes of emulsions related to turbulent energy dissipation rates," *Chemical Engineering Science*, vol. 40, no. 5, pp. 839–842, 1985.

16

Oxygen Absorption into Stirred Emulsions of n-Alkanes

Thanh Hai Ngo and Adrian Schumpe

Institut für Technische Chemie, Technische Universität Braunschweig, Hans-Sommer-Straße 10, 38106 Braunschweig, Germany

Correspondence should be addressed to Thanh Hai Ngo, t-h.ngo@tu-bs.de

Academic Editor: Jose Teixeira

Absorption of pure oxygen into aqueous emulsions of n-heptane, n-dodecane, and n-hexadecane, respectively, has been studied at 0 to 100% oil volume fraction in a stirred tank at the stirring speed of $1000\,min^{-1}$. The volumetric mass transfer coefficient, $k_L a$, was evaluated from the pressure decrease under isochoric and isothermal (298.2 K) conditions. The O/W emulsions of both n-dodecane and n-hexadecane show a $k_L a$ maximum at 1-2% oil fraction as reported in several previous studies. Much stronger effects never reported before were observed at high oil fractions. Particularly, all n-heptane emulsions showed higher mass-transfer coefficients than both of the pure phases. The increase is by upto a factor of 38 as compared to pure water at 50% n-heptane. The effect is tentatively interpreted by oil spreading on the bubble surface enabled by a high spreading coefficient. In W/O emulsions of n-heptane and n-dodecane $k_L a$ increases with the dispersed water volume fraction; the reason for this surprising trend is not clear.

1. Introduction

Oxygen absorption into emulsions is encountered, for example, in fermentations with an oil as the carbon source. Most studies have been carried out at low oil volume fractions typical for this application [1, 2]. Literature data on the effect of n-alkane addition on the volumetric mass transfer coefficient $k_L a$ for oxygen are illustrated in Figure 1. The $k_L a$ value has been reported to increase [3, 4], decrease [5, 6], or remain unaffected [7] compared to the one in water. Da Silva et al. [8] reported that 1% n-hexadecane or n-dodecane increased $k_L a$ in a stirred tank by factors of 1.68 and 1.36, respectively; Kundu et al. [3] found that addition of 1% n-dodecane or n-heptane could enhance oxygen transfer in a bubble column up to fourfold; Jia et al. [9] also found a fourfold increase by 2% soybean oil in an air-lift reactor. Among other factors, the oil spreading coefficient S (1) has been used to explain the differences between the oils as follows:

$$S = \sigma_{WG} - \sigma_{OG} - \sigma_{OW}. \tag{1}$$

The values of spreading coefficients S reported in the literature for the three n-alkanes used in this study, n-heptane, n-dodecane, and n-hexadecane, differ considerably (Table 1).

In the present work, absorption of pure oxygen into aqueous emulsions of these n-alkanes has been studied in the full range of oil volume fraction (0 to 100%) with a barometric technique. The main advantage of the pressure technique is that it can give information on both $k_L a$ and oxygen solubility. At high-oil volume fractions, not considered in the previous studies, the high oxygen solubilities in the oils (high driving force) should have a strong effect on the mass transfer characteristics.

2. Experimental

2.1. Chemicals. The oxygen gas had a purity of 99.999% (Westfalen AG, Germany). The emulsion (liquid phase) was composed of double-distilled water and n-heptane, n-dodecane, or n-hexadecane, respectively, with purities $\geq$99% (Merck, Germany).

2.2. Spreading Coefficient (S). The spreading coefficient S was calculated based on the values of the surface and interfacial tensions measured at 298.2 K with a tensiometer (K11, Krüss, Germany) using the ring method [10]. The measurements were carried out with mutually saturated liquid phases. The reported surface tensions are mean values

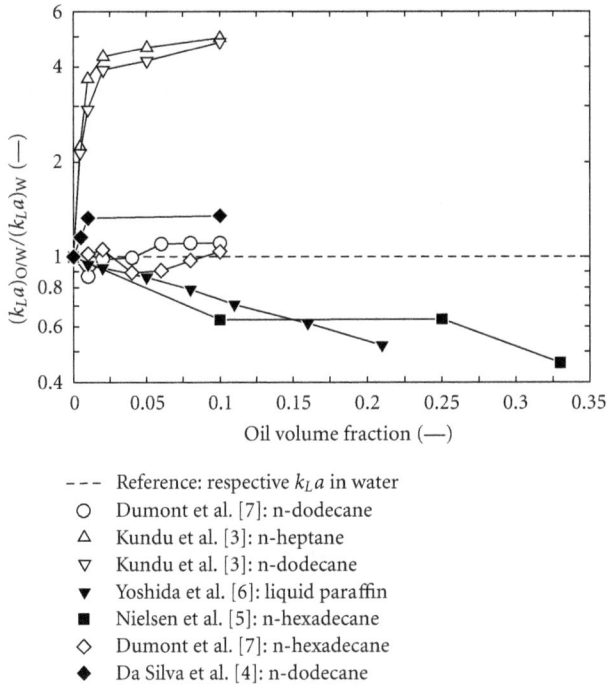

FIGURE 1: Variation of $k_L a$ for oxygen upon n-alkane addition reported in the literature for stirred tanks (filled symbols) and bubble columns (open symbols).

TABLE 1: Literature values of spreading coefficient S.

Oil	T (K)	S (mN m^{-1})	Authors
n-heptane	298.15	+1.2 to +2.3	Pinho and Alves [11]
	298.15	−5.3 to −5.9	Pinho and Alves [11]
	303.15	−2.6	Hassan and Robinson [12]
n-dodecane	298.15	+3.7	Oliveira et al. [13]
	?	+8.7	Wei and Liu [14]
	298.15	+0.6	Rols et al. [15]
n-hexadecane	303.15	−9.3	Hassan and Robinson [12]

of 3 measurements (maximum relative standard deviation 0.25%) and the interfacial tensions are mean values of 4 measurements (maximum relative standard deviation 4%). It should be noted that the aim was to understand the behavior of the oils used in this study; therefore, the oils were used without purification.

The evaluated spreading coefficients S are listed in Table 2. Different from the literature data (Table 1), all three oils have positive S values, but the value for n-heptane is noticeably higher than those for n-dodecane and n-hexadecane.

2.3. Volumetric Mass Transfer Coefficient ($k_L a$) and Oxygen Solubility.

The mass transfer experiments were carried out in a stirred glass vessel with a total volume of 1830 mL and an inner diameter of 100 mm. A schematic diagram of the experimental setup is shown in Figure 2. The stirrer shaft was equipped with two pitched-blade PTFE stirrers (50 mm diameter, 4 blades) mounted at 20 mm (liquid phase) and 140 mm height (gas phase). Four baffles with a width of

TABLE 2: Measured values of the surface tensions of the mutually saturated water (σ_{WG}) and oil (σ_{OG}) phase, interfacial tension (σ_{OW}), and spreading coefficient S at 298.2 K [10].

Oil	σ_{WG}	σ_{OG}	σ_{OW}	S
		(mN m^{-1})		
n-heptane	71.8	19.8	40.2	11.8
n-dodecane	68.2	24.6	41.6	2.0
n-hexadecane	71.2	26.7	42.3	2.2

10 mm were mounted to the walls. The reactor was jacketed and the temperature was controlled to 298.2 ± 0.1 K with a thermostat. All experiments were carried out under surface aeration with gas entrainment into bubbles at the same high stirring speed (1000 min^{-1}).

The volumetric mass transfer coefficient ($k_L a$) was evaluated from the pressure decrease measured during batchwise saturation of the initially gas-free emulsion under isothermal and isochoric conditions. To this end, the desired mixture of oil and water (total volume 700 mL) was placed in the reactor and degassed by applying vacuum under stirring. The liquid losses during degassing were determined using a cold trap with liquid nitrogen as the coolant. About 1.3 mL n-heptane loss was compensated by adding the respective volume in excess, whereas the losses were negligible for n-dodecane and n-hexadecane.

After degassing, the stirrer was stopped, and the reactor was pressurized in about 10 s to approximately 0.15 MPa with pure oxygen. After about 20 s, when the pressure reading had become constant, the absorption experiment was started by switching on the agitator. The time course of the pressure decrease in the head space was measured with a pressure transducer (0–0.5 MPa ± 0.05%). A typical experimental record is shown in Figure 3.

The volumetric mass transfer coefficient $k_L a$ was evaluated based on a pseudohomogenous model, that is, the oil-water emulsion was treated as a pseudohomogenous phase with mean physicochemical properties, specifically, mean gas solubility. This allowed determining the $k_L a$-value from the decrease of pressure P from initially P_i to finally P_f (see Figure 2) using (2) suggested by Albal et al. [16]:

$$-\frac{P_f - p_W - p_O}{P_i - p_W - p_O} \cdot \ln\left(P - P_f\right) = k_L a \cdot t + \text{const.} \quad (2)$$

Here p_W and p_O are the partial pressures of water and oil, respectively. The data at 15% to 90% saturation was used to evaluate $k_L a$ in order to exclude the initial re-emulsification phase as well as the final phase with low driving force. The reported $k_L a$ values are mean values of at least 5 measurements; the mean relative standard deviation was 8% for n-dodecane and n-heptane emulsions, and 7% for n-hexadecane emulsions.

From the total pressure decrease, ΔP, the overall oxygen solubility in the emulsion at a partial pressure of 101325 Pa can be calculated as follows:

$$C_{O_2}^* = \frac{V_G}{V_L} \cdot \frac{1}{RT} \cdot \Delta P \cdot \frac{101325\ Pa}{P_f - p_W - p_O}. \quad (3)$$

For the oxygen solubility, the mean relative standard deviation was always less than 5%.

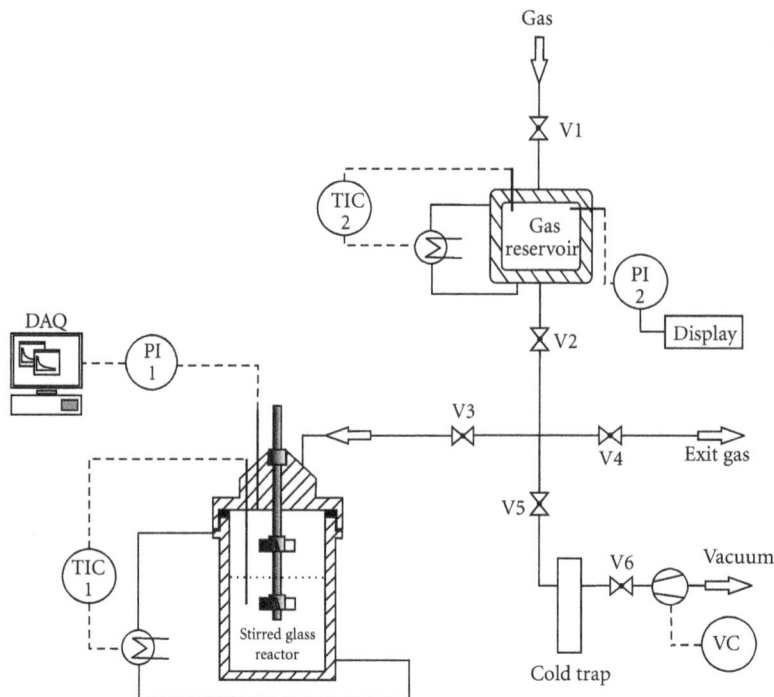

DAQ: acquisition computer

PI: pressure sensor

TIC: temperature controller

VC: vacuum controller

FIGURE 2: Experimental setup.

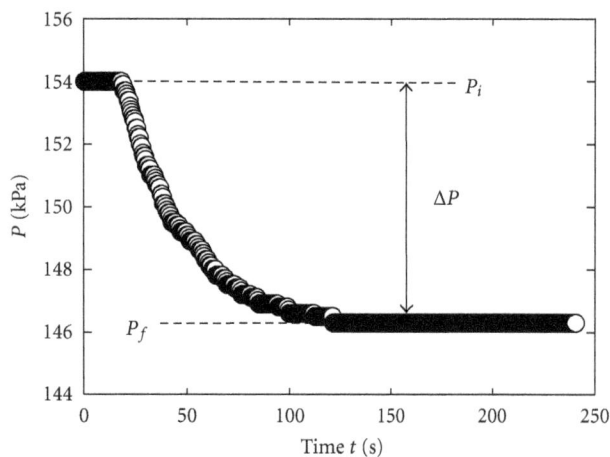

FIGURE 3: Pressure decrease during O_2 absorption into n-dodecane emulsion ($\Phi_{oil} = 30\%$).

3. Results and Discussion

3.1. Effect of Oil Volume Fraction on O_2 Solubility. The oxygen solubilities in the emulsions increases linearly with oil volume fraction (Figure 4). The oxygen solubility decreases with increasing chain length of the n-alkanes. The relative solubility (oxygen solubility in the pure oil compared to pure water) is 11.1, 7.7, and 6.5 for n-heptane, n-dodecane, and n-hexadecane, respectively.

3.2. Phase Inversion. In the previous study on CO_2 absorption [10], the electrical conductivity method showed the phase inversion taking place in the range of 60% to 65% oil volume fraction for n-dodecane as well as n-hexadecane. For n-heptane, the conductivity variation was rather continuous; observation of deemulsification suggested the inversion region to be at 50%–60% n-heptane.

3.3. Effect of Oil Volume Fraction on k_La. The volumetric mass transfer coefficient k_La in O/W emulsions of n-dodecane and n-hexadecane show similar trends (Figures 5 and 6): As the oil volume fraction increases, k_La first increases to a maximum at a small oil volume fraction and then decreases towards the phase inversion region. The increase in k_La at 1% n-dodecane and at 2% n-hexadecane is by 44% and 26%, respectively. Note that the standard deviation, indicated by error bars, is higher in this range. An increase in k_La at low oil concentrations (1-2%) was previously observed for oxygen by several authors. According to Clarke and Correia [1] hydrocarbons tend to decrease

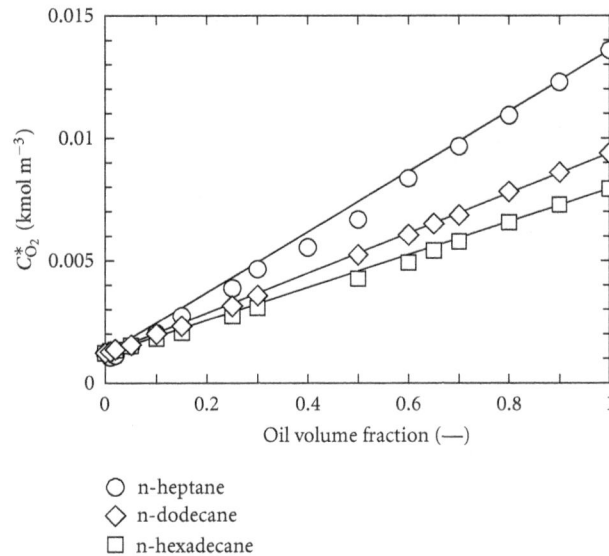

FIGURE 4: Effect of oil volume fraction on O_2 solubility at a partial pressure of 101325 Pa.

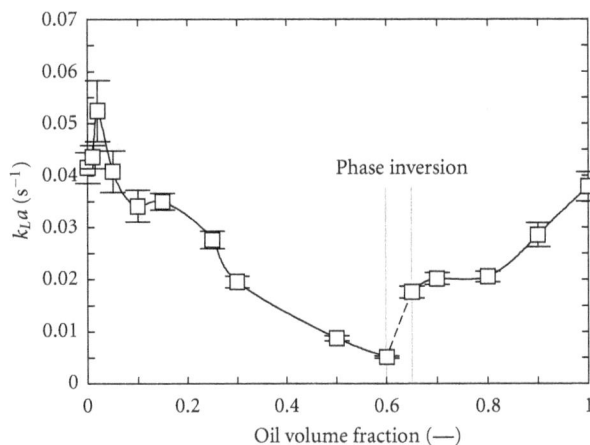

FIGURE 5: Effect of n-hexadecane volume fraction on $k_L a$.

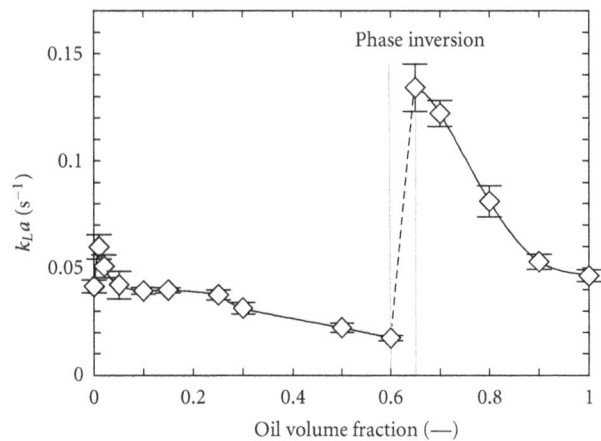

FIGURE 6: Effect of n-dodecane volume fraction on $k_L a$.

the surface tension of the water phase, resulting in smaller bubble size, therefore higher interfacial area (a), and hence higher $k_L a$. However, in this study the surface tension of water ($72.0 \, \text{mN} \, \text{m}^{-1}$) and that of water saturated with n-heptane ($71.8 \, \text{mN} \, \text{m}^{-1}$) or n-hexadecane ($71.2 \, \text{mN} \, \text{m}^{-1}$) were practically the same (cf. Table 2); only saturation with n-dodecane slightly reduced the surface tension ($68.2 \, \text{mN} \, \text{m}^{-1}$). This means that the "surface tension effect" is negligible for the systems studied. Bruining et al. [17] and Cents et al. [18] proposed that an additional transport mechanism might be responsible for the maximum at low oil fraction. Small oil droplets with higher solubility for the gas might enter the water film at the G/L interface, absorb the gas, and then return to the bulk water phase where the absorbed gas is discharged. After the maximum, the strong decrease of $k_L a$ is quite expected due to the increase of the emulsion viscosity. All trends are similar to those for CO_2 absorption in the same emulsions [10] but the effects are much more pronounced for oxygen. The main difference between the gases is that the

CO_2 solubilities in the organic liquids are only slightly higher than in water.

The O/W emulsions of n-heptane show quite different trends with increasing oil fraction (Figure 7): $k_L a$ strongly increases up to 15% oil fraction and then again from 40% oil towards the phase inversion region. All n-heptane emulsions have higher volumetric mass transfer coefficients than pure water.

The increase in $k_L a$ at 15% n-heptane volume fraction (+336% as compared to water) can probably be explained by the bubble-covering mechanism proposed by Rols et al. [19]. This mechanism is linked to the spreading coefficient S. Pinho and Alves [11] recently concluded that there is direct contact between gas and oil when $S > 0$, and very high mass transfer coefficients can be expected. The high S value for n-heptane (Table 2) might enable n-heptane to spread as a film on the oxygen bubbles. When a bubble finally breaks at the top surface, small oxygen-rich droplets may leave the oil film and release oxygen to the water phase. It should be noted that n-heptane has the highest oxygen solubility of the three alkanes used and the $k_L a$ value in pure n-heptane is remarkably higher than that in pure water (+358%), whereas the $k_L a$ values in pure n-dodecane and in pure n-hexadecane are similar to the one in pure water. This means that if n-dodecane or n-hexadecane could also spread as a thin film on the oxygen bubbles, the mechanism would be less effective. The differences of the $k_L a$ values in the pure oils mainly result from the viscosities [20] of only 0.39 mPa s for n-heptane and 1.38 mPa s and 3.03 mPa s for n-dodecane and n-hexadecane, respectively.

In W/O emulsions, from 100% n-hexadecane down to the phase inversion region, $k_L a$ monotonously decreases with increasing dispersed water fraction. This trend is expected due to the increase in viscosity of the W/O emulsions with increasing water content [21]. Very surprisingly, for both n-heptane and n-dodecane, from 100% oil volume fraction down to the phase inversion region, $k_L a$ does not decrease but increases substantially! The maximum increase in $k_L a$ compared to the pure oil is found at the phase inversion region where it is +101% for n-heptane and even +189% for n-dodecane. (For n-dodecane the $k_L a$ jump at the phase-inversion point is by a factor of 7.8). More extensive repetitions than usual have verified these results. To find an explanation, the structure of the W/O emulsion was studied with an endoscopic photographic technique [10, 22, 23]. The endoscopic photoprobe was installed at about half the emulsion height; after about 2.5 min. of stirring at 1000 rpm, photos were taken at 10 fps (frames per second). Figure 8 is a typical photo taken at 70% n-dodecane volume fraction. In this W/O emulsion, the water drops have a surface-to-volume mean diameter of 1.16 mm (standard deviation: 0.15 mm; count: 1142 droplets). There are some very tiny droplets but there is no multiemulsion (e.g., O/W/O emulsion).

It may be argued that the driving force should be based on the solubility $(C_{O_2}^*)_{cont}$ in the continuous phase rather than the mean solubility as assumed in (2). This will shift the $k_L a$ values in O/W emulsions upwards (lower driving force) while those in W/O emulsions will be shifted downwards (higher driving force). Nielsen et al. [5] additionally suggested a comparison with the volumetric mass transfer coefficient in the respective continuous phase $(k_L a)_{cont}$ in the form of the mass transfer enhancement factor E as follows:

$$E = \frac{k_L a}{(k_L a)_{cont.}} \cdot \frac{C_{O_2}^*}{\left(C_{O_2}^*\right)_{cont.}}. \qquad (4)$$

One might expect E to decrease from 1 on both sides to lower values at the phase inversion point due to the viscosity effect. Figures 9 and 10 present the effects of oil volume fraction on the mass transfer enhancement factor E. In O/W emulsions of both n-dodecane and n-hexadecane, as the oil fraction increases, at 25% oil fraction maxima are observed. There the mass transfer rate is enhanced by a factor of 2.3 for n-dodecane and a factor of 1.5 for n-hexadecane.

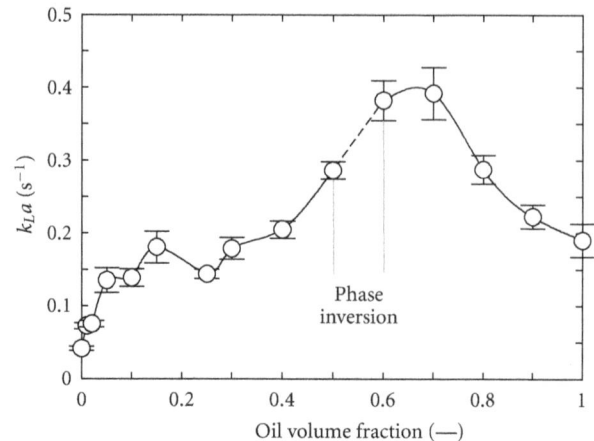

FIGURE 7: Effect of n-heptane volume fraction on $k_L a$.

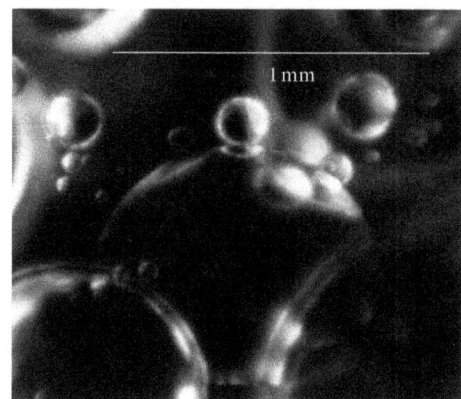

FIGURE 8: Photo of W/O emulsion at 70% n-dodecane taken with the SOPAT endoscope (SOPATec, Berlin, Germany).

At further increase in the oil fraction towards the phase-inversion region, the E value for n-dodecane decreases but still shows some mass transfer enhancement, whereas mass transfer retardation is observed above 40% n-hexadecane. This difference reflects the stronger $k_L a$ decrease in the case of n-hexadecane (Figure 5) as compared to n-dodecane (Figure 6), for example, at 60% oil fraction $k_L a$ is lower than in water by a factor of 0.12 for n-hexadecane and for n-dodecane the factor is only 0.42. In addition, the relative oxygen solubility in n-hexadecane as compared to water is 6.5 whereas it is 7.7 for n-dodecane. Both aspects lead to higher E values for n-dodecane (Figure 8).

In the case of O/W emulsions of n-heptane (Figure 10), as the oil volume fraction is increased towards the phase-inversion region, high mass transfer enhancement is always observed. This reflects the high $k_L a$ values (Figure 7) as well as the high relative oxygen solubility of 11.1 in n-heptane as compared to water. Mass transfer is enhanced by $E =$ 37.7 at the heptane volume fraction of 50%. The trend in O/W emulsions might be explained by the bubble-covering mechanism. However, further research is needed to explain the surprising increase in $k_L a$ and E with increasing water

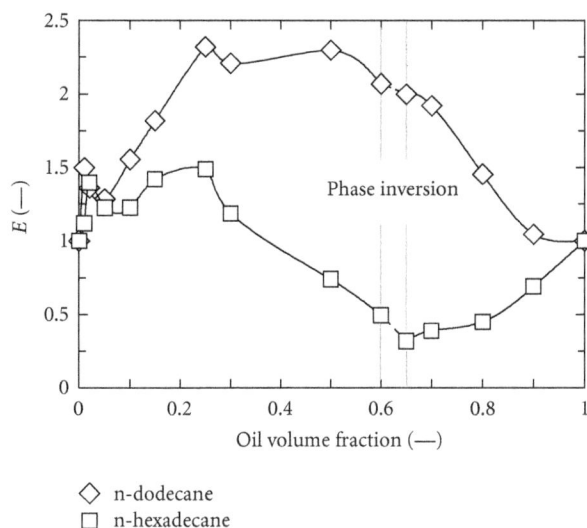

FIGURE 9: Variation of the mass transfer enhancement factor E with the n-dodecane and the n-hexadecane volume fraction.

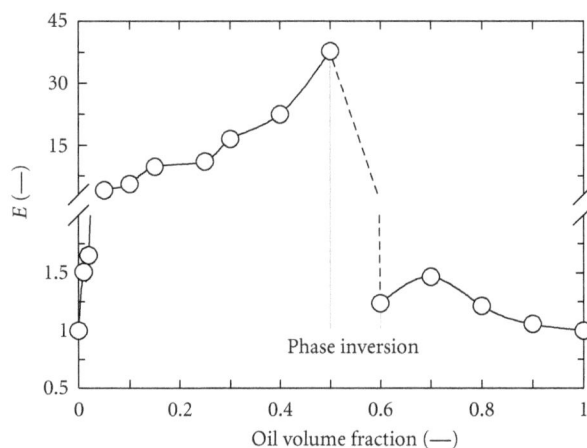

FIGURE 10: Variation of the mass transfer enhancement factor E with the n-heptane volume fraction.

content (from 100% oil down to the phase inversion region) for W/O emulsions of n-dodecane and n-heptane.

4. Conclusions

Oxygen absorption into aqueous emulsions of n-heptane, n-dodecane, and n-hexadecane has been studied in the full range of oil volume fraction. The oxygen solubility in emulsions of all three n-alkanes increases linearly with the oil volume fraction. All three oils have positive spreading coefficients S but a particularly high S value is found for n-heptane. In O/W emulsions, $k_L a$ maxima at low oil fraction (1-2%) are found for n-dodecane and n-hexadecane. Emulsions of n-heptane always show higher $k_L a$ than the pure liquids. This could be explained by a bubble-covering mechanism enabled by high spreading coefficient. In W/O emulsions of both n-heptane and n-dodecane, $k_L a$ increases with increasing dispersed water content towards a maximum

at the phase-inversion region. This is a quite unexpected trend that still needs to be explained.

Nomenclature

a:	Specific interfacial area $[\mathrm{m}^{-1}]$
$C_{O_2}^*$:	Oxygen solubility $[\mathrm{kmol\,m}^{-3}]$
$k_L a$:	Volumetric mass transfer coefficient $[\mathrm{s}^{-1}]$
k_L:	Mass transfer coefficient $[\mathrm{m\,s}^{-1}]$
ll$P_i (P_f)$:	Initial (final) total pressure $[\mathrm{Pa}]$
$p_O (p_W)$:	Partial pressure of oil (water) $[\mathrm{Pa}]$
R:	Gas constant $[\mathrm{Pa\,m}^3\,\mathrm{kmol}^{-1}\,\mathrm{K}^{-1}]$
S:	Spreading coefficient $[\mathrm{N\,m}^{-1}]$
T:	Temperature $[\mathrm{K}]$
t:	Time $[\mathrm{s}]$
V_G:	Gas phase volume $[\mathrm{m}^3]$
V_L:	Emulsion volume $[\mathrm{m}^3]$
$\sigma_{WG} (\sigma_{OG})$:	Surface tension of water (oil) $[\mathrm{N\,m}^{-1}]$
σ_{OW}:	Oil-water interfacial tension $[\mathrm{N\,m}^{-1}]$
Φoil:	Oil volume fraction $[-]$
E:	Mass transfer enhancement factor $[-]$.

Acknowledgment

The authors thank SOPATec (Berlin, Germany) for the chance to use its endoscopic photoprobe. This paper was presented as a poster at the 10th International Conference on Gas/Liquid and Gas/Liquid/Solid Reactor Engineering (GLS10), Braga, Portugal, 26–29 June 2011.

References

[1] K. G. Clarke and L. D. C. Correia, "Oxygen transfer in hydrocarbon-aqueous dispersions and its applicability to alkane bioprocesses: a review," *Biochemical Engineering Journal*, vol. 39, no. 3, pp. 405–429, 2008.

[2] E. Dumont and H. Delmas, "Mass transfer enhancement of gas absorption in oil-in-water systems: a review," *Chemical Engineering and Processing*, vol. 42, no. 6, pp. 419–438, 2003.

[3] A. Kundu, E. Dumont, A. M. Duquenne, and H. Delmas, "Mass transfer characteristics in gas-liquid-liquid system," *Canadian Journal of Chemical Engineering*, vol. 81, no. 3-4, pp. 640–646, 2003.

[4] T. L. Da Silva, A. Mendes, R. L. Mendes et al., "Effect of n-dodecane on *Crypthecodinium cohnii* fermentations and DHA production," *Journal of Industrial Microbiology and Biotechnology*, vol. 33, no. 6, pp. 408–416, 2006.

[5] D. R. Nielsen, A. J. Daugulis, and P. J. McLellan, "A novel method of simulating oxygen mass transfer in two-phase partitioning bioreactors," *Biotechnology and Bioengineering*, vol. 83, no. 6, pp. 735–742, 2003.

[6] F. Yoshida, T. Yamane, and Y. Miyamoto, "Oxygen absorption into oil-in-water emulsions: a study on hydrocarbon fermentors," *Industrial and Engineering Chemistry*, vol. 9, no. 4, pp. 570–577, 1970.

[7] E. Dumont, Y. Andrès, and P. Le Cloirec, "Effect of organic solvents on oxygen mass transfer in multiphase systems: application to bioreactors in environmental protection," *Biochemical Engineering Journal*, vol. 30, no. 3, pp. 245–252, 2006.

[8] T. L. da Silva, V. Calado, N. Silva et al., "Effects of hydrocarbon additions on gas-liquid mass transfer coefficients in biphasic

bioreactors," *Biotechnology and Bioprocess Engineering*, vol. 11, no. 3, pp. 245–250, 2006.

[9] S. Jia, G. Chen, P. Kahar, D. B. Choi, and M. Okabe, "Effect of soybean oil on oxygen transfer in the production of tetracycline with an airlift bioreactor," *Journal of Bioscience and Bioengineering*, vol. 87, no. 6, pp. 825–827, 1999.

[10] T. H. Ngo and A. Schumpe, "Absorption of CO_2 into alkane/water emulsions in a stirred tank," *Journal of Chemical Engineering of Japan*, vol. 45, 2012.

[11] H. J. O. Pinho and S. S. Alves, "Effect of spreading coefficient on gas-liquid mass transfer in gas-liquid-liquid dispersions in a stirred tank," *Chemical Engineering Communications*, vol. 197, no. 12, pp. 1515–1526, 2010.

[12] I. T. M. Hassan and C. W. Robinson, "Oxygen transfer in mechanically agitated aqueous systems containing dispersed hydrocarbon," *Biotechnology and Bioengineering*, vol. 19, no. 5, pp. 661–682, 1977.

[13] R. C. G. Oliveira, G. Gonzalez, and J. F. Oliveira, "Interfacial studies on dissolved gas flotation of oil droplets for water purification," *Colloids and Surfaces A*, vol. 154, no. 1-2, pp. 127–135, 1999.

[14] D. Z. Wei and H. Liu, "Promotion of L-asparaginase production by using n-dodecane," *Biotechnology Techniques*, vol. 12, no. 2, pp. 129–131, 1998.

[15] J. L. Rols, J. S. Condoret, C. Fonade, and G. Goma, "Mechanism of enhanced oxygen transfer in fermentation using emulsified oxygen-vectors," *Biotechnology and Bioengineering*, vol. 35, no. 4, pp. 427–435, 1990.

[16] R. S. Albal, Y. T. Shah, A. Schumpe, and N. L. Carr, "Mass transfer in multiphase agitated contactors," *Chemical Engineering Journal*, vol. 27, no. 2, pp. 61–80, 1983.

[17] W. J. Bruining, G. E. H. Joosten, A. A. C. M. Beenackers, and H. Hofman, "Enhancement of gas-liquid mass transfer by a dispersed second liquid phae," *Chemical Engineering Science*, vol. 41, no. 7, pp. 1873–1877, 1986.

[18] A. H. G. Cents, D. W. F. Brilman, and G. F. Versteeg, "Gas absorption in an agitated gas-liquid-liquid system," *Chemical Engineering Science*, vol. 56, no. 3, pp. 1075–1083, 2001.

[19] J. L. Rols, J. S. Condoret, C. Fonade, and G. Goma, "Modeling of oxygen transfer in water through emulsified organic liquids," *Chemical Engineering Science*, vol. 46, no. 7, pp. 1869–1873, 1991.

[20] D. R. Lide and H. V. Kehiaian, *CRC Handbook of Thermophysical and Thermochemical Data*, CRC Press, Boca Raton, Fla, USA, 1994.

[21] K. H. Ngan, K. Ioannou, L. D. Rhyne, W. Wang, and P. Angeli, "A methodology for predicting phase inversion during liquid-liquid dispersed pipeline flow," *Chemical Engineering Research and Design*, vol. 87, no. 3, pp. 318–324, 2009.

[22] S. Maaß, S. Wollny, A. Voigt, and M. Kraume, "Experimental comparison of measurement techniques for drop size distributions in liquid/liquid dispersions," *Experiments in Fluids*, vol. 50, no. 2, pp. 259–269, 2011.

[23] S. Maaß, J. Rojahn R. Hänsch, and M. Kraume, "Automated drop detection using image analysis for online particle size monitoring in multiphase systems," *Computers & Chemical Engineering*, vol. 87, pp. 27–37, 2012.

17

Effect of Temperature Change on Geometric Structure of Isolated Mixing Regions in Stirred Vessel

Nor Hanizah Shahirudin, Alatengtuya, Norihisa Kumagai, Takafumi Horie, and Naoto Ohmura

Department of Chemical Science and Engineering, Kobe University, 1-1 Rokkodai, Nada, Hyogo, Kobe 657-8501, Japan

Correspondence should be addressed to Naoto Ohmura, ohmura@kobe-u.ac.jp

Academic Editor: See-Jo Kim

The present work experimentally investigated the effect of temperature change on the geometric structure of isolated mixing regions (IMRs) in a stirred vessel by the decolorization of fluorescent green dye by acid-base neutralization. A four-bladed Rushton turbine was installed in an unbaffled stirred vessel filled with glycerin as a working fluid. The temperature of working fluid was changed in a stepwise manner from 30°C to a certain fixed value by changing the temperature of the water jacket that the vessel was equipped with. The step temperature change can dramatically reduce the elimination time of IMRs, as compared with a steady temperature operation. During the transient process from an initial state to disappearance of IMR, the IMR showed interesting three-dimensional geometrical changes, that are, simple torus with single filament, simple torus without filaments, a combination of crescent shape and circular tori, and doubly entangled torus.

1. Introduction

Stirred vessels are frequently used to homogenize different substances, conduct chemical reactions, and enhance mass transfer between different phases. These vessels are versatile and they are available in a wide variety of sizes and impeller configurations for use in industrial processes. Although turbulent flow is efficient for mixing, laminar mixing is required in some cases such as for high-viscosity fluids and shear-sensitive materials. Koiranen et al. [1] proposed specific principles for effective mixing of highly viscous liquids or shear-sensitive materials in laminar flow mixing regimes. In these regimes, global mixing is inefficient due to the existence of isolated mixing regions (IMRs). Makino et al. [2] characterized IMRs in a stirred vessel using radial flow impellers and found that IMRs consisted of various Kolmogorov-Arnold-Moser (KAM) tori. Ohmura et al. [3] reported the existence of KAM tori as island structures in a phase-locked orbit that has a rational relation of the time period between the primary and secondary circulation flows. Noui-Mehidi et al. [4] found that the mechanism of IMR disappearance could be described by the formation of a period-doubling

locus in the physical space when using a six-blade Rushton turbine impeller. Hashimoto et al. [5] successfully visualized a three-dimensional structure of thin filaments spirally wrapping around the core of toroidal region. They formulated and estimated the relation between mixing conditions and filament numbers and/or wire turns.

The elimination of IMR at low Reynolds numbers has also been studied extensively. Lamberto et al. [6] and Yao et al. [7] demonstrated that IMRs could be eliminated by using an unsteady rotation method. Takahashi and Motoda [8] proposed a method in which relatively large objects are introduced to the system in order to improve the mixing performance. Nishioka et al. [9] also found that small particles released at the liquid surface were captured within IMRs. Alatengtuya et al. [10] showed that the motion of small particles improves the material transfer between IMRs and the surrounding active mixing region (AMR).

Although research studies on IMRs in stirred vessels have been conducted extensively, few have paid attention to the effect of temperature change on laminar mixing. As industrial batch processes using stirred vessels are often conducted under nonisothermal conditions, it is crucial to

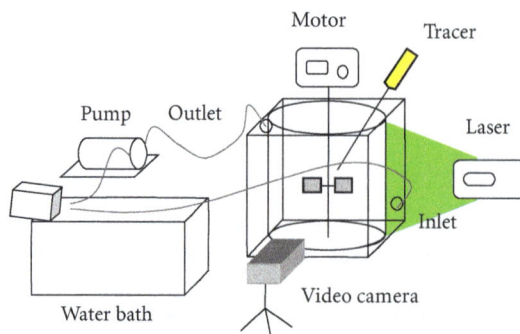

FIGURE 1: Experimental apparatus for flow visualization.

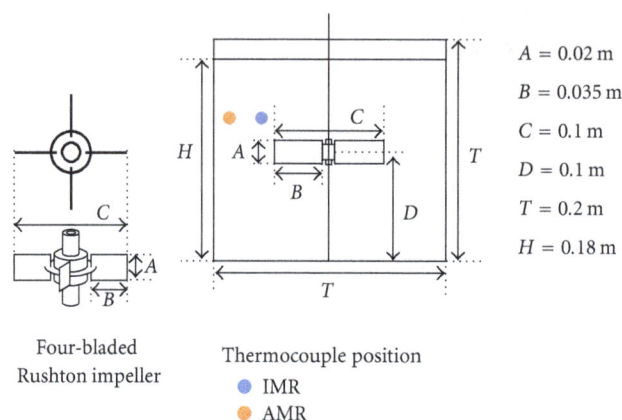

$A = 0.02\,\mathrm{m}$

$B = 0.035\,\mathrm{m}$

$C = 0.1\,\mathrm{m}$

$D = 0.1\,\mathrm{m}$

$T = 0.2\,\mathrm{m}$

$H = 0.18\,\mathrm{m}$

Four-bladed
Rushton impeller

Thermocouple position
● IMR
● AMR

FIGURE 2: Configuration and location of a four bladed turbine impeller and position of thermocouples.

obtain knowledge of the effect of thermal unsteadiness on laminar mixing. This study, therefore, investigated transient behaviors of isolated mixing regions undergoing change of temperature in a stirred vessel.

2. Experimental

The mixing system consisted of a cylindrical flat-bottom vessel with the diameter of 0.1 m without a baffle and a four-bladed Rushton turbine, as shown in Figure 1. The cylindrical vessel was immersed in a square vessel of acrylic resin filled with water so that the temperature of glycerin solution was controlled by feeding water from a water bath, and photographic distortion was reduced. The dimensions and location of the impeller are shown in Figure 2. The impeller was installed 0.1 m from the bottom of the vessel. Hence, the ratio of impeller off-bottom clearance to the tank diameter, D/T, was 0.5.

The working fluid was glycerine ($\rho = 1260\,\mathrm{kg/m^3}$, where $\mu = 0.53\,\mathrm{Pa\ s}$ at 30°C). A fluorescent, pH sensitive and neutrally buoyant green dye, uranine, was used as a passive tracer to observe the mixing process. A small amount of basic solution made of 5 mL 1 N NaOH and 20 mL glycerin was added to the working fluid. After the impeller reached 67 rpm of rotational speed and glycerin temperature reached 30°C (the initial Reynolds number can be estimated at $Re = 27$), a small amount of acidic solution made of 5 mL 1 N HCl and

20 mL glycerin was injected approximately 1 cm away from the shaft and near to the center of impeller to decolorize the green dye by neutralization reaction. Then the feed of water was stopped for 30 minutes to maintain a stable IMR structure. While the glycerin temperature was kept at 30°C, the water bath temperature was changed to a fixed higher temperature (40, 50, 60°C). Then, the water was fed into the jacket after 30 minutes. The cross-sectional view of IMRs was visualized using a plane sheet of semiconductor laser light. The images of decolorization process were recorded by a digital video camera.

To measure the temperature in the upper IMR and its surrounding AMR, an experiment independent of the above flow visualization was conducted using two thermocouples. The positions of the two thermocouples were determined by the flow visualization experiment as shown in Figure 2 where the dots represent the thermocouples. One thermocouple was set in the upper IMR, and the other one was set in AMR.

3. Results and Discussion

3.1. Transition of IMR Structure. Figures 3 and 4 show overviews and cross-sectional views of IMR transition, respectively. One similarity in all of the experiments is that the upper torus is eliminated before the lower torus. Influence of the free liquid surface at the top of the stirred tank makes the volume of the upper torus structure smaller

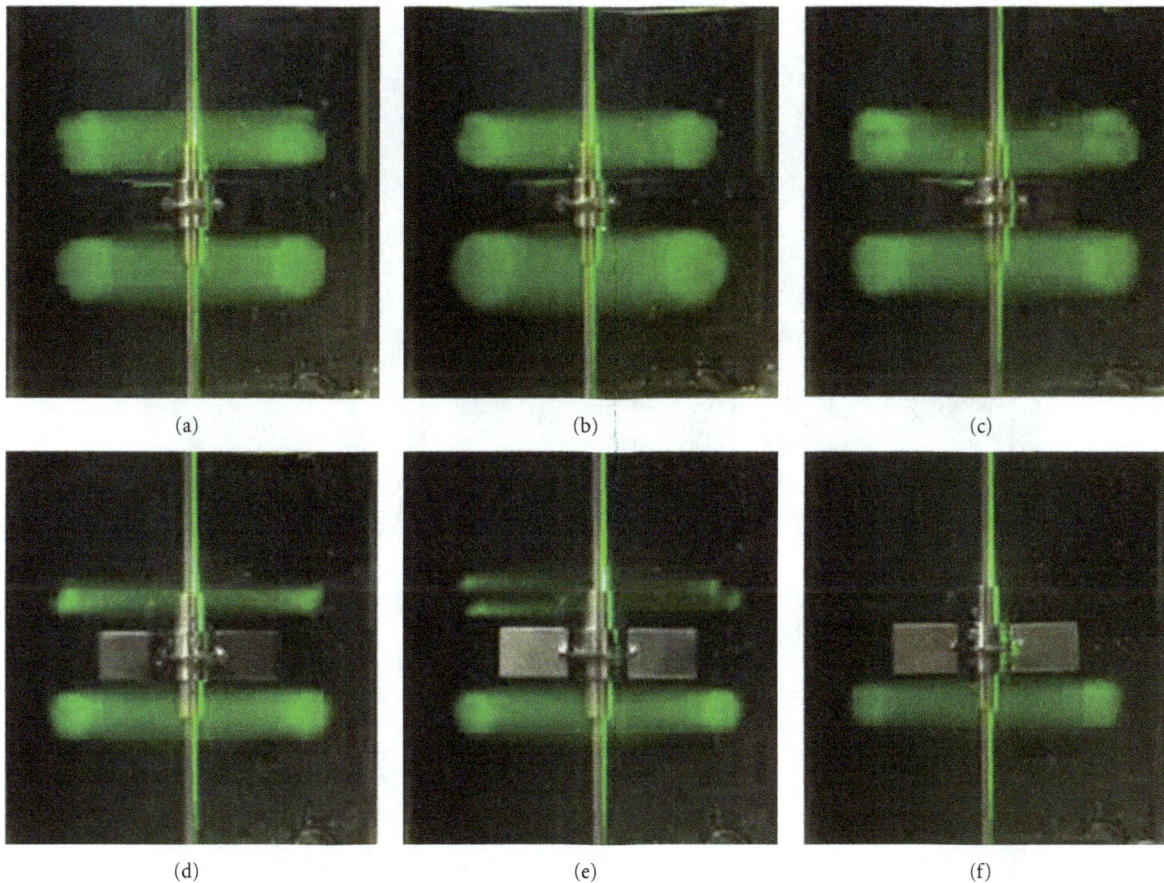

FIGURE 3: Overviews of IMR transition, $\Delta T = 10$ K.

than volume of the lower torus structure resulting in the elimination of the upper IMR first and later the lower IMR. Moreover, a larger temperature difference enhances the diffusion mechanism between IMR and AMR more easily. As illustrated in Figures 3(a) and 4(a), when the glycerin solution is maintained at 30°C, IMR is created above and below the impeller while it is preserved for 30 minutes to obtain a stable IMR structure. Since more acid is used compared to base, mixed regions of the stirred tank contain excess acid, causing the indicator in these regions to appear colorless. Segregated regions, on the other hand, contain unreacted base and display a green color. Lamberto et al. [6] explain that these regions remain segregated from the rest of the system and are not mixed by convective flow mechanisms. As time increases, single filament wrapped around the upper torus where the filament whose cross-section appears as islands formed small tori.

As temperature increases, the filament simultaneously disappear by diffusion and leave only the core torus as shown in Figures 3(b) and 4(b). After a few minutes, a large amount of fluid from the AMR penetrates into the coherent structure of IMRs resulting in combination of torus and crescent shape from the cross-section view. These shapes remain stable while the fluids inside both shapes rotate counterclockwise. Influence from the temperature difference between AMR and IMR enhances the erosion at the "outer" shells of the

toroidal regions, corresponding to Figures 3(c) and 4(c). This phenomenon continues for several minutes and returns to a stable toroidal shape. This toroidal shape becomes narrow, forming an ellipse as in Figures 3(d) and 4(d). Later, it continues to form a spiral structure of string IMR [11] that appeared as two islands from the cross-sectional view as in Figures 3(e) and 4(e). These islands' cross-sectional area decreases slowly and finally the string IMR gradually disappears, as shown in Figures 3(f) and 4(f). The above-mentioned results indicate that changing the temperature by steps can dramatically reduce the elimination time of IMRs, as compared with a steady temperature operation. This mixing enhancement might be attributed to dynamic change of flow structure due to decrease of viscosity when temperature increases. The changes of IMR structures occurred in both upper and lower IMR for different temperature variations.

3.2. Time Variations of Temperature of IMR and AMR. Figure 5 shows the time variation of temperature of IMR and AMR. Points notated from (c) to (f) in Figure 5 correspond to Figures 3 and 4. As can be seen from the graph, there is slight temperature difference between IMR and AMR for all of the temperature variations. With only 1°C to 2°C of temperature increase from the initial temperature, the changes of IMR structure started to occur. The curve in every graph for the upper IMR temperature is identical, where there is

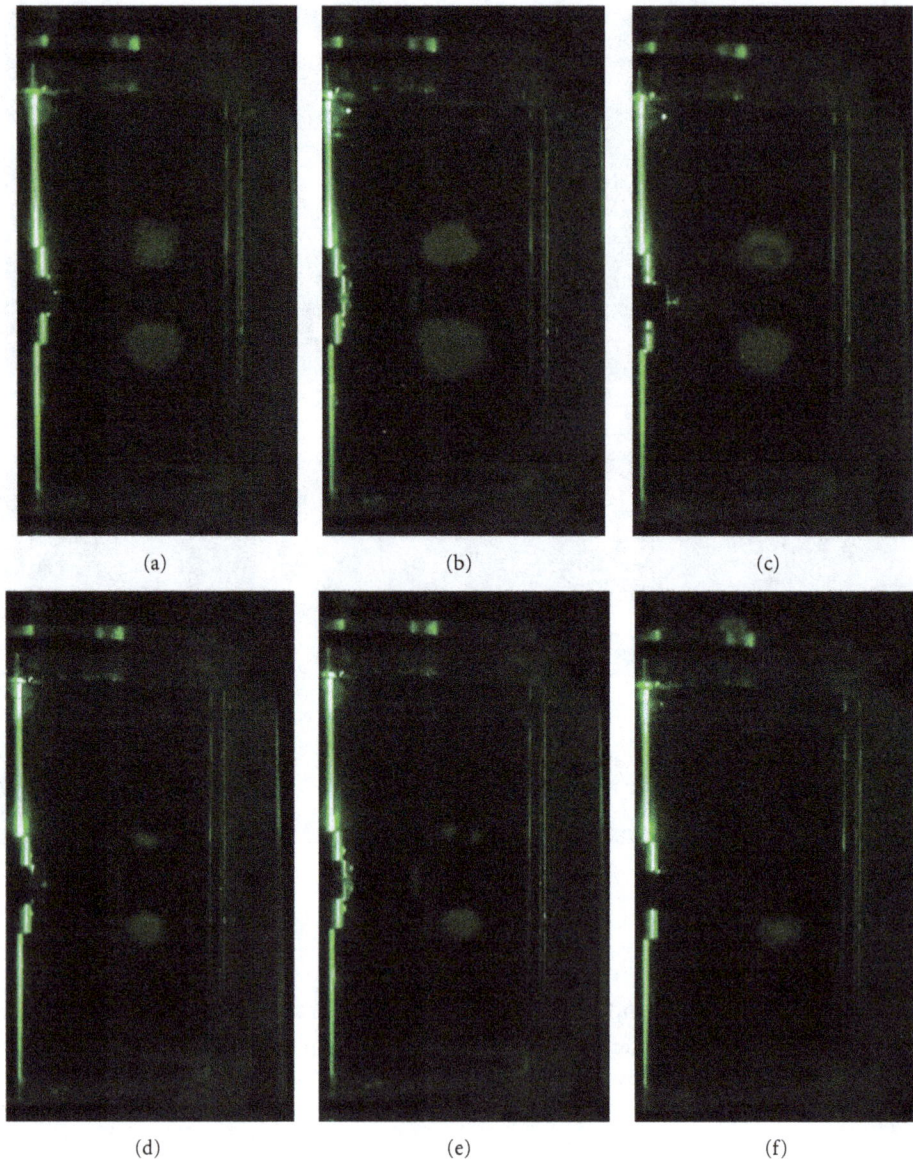

FIGURE 4: Cross-sectional views of IMR transition, $\Delta T = 10$ K.

a curve-step-like change. There is a unique characteristic indicated at (c) where the time is equivalent to the time during formation of torus and crescent shape. At point (d), as temperature increased continuously the formation of a narrow torus occurred and later a spiral structure of string IMR formed, at (e). During the elimination of the segregated region, there is no temperature difference between the two positions, as indicated at (f). Although the points in Figure 5 do not perfectly match the flow structures shown in Figures 3 and 4 because of the disturbance caused by a thermocouple in the IMR structure, nonetheless it can be seen that influence from the nonuniformity of temperature in the stirred tank allowed more interaction between IMR structure and AMR.

3.3. Dimensionless Time for Every Step of IMR Structure Transformation. Figure 6 displays the relation between dimensionless time against initial temperature difference for the upper and lower IMRs. The dimensionless time is defined as $t^* = t/t_e$, where t_e is the elimination time of IMRs, which is one criterion for mixing time. The t^* value when the combination of torus and crescent shape is formed can be considered constant while the t^* value during the formation of narrow torus and string IMR is proportional to the initial temperature difference. The dimensionless time in each step of IMR changes for 10 K temperature difference is the smallest compared to other temperature differences. The temperature step changes for 10 K temperature difference increase very slowly, which decreases the diffusive

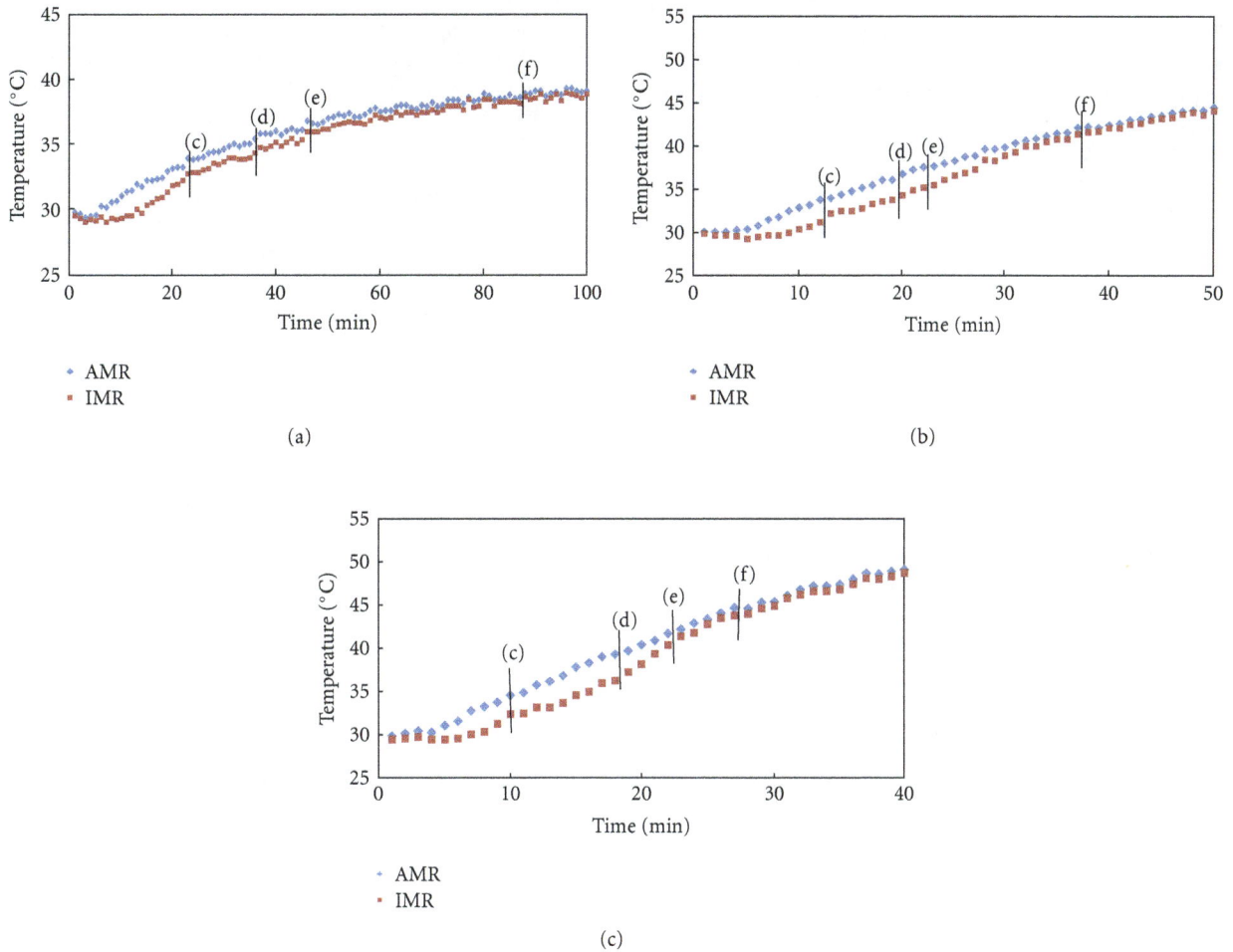

Figure 5: Time variation of temperature in the upper IMR and AMR: (a) $\Delta T = 10\,K$, (b) $\Delta T = 20\,K$, and (c) $\Delta T = 30\,K$.

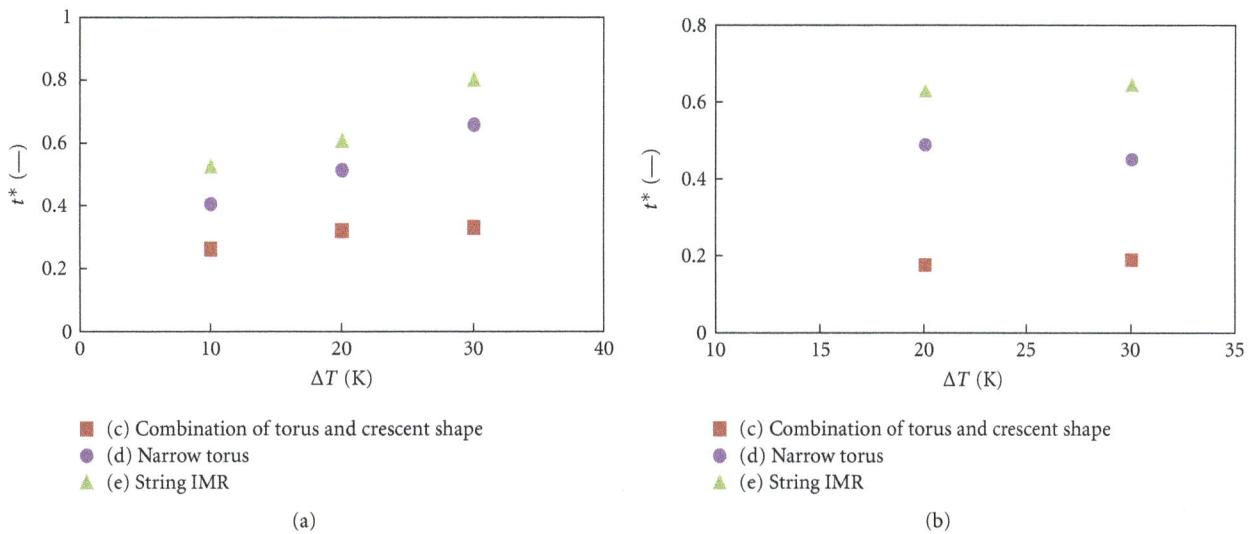

Figure 6: Dimensionless time as geometric structures of IMRs against temperature difference: (a) the upper IMR and (b) lower IMR.

mechanism between IMR and AMR. Therefore, the time for each process of IMR transformation needs to be lengthened. Moreover, it can also be considered that influence from the free surface liquid causes different trends in the results of upper IMR. Compared to the dimensionless time for upper IMR, dimensionless time for lower IMR is nearly constant for all steps of IMR structure changes. This is because the flow circulation in the lower region becomes weaker due to the existence of the bottom.

4. Conclusion

Presence of IMRs in a stirred vessel interrupts the process of achieving uniform mixing. Temperature changes from the jacket vessel can impact the structure of IMR. In this research, as the temperature started to change, the IMR structure transformed from a combination of crescent shape and torus, to narrow torus, to a spiral structure of string IMR, and finally the IMR disappeared. Influence from the step temperature change of the outer jacket enhances the elimination of IMR. Moreover, as the temperature difference increases, the time to eliminate IMR decreases by enhancing global mixing in the stirred vessel because a larger temperature difference enhances diffusion mechanism between IMR and AMR more easily.

Nomenclature

A: Height of turbine blade [m]
B: Length of turbine blade [m]
C: Impeller diameter [m]
D: Impeller off-bottom clearance [m]
H: Liquid height [m]
T: Diameter and height of vessel [m]
T: Temperature [K]
t: Time [min]
t_e: Elimination time of IMRs [min]
t^*: Dimensionless time ($=t/t_e$) [−].

Greek Letters

μ: Viscosity [Pa · s]
ρ: Density [kg · m^{-3}].

Acknowledgments

This research was supported by the Ministry of Education, Science, Sports, and Culture of Japan, Grant-in-Aid for Scientific Research (A) (no. 20246115) and Grant-in-Aid for Challenging Exploratory Research (no. 23656492) from the Japan Society for the Promotion of Science (JSPS).

References

[1] T. Koiranen, A. Kraslawski, and L. Nyström, "Knowledge-based system for the preliminary design of mixing equipment," *Industrial and Engineering Chemistry Research*, vol. 34, no. 9, pp. 3059–3067, 1995.

[2] T. Makino, N. Ohmura, and K. Kataoka, "Observation of isolated mixing regions in a stirred vessel," *Journal of Chemical Engineering of Japan*, vol. 34, no. 5, pp. 574–578, 2001.

[3] N. Ohmura, T. Makino, T. Kaise, and K. Kataoka, "Transition of organized flow structure in a stirred vessel at low Reynolds numbers," *Journal of Chemical Engineering of Japan*, vol. 36, no. 12, pp. 1458–1463, 2003.

[4] M. N. Noui-Mehidi, N. Ohmura, J. Wu, B. Van Nguyen, N. Nishioka, and T. Takigawa, "Characterisation of isolated mixing regions in a stirred vessel," *International Journal of Chemical Reactor Engineering*, vol. 6, article no. A25, 2008.

[5] S. Hashimoto, H. Ito, and Y. Inoue, "Experimental study on geometric structure of isolated mixing region in impeller agitated vessel," *Chemical Engineering Science*, vol. 64, no. 24, pp. 5173–5181, 2009.

[6] D. J. Lamberto, F. J. Muzzio, P. D. Swanson, and A. L. Tonkovich, "Using time-dependent RPM to enhance mixing in stirred vessels," *Chemical Engineering Science*, vol. 51, no. 5, pp. 733–741, 1996.

[7] W. G. Yao, H. Sato, K. Takahashi, and K. Koyama, "Mixing performance experiments in impeller stirred tanks subjected to unsteady rotational speeds," *Chemical Engineering Science*, vol. 53, no. 17, pp. 3031–3040, 1998.

[8] K. Takahashi and M. Motoda, "Chaotic mixing created by object inserted in a vessel agitated by an impeller," *Chemical Engineering Research and Design*, vol. 87, no. 4, pp. 386–390, 2009.

[9] N. Nishioka, Y. Tago, T. Takigawa, N. M. Noui-Mehidi, J. Wu, and N. Ohmura, "Particle migration in a stirred vessel at low reynolds numbers," in *Proceedings of the 8th Italian Conference on Chemical and Process Engineering (AIDIC Conference Series)*, vol. 8, pp. 243–247, Milano, Italy, 2007.

[10] Alatengtuya, N. Nishioka, T. Horie, M. N. Noui-Mehidi, and N. Ohmura, "Effect of particle motion in isolated mixing regions on mixing in stirred vessel," *Journal of Chemical Engineering of Japan*, vol. 42, no. 7, pp. 459–463, 2009.

[11] Y. Inoue, H. Ito, Y. Nakata, and S. Hashimoto, "Theoretical analysis of isolated mixing regions in stirred vessels," *Kagaku Kogaku Ronbunshu*, vol. 36, no. 1, pp. 1–16, 2010.

18

MHD Heat and Mass Transfer of Chemical Reaction Fluid Flow over a Moving Vertical Plate in Presence of Heat Source with Convective Surface Boundary Condition

B. R. Rout,[1] **S. K. Parida,**[2] **and S. Panda**[3]

[1] *Department of Mathematics, Krupajal Engineering College, Prasanti Vihar, Pubasasan, Kausalyaganga, Bhubaneswar, Odisha 751002, India*
[2] *Department of Mathematics, Institute of Technical Education and Research (ITER), SOA University, Bhubaneswar, Odisha 751019, India*
[3] *Department of Mathematics and Civil Engineering, National Institute of Technology (NIT) Calicut, Calicut 673601, India*

Correspondence should be addressed to S. K. Parida; sparidamath2007@rediffmail.com

Academic Editor: Jose C. Merchuk

This paper aims to investigate the influence of chemical reaction and the combined effects of internal heat generation and a convective boundary condition on the laminar boundary layer MHD heat and mass transfer flow over a moving vertical flat plate. The lower surface of the plate is in contact with a hot fluid while the stream of cold fluid flows over the upper surface with heat source and chemical reaction. The basic equations governing the flow, heat transfer, and concentration are reduced to a set of ordinary differential equations by using appropriate transformation for variables and solved numerically by Runge-Kutta fourth-order integration scheme in association with shooting method. The effects of physical parameters on the velocity, temperature, and concentration profiles are illustrated graphically. A table recording the values of skin friction, heat transfer, and mass transfer at the plate is also presented. The discussion focuses on the physical interpretation of the results as well as their comparison with previous studies which shows good agreement as a special case of the problem.

1. Introduction

The study of convective flow with heat and mass transfer under the influence of magnetic field and chemical reaction with heat source has practical applications in many areas of science and engineering. This phenomenon plays an important role in chemical industry, petroleum industry, cooling of nuclear reactors, and packed-bed catalytic reactors. Natural convection flows occur frequently in nature due to temperature differences, concentration differences, and also due to combined effects. The concentration difference may sometimes produce qualitative changes to the rate of heat transfer. The study of heat generation in many fluids due to exothermic and endothermic chemical reactions and natural convection with heat generenation can be added to combustion modeling. In this direction Vajrvelu and Nayfeh [1] studied the hydromagnetic convection at a cone and at a wedge

in presence of temperature-dependent heat generation and absorption effect. Chamkha [2] later examined the effect of heat generation or absorption on hydromagnetic three-dimensional free convection flow over a vertical stretching surface. The flow through porous media is a subject of most common interest and has emerged as a separate intensive research area because heat and mass transfer in porous medium is very much prevalent in nature and can also be encountered in many technological processes. In this context the effect of temperature-dependent heat sources has been studied by Moalem [3] taking into account the steady state heat transfer within porous medium. Rahman and Sattar [4] have investigated the effect of heat generation or absorption on convective flow of a micropolar fluid past a continuously moving vertical porous plate in presence of a magnetic field. Analysis of transport processes and their interaction with chemical reaction has the greatest contributions to many areas of chemical science. The effect of chemical reaction on different geometry of the problem has been investigated by many authors. Das et al. [5] have studied the effect of mass transfer flow past an impulsively started infinite vertical plate with heat flux and chemical reaction. The chemical reaction effect on heat and mass transfer flow along a semi-infinite horizontal plate has been studied by Anjalidevi and Kandaswamy [6] and later it was extended for Hiemenz flow by Seddeek et al. [7] and for polar fluid by Patil and Kulkarni [8]. Salem and Abd El-Aziz [9] have reported the effect of hall currents and chemical reaction on hydromagnetic flow of a stretching vertical surface with internal heat generation or absorption. Ibrahim et al. [10] studied the effect of chemical reaction and radiation absorption on the unsteady MHD free convection flow past a semi-infinite vertical permeable moving plate with heat source and suction. A detailed numerical study has been carried out for unsteady hydromagnetic natural convection heat and mass transfer with chemical reaction over a vertical plate in rotating system with periodic suction by Parida et al. [11]. Rajeswari et al. [12] have investigated chemical reaction, heat and mass transfer on nonlinear MHD boundary layer flow through a vertical porous surface in presence of suction. Mahdy [13] has studied the effect of chemical reaction and heat generation or absorption on double diffusive convection from vertical truncated cone in a porous media with variable viscosity. Pal and Talukdar [14] have studied perturbation analysis of unsteady magnetohydrodynamic convective heat mass transfer in boundary layer slip flow past a vertical permeable plate with a thermal radiation and chemical reaction. Further the effect of thermal radiation, heat and mass transfer flow of a variable viscosity fluid past a vertical porous plate in presence of transverse magnetic field was investigated by Makinde and Ogulu [15]. The analysis of MHD mixed-convection interaction with thermal radiation and higher order chemical reaction is carried out by Makinde [16]. Aziz [17] theoretically examined a similarity solution for a laminar thermal boundary layer over a flat plate with a convective surface boundary condition. He found an interesting result that a similarity solution is possible if the convective heat transfer along with the hot fluid on the lower surface of the plate is inversely proportional to the square root of the axial distance. Recently, the combined effects of an exponentially decaying internal heat generation and a convective boundary condition on the thermal boundary layer over a flat plate are investigated by Olanrewaju et al. [18]. In their study authors have neglected the Sherwood effect. Similar analysis has been carried out by Makinde [19, 20] without heat source and with heat source [21], neglecting chemical reaction effect. There has been considerable interest in studying the effect of chemical reaction [22] and heat source effect on the boundary layer flow problem with heat and mass transfer of an electrically conducting fluid in different geometry [23–25].

Heat source and chemical reaction effects are crucial in controlling the heat and mass transfer. The present paper attempts to investigate the influence of chemical reaction and the combined effects of internal heat generation and the convective boundary condition on the MHD heat and mass transfer flow. To the best of the authors' knowledge, so far no one has considered the combined effect of chemical reaction and heat source along with convective surface boundary condition on MHD flow and the heat and mass transfer over a moving vertical plate. This fact motivated us to propose the similar study. We extend the recent work of Makinde [19], Olanrewaju et al. [18], and Gangadhar et al. [22] to expose the effect of chemical reaction on MHD heat and mass transfer over a moving vertical plate in presence of heat source along with convective surface boundary condition. The coupled nonlinear partial differential equations governing the flow, heat and mass transfer have been reduced to a set of coupled nonlinear ordinary differential equations by using similarity transformation. Following [19] the similarity solutions exist, if the convective heat transfer associated with the hot fluid on lower surface of the plate is proportional to the inverse square root of the axial distance. The reduced equations are solved numerically using Runge-Kutta fourth-order integration scheme together with shooting method. The effect of various physical parameters on the velocity, temperature, and concentration fields is studied.

2. Mathematical Formulation

A typical flow scenario is illustrated in Figure 1; it shows a steady two-dimensional boundary layer flow of a stream of cold incompressible electrically conducting fluid over a moving vertical flat plate at temperature T_∞ in presence of heat source and chemical reaction. The left surface of the plate is being heated by convection from a hot fluid at temperature T_f that gives a heat transfer coefficient h_f, and T_∞ is the temperature of the fluid away from the plate. The cold fluid in contact with the upper surface of the plate generates heat internally at the volumetric rate Q_0. Here the x-axis is taken along the direction of plate and y-axis is normal to it. A magnetic field of uniform field strength B_0 is applied in the negative direction of y-axis.

The continuity, momentum, energy, and concentration equations describing the flow under the Boussinesq approximation can be written as

$$\frac{\partial u}{\partial x} + \frac{\partial v}{\partial y} = 0, \tag{1}$$

$$u\frac{\partial u}{\partial x} + v\frac{\partial u}{\partial y} = \nu\frac{\partial^2 u}{\partial y^2} - \frac{\sigma B_0^2}{\rho}u + g\beta\left(T - T_\infty\right) + g\beta^*\left(C - C_\infty\right), \tag{2}$$

$$u\frac{\partial T}{\partial x} + v\frac{\partial T}{\partial y} = \alpha\frac{\partial^2 T}{\partial y^2} + \frac{Q_0}{\rho C_p}\left(T - T_\infty\right), \tag{3}$$

$$u\frac{\partial C}{\partial x} + v\frac{\partial C}{\partial y} = D\frac{\partial^2 C}{\partial y^2} - \mathrm{Kr}'C. \tag{4}$$

The symbols u and v denote the fluid velocity in the x- and y-direction. Here T and C are the temperature and concentration variables, ν is the kinematic viscosity, α is the thermal diffusivity, D is the mass diffusivity, β is the thermal expansion coefficient, β^* is the solutal expansion coefficient,

MHD Heat and Mass Transfer of Chemical Reaction Fluid Flow over a Moving Vertical Plate in Presence of Heat Source with Convective Surface Boundary Condition

157

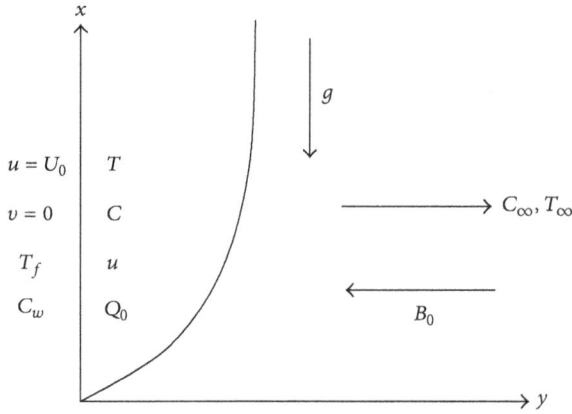

FIGURE 1: Sketch of flow geometry.

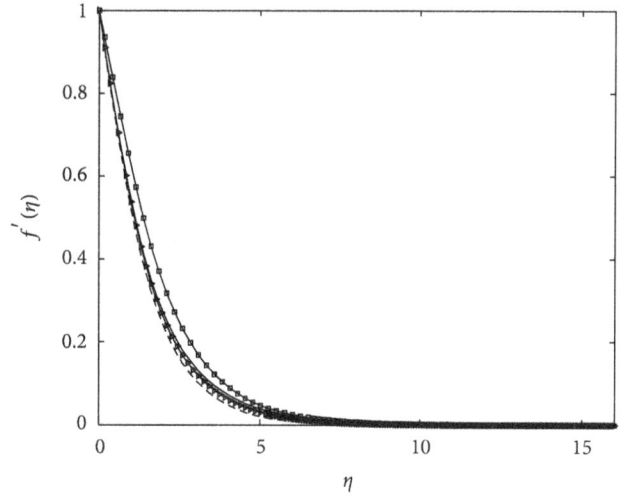

- Ha = 0.1, S = 0, Kr = 0
- Ha = 0.3, S = 0, Kr = 0
- Ha = 0.3, S = 0.02, Kr = 0.1
- Ha = 0.3, S = 0.02, Kr = 0.5

FIGURE 2: Velocity profiles for different values of Ha, S, and Kr for Gr = 0.1, Gc = 0.1, Sc = 0.62, Bi = 0.1, Nc = 0.01, and Pr = 0.72.

ρ is the fluid density, g is the gravitational acceleration, σ is the electrical conductivity, Q_0 is the heat source, C_p is the specific heat at constant pressure, and Kr' is the chemical reaction rate on the species concentration. In the above equations, several assumptions have been made. First, the plate is nonconducting, and the effects of radiant heating, viscous dissipation, Hall effects, and induced fields are neglected. Second, the physical properties, that is, viscosity, heat capacity, thermal diffusivity, and the mass diffusivity of the fluid remain invariant throughout the fluid.

The appropriate boundary conditions at the plate surface and far into the cold fluid are

$$u(x, 0) = U_0, \qquad v(x, 0) = 0,$$

$$-k \frac{\partial T}{\partial y}(x, 0) = h_f \left[T_f - T(x, 0) \right],$$

$$C_w(x, 0) = A x^\lambda + C_\infty,$$

$$u(x, \infty) = 0, \qquad T(x, \infty) = T_\infty, \qquad C(x, \infty) = C_\infty, \tag{5}$$

where C_w is the species concentration at the plate surface, A is the constant, λ is the power index of the concentration, U_0 is the plate velocity, k is the thermal conductivity coefficient, and C_∞ is the concentration of the fluid away from the plate. The boundary layer equations presented are nonlinear partial differential equations and, are in general, difficult to solve. However, the equations admit of a self-similar solution. Therefore transformation allows them to be reduced to a system of ordinary differential equations that are relatively easy to solve numerically. We look for solution compatible with (1) of the form

$$u = U_0 f'(\eta), \qquad v = -\frac{1}{2}\sqrt{\frac{\nu U_0}{x}} \, f(\eta) + \frac{U_0 y}{2 \, x} f'(\eta), \tag{6}$$

where $\eta = y\sqrt{U_0/(\nu x)}$, and prime denotes the differentiation with respect to η.

Let us introduce the dimensionless quantities, that is,

$$\theta(\eta) = \frac{T - T_\infty}{T_f - T_\infty}, \qquad \phi(\eta) = \frac{C - C_\infty}{C_w - C_\infty},$$

$$\mathrm{Ha}_x = \frac{\sigma B_0^2 x}{\rho U_0}, \qquad \mathrm{Gr}_x = \frac{g\beta \left(T_f - T_\infty\right) x}{U_0^2},$$

$$\mathrm{Gc}_x = \frac{g\beta^* \left(C_w - C_\infty\right) x}{U_0^2},$$

$$\mathrm{Bi}_x = \frac{h_f}{k}\sqrt{\frac{\nu x}{U_0}}, \qquad \mathrm{Pr} = \frac{\nu}{\alpha}, \qquad \mathrm{Sc} = \frac{\nu}{D},$$

$$S_x = \frac{Q_0 x}{U_0 \rho C_p}, \qquad \mathrm{Kr}_x = \frac{\mathrm{Kr}' x}{U_0}, \qquad \mathrm{Nc} = \frac{C_\infty}{C_w - C_\infty}. \tag{7}$$

Here Ha_x is the local magnetic field parameter, Gr_x is the local thermal Grashof number, Gc_x is the modified Grashof number, Bi_x is the local convective heat transfer parameter, Pr is the Prandtl number, Sc is the Schmidt number, S_x is the local heat source parameter, Kr_x is the local chemical reaction parameter, and Nc is the concentration difference parameter. Using (6) and (7) in (2)–(4), we get the following equations:

$$f''' + \frac{1}{2} ff'' - \mathrm{Ha}_x \, f' + \mathrm{Gr}_x \theta + \mathrm{Gc}_x \phi = 0, \tag{8}$$

$$\theta'' + \frac{1}{2}\mathrm{Pr} f \, \theta' + \mathrm{Pr} S_x \theta = 0, \tag{9}$$

$$\phi'' + \frac{1}{2}\,\mathrm{Sc} f\phi' - \mathrm{Sc} \mathrm{Kr}_x \left(\phi + \mathrm{Nc}\right) = 0. \tag{10}$$

The corresponding boundary conditions equation (5) for velocity, temperature, and concentration fields in terms of nondimensional variables are

$$f(0) = 0, \qquad f'(0) = 1,$$

$$\theta'(0) = \text{Bi}_x\left[\theta(0) - 1\right], \qquad \phi(0) = 1, \qquad (11)$$

$$f'(\infty) = 0, \qquad \theta(\infty) = 0, \qquad \phi(\infty) = 0.$$

It is observed that in the absence of local source parameter and chemical reaction parameter, that is, for $S_x = 0$ and $\text{Kr}_x = 0$; (8), (9), and (10) together with boundary condition (11) are the same as those obtained by Makinde [19]. It is noticed that the concentration equation (10) in presence of the chemical reaction parameter (Kr_x) in the fluid yields nonhomogeneous differential equation which is coupled with momentum equation (8), and in general, difficult to solve analytically. In order to overcome this difficulty, we solve these equations numerically by fourth-order Runge-Kutta method in association with shooting technique. Firstly, these equations together with associated boundary conditions are reduced to first-order differential equations. Since equations to be solved are the third order for the velocity and second order for the temperature and concentration, the values of f', θ', and ϕ' are needed at $\eta = 0$. Therefore, the shooting method is used to solve this boundary value problem. The local skin friction coefficient, the local Nusselt number, the local Sherwood number, and the plate surface temperature are computed in terms of $f''(0)$, $-\theta'(0)$, $-\phi'(0)$ and $\theta(0)$, respectively. It can be noted that the local parameters Ha_x, Gr_x, Gc_x, Bi_x, S_x, and Kr_x in (8)–(10) are functions of x and generate local similarity solution. In order to have a true similarity solution we assume the following relation [19]:

$$h_f = \frac{a}{\sqrt{x}}, \qquad \sigma = \frac{b}{x}, \qquad \beta = \frac{c}{x},$$

$$\beta^* = \frac{d}{x}, \qquad Q_0 = \frac{e}{x}, \qquad \text{Kr}' = \frac{m}{x}, \qquad (12)$$

where a, b, c, d, e, and m are the constants with appropriate dimensions. In view of relation (12) the parameters Ha_x, Gr_x, Gc_x, Bi_x, S_x, and Kr_x are now independent of x and henceforth, we drop the index "x" for simplicity.

3. Result Discussion

The numerical solutions of the boundary value problem for system of ordinary differential equations were obtained by Runge-Kutta method along with shooting technique. Since the physical domain of the underlying problem is unbounded, the computational domain is chosen sufficiently large in order to meet the far field boundary condition at infinity. Here the transverse distance is fixed to 10 and suitably more than 10 depending upon the choice of the parameters. To demonstrate successful implementation of the numerical scheme, the numerical results are compared to those obtained by a previous published paper (see [19]) for the local skin friction coefficient, plate surface temperature, and the local

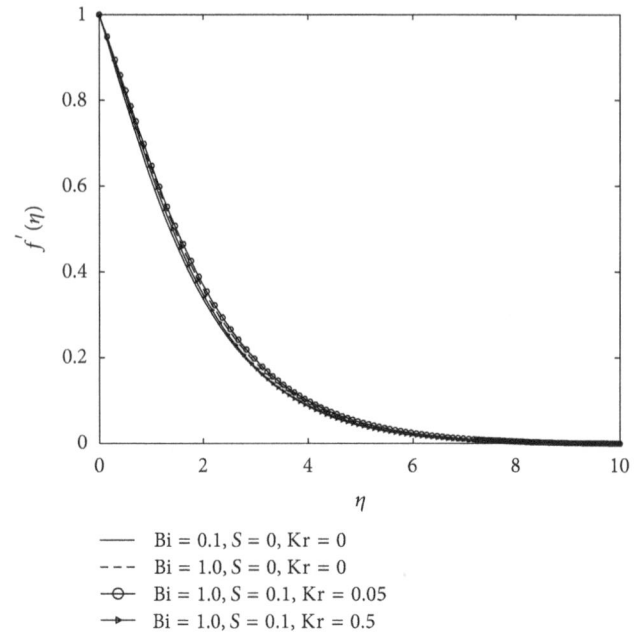

FIGURE 3: Velocity profiles for different values of Bi, S, and Kr for Ha = 0.1, Gr = 0.1, Gc = 0.1, Sc = 0.62, Nc = 0.01, and Pr = 0.72.

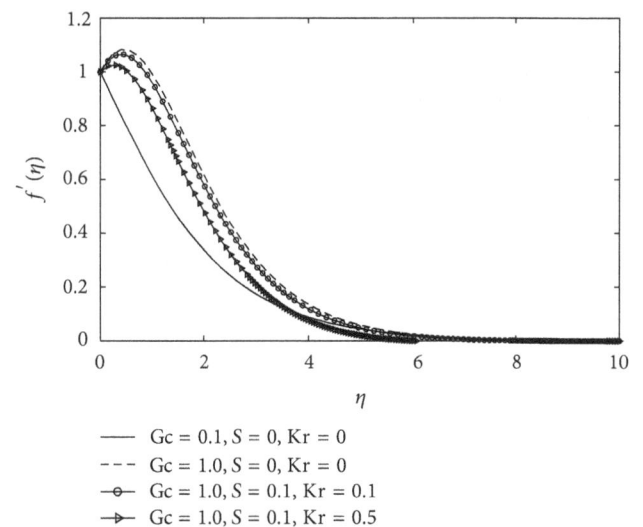

FIGURE 4: Velocity profiles for different values of Gc, S, and Kr for Ha = 0.1, Gr = 0.1, Sc = 0.62, Bi = 0.1, Nc = 0.01, and Pr = 0.72.

Sherwood and Nusselt numbers in Table 1 for the parameters embedded in absence of local chemical reaction parameter (Kr) and local heat source parameter (S). Table 1 presents a comparison of $f''(0)$, $-\theta'(0)$, $\theta(0)$, and $-\phi'(0)$ between the present results and the results obtained by Makinde [19] for various values of Ha, Gr, Gc, Bi, Pr, and Sc when $S = \text{Kr} = 0$. The results are found to be in excellent agreement. It is important to note that the momentum equation is coupled with heat and mass transfer equations and hence the Prandtl number, Schmidt number, chemical reaction parameter, and source term have an influence on skin friction in our present

TABLE 1: Comparison of $f''(0)$, $-\theta'(0)$, $\theta(0)$, and $-\phi'(0)$ for various values of Bi, Ha, Gr, Gc, Pr, and Sc when Kr $= S = 0$.

Bi	Gr	Gc	Ha	Pr	Sc	Makinde [19]				Present study			
						$f''(0)$	$-\theta'(0)$	$\theta(0)$	$-\phi'(0)$	$f''(0)$	$-\theta'(0)$	$\theta(0)$	$-\phi'(0)$
0.1	0.1	0.1	0.1	0.72	0.62	−0.402271	0.078635	0.213643	0.3337425	−0.402271	0.078635	0.213643	0.333742
1.0	0.1	0.1	0.1	0.72	0.62	−0.352136	0.273153	0.726846	0.3410294	−0.352136	0.273153	0.726846	0.341029
0.1	0.5	0.1	0.1	0.72	0.62	−0.322212	0.079173	0.208264	0.3451301	−0.322212	0.079173	0.208264	0.345130
0.1	1.0	0.1	0.1	0.72	0.62	−0.231251	0.079691	0.203088	0.3566654	−0.231251	0.079691	0.203088	0.356665
0.1	0.1	0.5	0.1	0.72	0.62	−0.026410	0.080711	0.192889	0.3813954	−0.026410	0.080711	0.192889	0.381395
0.1	0.1	1.0	0.1	0.72	0.62	0.3799184	0.082040	0.179592	0.4176699	0.379918	0.082040	0.179592	0.417669
0.1	0.1	0.1	5.0	0.72	0.62	−2.217928	0.066156	0.338435	0.1806634	−2.217928	0.066156	0.338435	0.180664
0.1	0.1	0.1	0.1	1.0	0.62	−0.407908	0.081935	0.180640	0.3325180	−0.407908	0.081935	0.180640	0.332518
0.1	0.1	0.1	0.1	7.10	0.62	−0.421228	0.093348	0.066513	0.3305618	−0.421431	0.093348	0.066515	0.330843
0.1	0.1	0.1	0.1	0.72	0.78	−0.411704	0.078484	0.215159	0.3844559	−0.411704	0.078484	0.215159	0.384455
0.1	0.1	0.1	0.1	0.72	2.63	−0.453094	0.077915	0.220841	0.7981454	−0.453094	0.077915	0.220841	0.798146

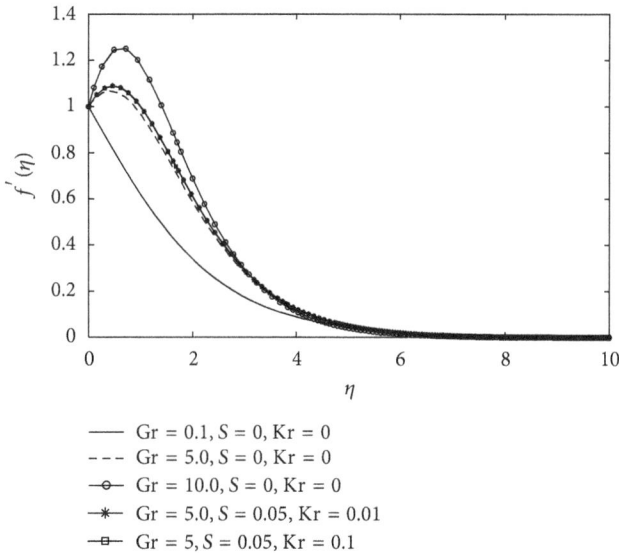

— Gr = 0.1, S = 0, Kr = 0
- - - Gr = 5.0, S = 0, Kr = 0
—○— Gr = 10.0, S = 0, Kr = 0
—✳— Gr = 5.0, S = 0.05, Kr = 0.01
—□— Gr = 5, S = 0.05, Kr = 0.1

FIGURE 5: Velocity profiles for different values of Gr, S, and Kr for Ha = 0.1, Gc = 0.1, Sc = 0.62, Bi = 0.1, Nc = 0.01, and Pr = 0.72.

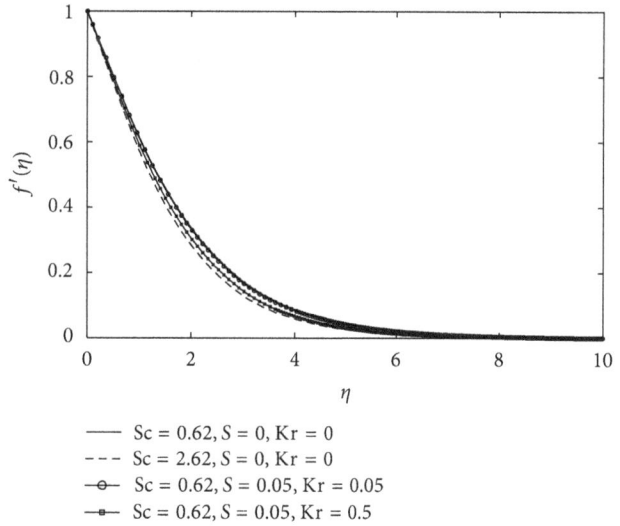

— Sc = 0.62, S = 0, Kr = 0
- - - Sc = 2.62, S = 0, Kr = 0
—○— Sc = 0.62, S = 0.05, Kr = 0.05
—■— Sc = 0.62, S = 0.05, Kr = 0.5

FIGURE 6: Velocity profiles for different values of Sc, S, and Kr for Ha = 0.1, Gr = 0.1, Gc = 0.1, Bi = 0.1, Nc = 0.01, and Pr = 0.72.

problem. The influence of heat source and chemical reaction parameters on local skin friction, local Nusselt number, plate surface temperature, and Sherwood number are highlighted in Tables 2 and 3 for various nondimensional flow parameters. It is clearly seen from Table 2 that the magnitude of skin friction and Nusselt number increase whereas the plate surface temperature, and Sherwood number decrease with the increase of source parameter. Furthermore, increasing the strength of chemical reacting substances is to increase the local skin friction, the plate surface temperature and Sherwood number but opposite behavior is seen for local Nusselt number. The obvious observation from Table 2 is that the fluid low Prandtl number increases the magnitude of local skin friction and local Nusselt number while decreasing the plate surface temperature and local Sherwood number. Again the fluid with low convective resistance (or external resistance) decreases the magnitude of local skin friction while the Nusselt number, the plate surface temperature, and local Sherwood number increase. It is also observed

TABLE 2: Computation of skin friction $(f''(0))$, Nusselt number $(-\theta'(0))$, plate surface temperature $(\theta(0))$, and Sherwood number $(-\phi'(0))$ for different values of Bi, Pr, S, and Kr. The other parameters are Ha = 0.1, Gr = 0.1, Gc = 0.1, Sc = 0.62, and Nc = 0.01.

Bi	Pr	S	Kr	$f''(0)$	$-\theta'(0)$	$\theta(0)$	$-\phi'(0)$
0.1	0.72	0.1	0.1	−0.604781	0.154830	0.548301	0.376062
0.5	0.72	0.1	0.1	−0.358517	0.167824	0.664351	0.428429
1.0	0.72	0.1	0.1	−0.344220	0.204287	0.795712	0.430013
1.0	0.72	0.3	0.1	−0.410247	0.283270	0.716729	0.413375
1.0	0.72	0.3	0.3	−0.419822	0.265179	0.734820	0.552735
1.0	0.72	0.3	0.6	−0.428749	0.253741	0.746258	0.708364
1.0	1.0	0.3	0.6	−0.412617	0.225282	0.774717	0.709739

from Table 3 that the increase of Schimdt number results in increase of the magnitude of the local skin friction but opposite behavior is marked in case of Nusselt number and the plate surface temperature.

TABLE 3: Computation of skin friction ($f''(0)$), Nusselt number ($-\theta'(0)$), plate surface temperature ($\theta(0)$), and Sherwood number ($-\phi'(0)$) for different values of Sc, S, and Kr. The other parameters are fixed at Bi = Gc = Gr = 0.1, Pr = 0.72, and Nc = 0.01.

Sc	S	Kr	$f''(0)$	$-\theta'(0)$	$\theta(0)$	$-\phi'(0)$
0.62	0.05	0.05	−0.403557	0.075603	0.243969	0.381077
0.62	0.1	0.05	−0.395738	0.070777	0.292201	0.382425
0.62	0.3	0.05	−0.433542	0.087237	0.127620	0.373824
0.62	0.1	0.5	−0.606262	0.141007	−0.410071	0.648094
0.62	0.1	1.0	−0.612201	0.137188	−0.371886	0.864236
0.78	0.1	1.0	−0.615507	0.135984	−0.359841	0.972618
2.63	0.1	1.0	−0.634123	0.132985	−0.329852	1.811414

3.1. Velocity Profiles. Figures 2–7 exhibit the velocity profiles obtained by the numerical simulations for various flow parameters involved in the problem. The simulated parameters are reported in the figure caption. The effects of magnetic parameter on the velocity field in presence and absence of source and chemical reaction parameter are shown in Figure 2. It illustrates that the velocity profile decreases with the increase of magnetic parameter in absence of source and chemical reaction parameter because Lorentz force acts against the flow if the magnetic field is applied in the normal direction. In presence of source and chemical reaction parameter no such appreciable change is observed in Figure 2. This corresponds to Figure (2) in [19] and thereby again validating our numerical scheme. A little increase in the velocity profile near the boundary layer is marked in Figure 3 with the increase in the convective heat parameter because the fluid adjacent to the right surface of the plate becomes lighter by hot fluid and rises faster. The boundary layer flows develop adjacent to vertical surface and velocity reaches a maximum in the boundary layer. It is evident from Figures 4 and 5 that greater cooling of surface, an increase in Gc, and increase in Gr result in an increase in the velocity. It is due to the fact that the increase in the values of Grashof number and modified Grashof number has the tendency to increase the thermal and mass buoyancy effect. The increase is also evident due to the presence of source and chemical reaction parameters. Furthermore the velocity increases rapidly and suddenly falls near the boundary and then approaches the far field boundary condition due to favorable buoyancy force with the increase of both Gr and Gc. It can be seen that the increase in the Prandtl number and Schmidt number leads to a fall in the velocity as shown in Figures 6 and 7.

3.2. Temperature Profiles. Figures 8–13 show the temperature profiles obtained by the numerical simulations for various values of flow parameters. Figure 8 clearly demonstrates that the temperature profiles increase with the increase of the magnetic field parameter, which implies that the applied magnetic field tends to heat the fluid, and thus reduces the heat transfer from the wall. Further it can be seen that temperature profile increases due to increase of heat source as well as chemical reaction parameter. The thermal boundary layer thickness increases with an increase in the

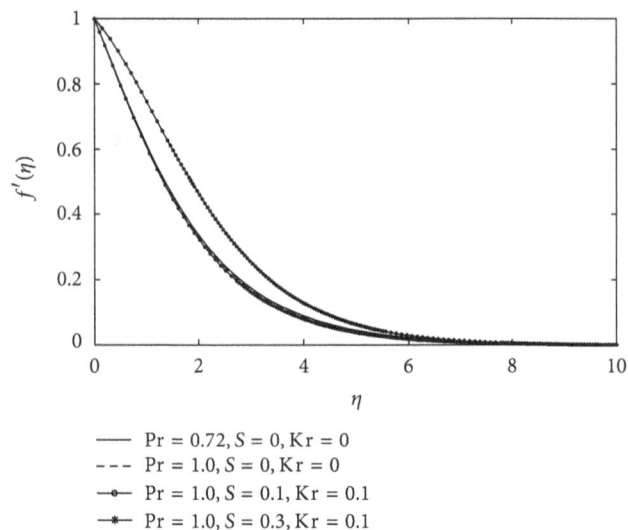

FIGURE 7: Velocity profiles for different values of Pr, S, and Kr for Ha = 0.1, Gr = 0.1, Gc = 0.1, Sc = 0.62, Bi = 0.1, and Nc = 0.01.

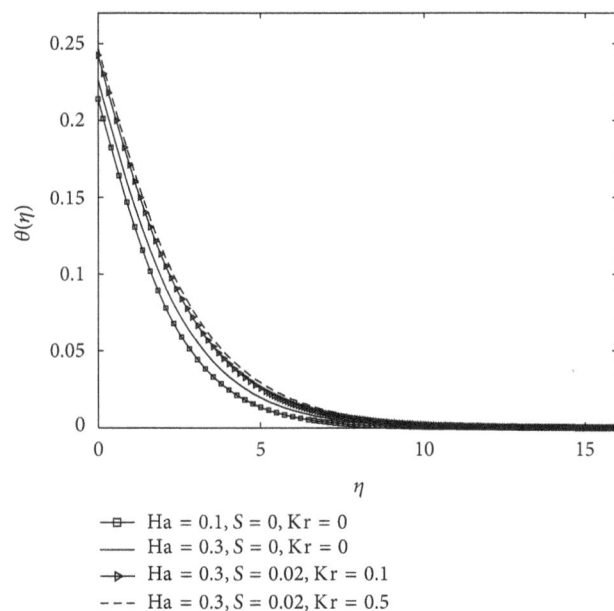

FIGURE 8: Temperature profiles for different values of Ha, S, and Kr for Gr = 0.1, Gc = 0.1, Sc = 0.62, Bi = 0.1, Nc = 0.01, and Pr = 0.72.

plate surface convective heat parameter (Figure 9), and a similar effect is also observed in Figure 12 with the increase of Schmidt number. It can be observed that the amplitude of fluid temperature in presence of heat source and chemical reacting substances is more in comparison to in absence of these parameters. The steady state temperatures for different Grashof number, modified Grashof number, internal heat source, and chemical reaction parameters are shown in Figures 10 and 11. The thermal boundary layer decreases with increasing Grashof number and modified Grashof number, but reverse effect is observed with the presence of chemical reaction parameter. This is also revealed in Figure 13, which

MHD Heat and Mass Transfer of Chemical Reaction Fluid Flow over a Moving Vertical Plate in Presence of Heat
Source with Convective Surface Boundary Condition

161

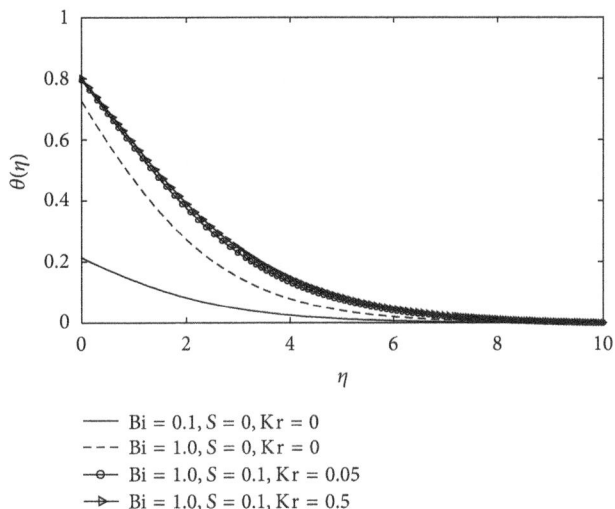

FIGURE 9: Temperature profiles for different values of Bi, S, and Kr for Ha = 0.1, Gr = 0.1, Gc = 0.1, Sc = 0.62, Nc = 0.01, and Pr = 0.72.

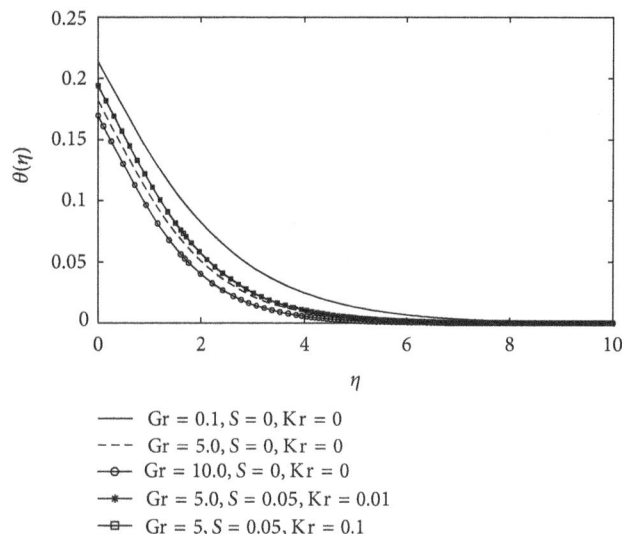

FIGURE 11: Temperature profiles for different values of Gr, S, and Kr for Ha = 0.1, Gc = 0.1, Sc = 0.62, Bi = 0.1, Nc = 0.01, and Pr = 0.72.

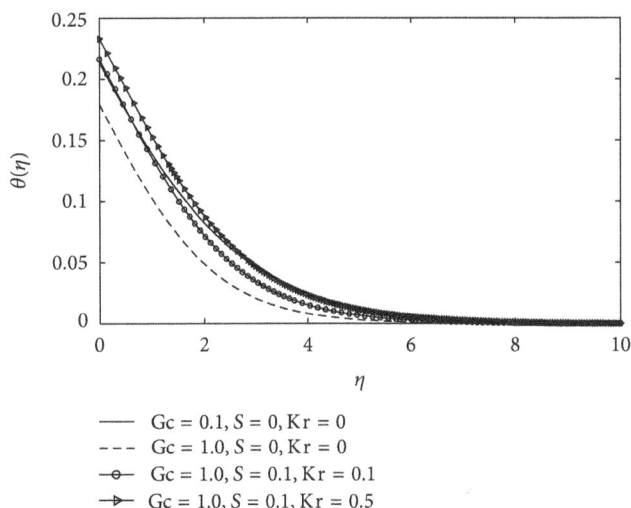

FIGURE 10: Temperature profiles for different values of Gc, S, and Kr for Ha = 0.1, Gr = 0.1, Sc = 0.62, Bi = 0.1, Nc = 0.01, and Pr = 0.72.

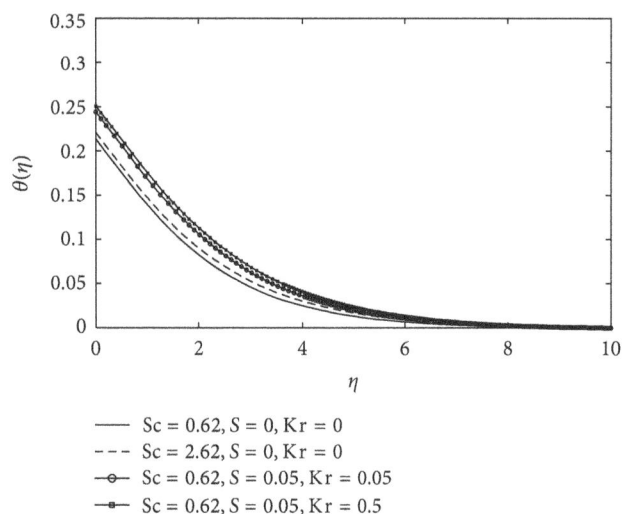

FIGURE 12: Temperature profiles for different values of Sc, S, and Kr for Ha = 0.1, Gr = 0.1, Gc = 0.1, Bi = 0.1, Nc = 0.01, and Pr = 0.72.

shows that thermal boundary layer thickness decreases as the Prandtl number increases implying higher heat transfer. It is due to that smaller values of Pr means increasing the thermal conductivity, and therefore heat is able to diffuse away from the plate more quickly than higher values of Pr, hence the rate of heat transfer is reduced. It is noted that, owing to the presence of heat source effect ($S > 0$) and chemical reaction parameter (Kr > 0), the thermal state of the fluid increases. Hence, the temperature of the fluid increases within the boundary layers. In the event that the strength of the heat source and chemical reaction parameters are relatively large, a remarkable change is observed in the temperature profiles within the thermal boundary layer as can be seen in Figure 13. Further, the effect of heat generation

is more pronounced on temperature profiles for high Prandtl number fluids.

3.3. Concentration Profile.

Figures 14–19 show the concentration profiles obtained by the numerical simulations for various values of nondimensional parameters Bi, Ha, Gr, Gc, Kr, Pr, Nc, Sc, and S. In Figure 14, the effect of an applied magnetic field is found to increase the concentration boundary layer. However, it is interesting to note that the concentration profiles decrease with the increase of both heat source and chemical reaction parameters (Figures 14 and 15). Figures 16 and 17 reveal the concentration variations with Gr, and Gc respectively. It is due to the fact that an increase in the values of Grashof number and modified Grashof number has the tendency to increase the mass buoyancy effect. This gives

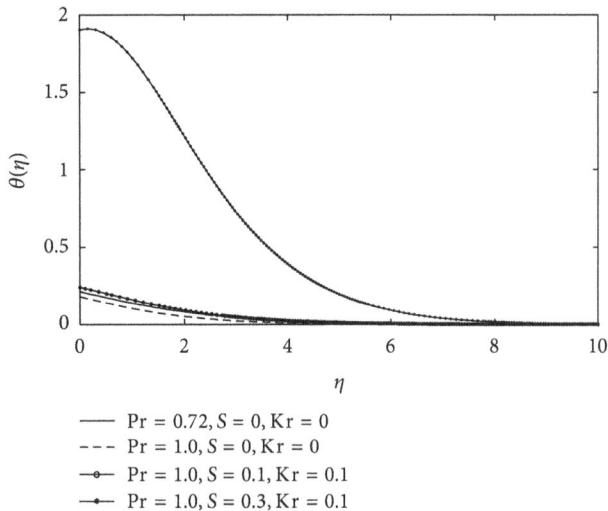

FIGURE 13: Temperature profiles for different values of Pr, S, and Kr for Ha = 0.1, Gr = 0.1, Gc = 0.1, Sc = 0.62, Bi = 0.1, and Nc = 0.01.

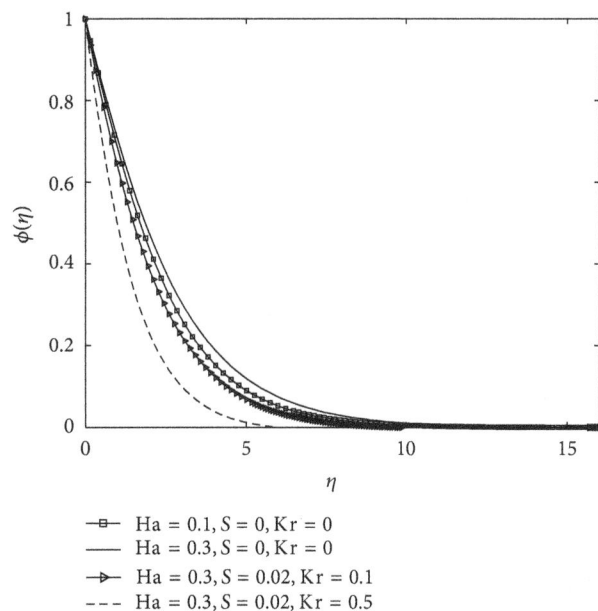

FIGURE 15: Concentration profiles for different values of Bi, S, and Kr for Ha = 0.1, Gr = 0.1, Gc = 0.1, Sc = 0.62, Nc = 0.01, and Pr = 0.72.

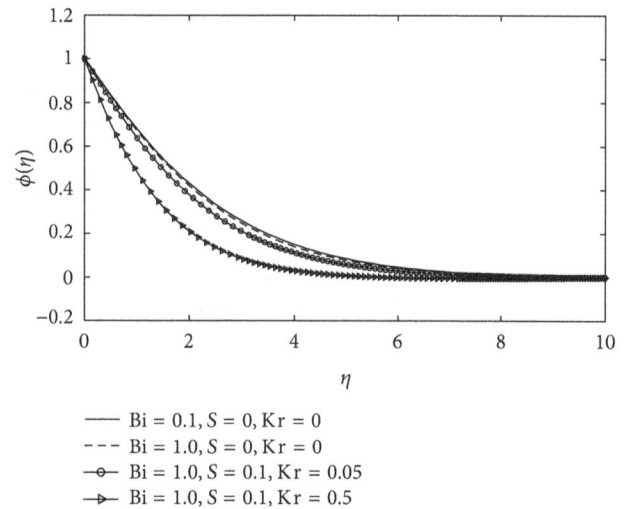

FIGURE 14: Concentration profiles for different values of Ha, S, and Kr for Gr = 0.1, Gc = 0.1, Sc = 0.62, Bi = 0.1, Nc = 0.01, and Pr = 0.72.

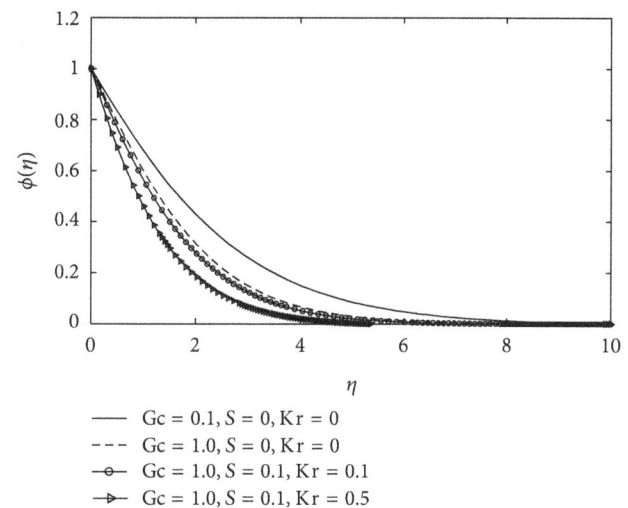

FIGURE 16: Concentration profiles for different values of Gc, S, and Kr for Ha = 0.1, Gr = 0.1, Sc = 0.62, Bi = 0.1, Nc = 0.01, and Pr = 0.72.

resulting a decrease in thermal boundary layer thickness, and source term further influences the decrease of concentration.

4. Conclusions

The present numerical study has been carried out for heat and mass transfer of MHD flow over a moving vertical plate in presence of heat source and chemical reaction along with convective surface boundary condition. The shooting method with Runge-Kutta fourth-order iteration scheme has been implemented to solve the dimensionless velocity, thermal, and mass boundary layer equations. It has been shown that the magnitude of local skin friction and local Nusselt number increase whereas the plate surface temperature and

rise to an increase in the induced flow and thereby decreases concentration. Figure 18 depicts the effect of Schmidt number on the concentration. Like temperature, the concentration value is higher at the surface and falls exponentially. The concentration decreases with an increase in Sc and the decrease is more with the increase in concentration parameter. Figure 19 displays the effect of Pr on concentration profile against η with the variation of source and chemical reaction parameters. The magnitude of concentration is higher at the plate and then decays to zero asymptotically; this is due to the fact that thermal conductivity of fluid decreases with the increase of Pr

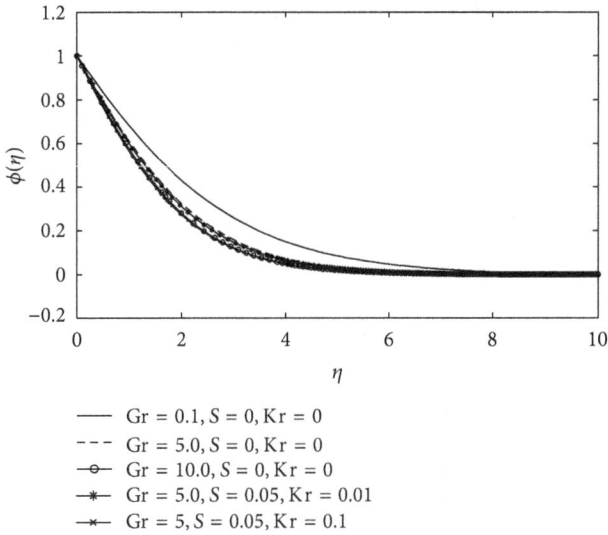

FIGURE 17: Concentration profiles for different values of Gr, S, and Kr for Ha = 0.1, Gc = 0.1, Sc = 0.62, Bi = 0.1, Nc = 0.01, and Pr = 0.72.

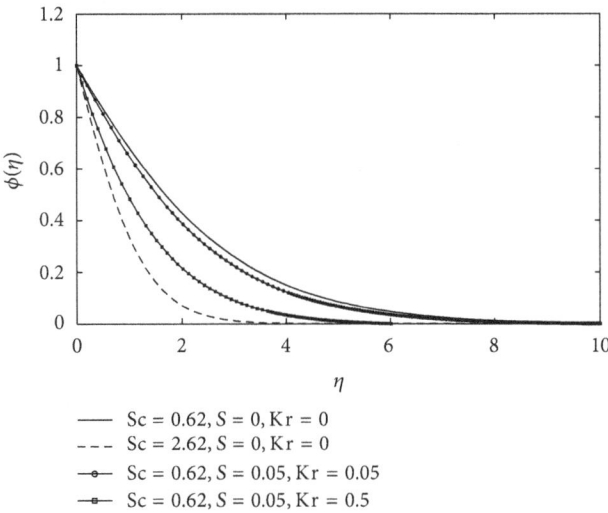

FIGURE 19: Concentration profiles for different values of Pr, S, and Kr for Ha = 0.1, Gr = 0.1, Gc = 0.1, Sc = 0.62, Bi = 0.1, Nc = 0.01.

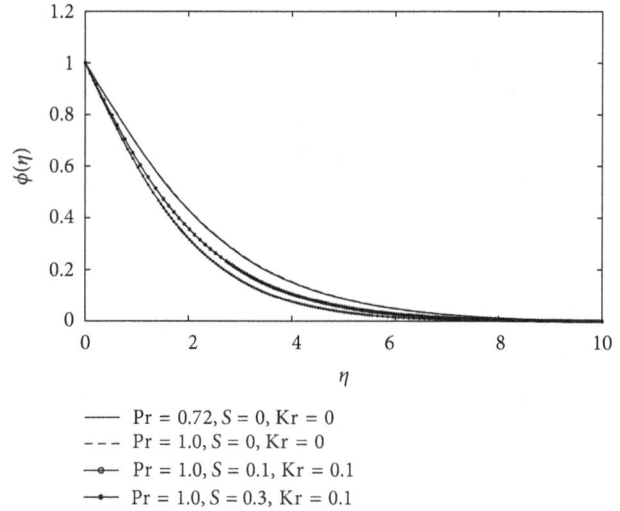

FIGURE 18: Concentration profiles for different values of Sc, S, and Kr for Ha = 0.1, Gr = 0.1, Gc = 0.1, Bi = 0.1, Nc = 0.01, and Pr = 0.72.

Sherwood number decreases with an increase in source parameter. The increase in the strength of chemical reacting substances causes an increase in the magnitude of local skin friction, the plate surface temperature, and Sherwood number, but opposite behavior is seen for local Nusselt number. The velocity profile decreases by increasing the magnetic parameter and even the increase is more prominent with the increase in source and chemical reaction parameter. The thermal boundary layer thickness increases with the increase of source, chemical reaction parameter, plate surface convective heat parameter, and Schmidt number while the mass flux boundary layer thickness decreases. Moreover, the thermal boundary layer thickness, the mass boundary layer, and velocity decrease as the Prandtl number increases.

Nomenclature

u,v:	Velocity components along x- and y-axis direction, respectively
g:	Acceleration due to gravity
Q_0:	Heat source parameter
μ:	Dynamic viscosity
v:	Kinematic viscosity
U_0:	Characteristic velocity at the plate
h_f:	Heat transfer coefficient
T_f:	Temperature of hot fluid at the wall
C_w:	Plate surface concentration
C_∞:	Free stream concentration
β_c:	Concentration expansion coefficient
Pr:	Prandtl number
D:	Mass diffusivity
α:	Thermal diffusivity
T:	Temperature
C:	Concentration
B_0:	Magnetic field strength
Ha_x:	Local magnetic field parameter
Bi_x:	Local convective heat transfer parameter
T_∞:	Temperature of the fluid away from the plate
Gr_x:	Local thermal Grashof number
Gc_x:	Local modified Grashof number
Kr_x:	Local chemical reaction parameter
S_x:	Local heat source parameter
N_c:	Concentration difference parameter
η:	Similarity variable
ρ:	Fluid density
σ:	Fluid electrical conductivity
f:	Dimensionless velocity
θ:	Dimensionless temperature
ϕ:	Dimensionless concentration.

References

[1] K. Vajrevelu and J. Nayfeh, "Hydromagnetic convection at a cone and a wedge," *International Communications in Heat and Mass Transfer*, vol. 19, pp. 701–710, 1992.

[2] A. J. Chamkha, "Hydromagnetic three-dimensional free convection on a vertical stretching surface with heat generation or absorption," *International Journal of Heat and Fluid Flow*, vol. 20, no. 1, pp. 84–92, 1999.

[3] D. Moalem, "Steady state heat transfer within porous medium with temperature dependent heat generation," *International Journal of Heat and Mass Transfer*, vol. 19, no. 5, pp. 529–537, 1976.

[4] M. M. Rahman and M. A. Sattar, "Magnetohydrodynamic convective flow of a micropolar fluid past a continuously moving vertical porous plate in the presence of heat generation/absorption," *Journal of Heat Transfer*, vol. 128, no. 2, pp. 142–152, 2006.

[5] U. N. Das, R. Deka, and V. M. Soundalgekar, "Effects of mass transfer on flow past an impulsively started infinite vertical plate with constant heat flux and chemical reaction," *Forschung im Ingenieurwesen/Engineering Research*, vol. 60, no. 10, pp. 284–287, 1994.

[6] S. P. Anjalidevi and R. Kandasamy, "Effects of chemical reaction, heat and mass transfer on laminar flow along a semi infinite horizontal plate," *Heat and Mass Transfer*, vol. 35, no. 6, pp. 465–467, 1999.

[7] M. A. Seddeek, A. A. Darwish, and M. S. Abdelmeguid, "Effects of chemical reaction and variable viscosity on hydromagnetic mixed convection heat and mass transfer for Hiemenz flow through porous media with radiation," *Communications in Nonlinear Science and Numerical Simulation*, vol. 12, no. 2, pp. 195–213, 2007.

[8] P. M. Patil and P. S. Kulkarni, "Effects of chemical reaction on free convective flow of a polar fluid through a porous medium in the presence of internal heat generation," *International Journal of Thermal Sciences*, vol. 47, no. 8, pp. 1043–1054, 2008.

[9] A. M. Salem and M. Abd El-Aziz, "Effect of Hall currents and chemical reaction on hydromagnetic flow of a stretching vertical surface with internal heat generation/absorption," *Applied Mathematical Modelling*, vol. 32, no. 7, pp. 1236–1254, 2008.

[10] F. S. Ibrahim, A. M. Elaiw, and A. A. Bakr, "Effect of the chemical reaction and radiation absorption on the unsteady MHD free convection flow past a semi infinite vertical permeable moving plate with heat source and suction," *Communications in Nonlinear Science and Numerical Simulation*, vol. 13, no. 6, pp. 1056–1066, 2008.

[11] S. K. Parida, M. Acharya, G. C. Dash, and S. Panda, "MHD heat and mass transfer in a rotating system with periodic suction," *Arabian Journal for Science and Engineering*, vol. 36, no. 6, pp. 1139–1151, 2011.

[12] R. Rajeswari, B. Jothiram, and V. K. Nelson, "Chemical reaction, heat and mass transfer on nonlinear MHD boundary layer flow through a vertical porous surface in the presence of suction," *Applied Mathematical Sciences*, vol. 3, no. 49-52, pp. 2469–2480, 2009.

[13] A. Mahdy, "Effect of chemical reaction and heat generation or absorption on double-diffusive convection from a vertical truncated cone in porous media with variable viscosity," *International Communications in Heat and Mass Transfer*, vol. 37, no. 5, pp. 548–554, 2010.

[14] D. Pal and B. Talukdar, "Perturbation analysis of unsteady magnetohydrodynamic convective heat and mass transfer in a boundary layer slip flow past a vertical permeable plate with thermal radiation and chemical reaction," *Communications in Nonlinear Science and Numerical Simulation*, vol. 15, no. 7, pp. 1813–1830, 2010.

[15] O. D. Makinde and A. Ogulu, "The effect of thermal radiation on the heat and mass transfer flow of a variable viscosity fluid past a vertical porous plate permeated by a transverse magnetic field," *Chemical Engineering Communications*, vol. 195, no. 12, pp. 1575–1584, 2008.

[16] O. D. Makinde, "MHD mixed-convection interaction with thermal radiation and nth order chemical reaction past a vertical porous plate embedded in a porous medium," *Chemical Engineering Communications*, vol. 198, no. 4, pp. 590–608, 2011.

[17] A. Aziz, "A similarity solution for laminar thermal boundary layer over a flat plate with a convective surface boundary condition," *Communications in Nonlinear Science and Numerical Simulation*, vol. 14, no. 4, pp. 1064–1068, 2009.

[18] P. O. Olanrewaju, O. T. Arulogun, and K. Adebimpe, "Internal heat generation effect on thermal boundary layer with a convective surface boundary condition," *American Journal of Fluid Dynamics*, vol. 2, no. 1, pp. 1–4, 2012.

[19] O. D. Makinde, "On MHD heat and mass transfer over a moving vertical plate with a convective surface boundary condition," *Canadian Journal of Chemical Engineering*, vol. 88, no. 6, pp. 983–990, 2010.

[20] O. D. Makinde, "Similarity solution of hydromagnetic heat and mass transfer over a vertical plate with a convective surface boundary condition," *International Journal of Physical Sciences*, vol. 5, no. 6, pp. 700–710, 2010.

[21] O. D. Makinde, "Similarity solution for natural convection from a moving vertical plate with internal heat generation and a convective boundary condition," *Thermal Science*, vol. 15, supplement 1, pp. S137–S143, 2011.

[22] K. Gangadhar, N. B. Reddy, and P. K. Kameswaran, "Similarity solution of hydromagnetic heat and mass transfer over a vertical plate with convective surface boundary condition and chemical reaction," *International Journal of Nonlinear Science*, vol. 3, no. 3, pp. 298–307, 2012.

[23] N.F.M. Noor, S. Abbabandy, and I. Hasim, "Heat and Mass Transfer of thermophoretic MHD flow over an inclined radiate isothermal permeable surface in presence of heat source /sink," *International Journal of Heat and Mass Transfer*, vol. 55, no. 7, pp. 2122–2128, 2012.

[24] V. Bisht, M. Kumar, and Z. Uddin, "Effect of variable thermal conductivity and chemical reaction on steady mixed convection boundary layer flow with heat and mass transfer inside a cone due to a point," *Journal of Applied Fluid Mechanics*, vol. 4, no. 4, pp. 59–63, 2011.

[25] A. A. Bakr, "Effects of chemical reaction on MHD free convection and mass transfer flow of a micropolar fluid with oscillatory plate velocity and constant heat source in a rotating frame of reference," *Communications in Nonlinear Science and Numerical Simulation*, vol. 16, no. 2, pp. 698–710, 2011.

19

Effect of Operating Conditions on Catalytic Gasification of Bamboo in a Fluidized Bed

Thanasit Wongsiriamnuay,[1] Nattakarn Kannang,[2] and Nakorn Tippayawong[2]

[1] *Division of Agricultural Engineering, Faculty of Engineering and Agro-Industry, Maejo University, Chiang Mai 50290, Thailand*
[2] *Department of Mechanical Engineering, Faculty of Engineering, Chiang Mai University, Chiang Mai 50200, Thailand*

Correspondence should be addressed to Thanasit Wongsiriamnuay; w_thanasit@hotmail.com

Academic Editor: Deepak Kunzru

Catalytic gasification of bamboo in a laboratory-scale, fluidized bed reactor was investigated. Experiments were performed to determine the effects of reactor temperature (400, 500, and 600°C), gasifying medium (air and air/steam), and catalyst to biomass ratio (0:1, 1:1, and 1.5:1) on product gas composition, H_2/CO ratio, carbon conversion efficiency, heating value, and tar conversion. From the results obtained, it was shown that at 400°C with air/steam gasification, maximum hydrogen content of 16.5% v/v, carbon conversion efficiency of 98.5%, and tar conversion of 80% were obtained. The presence of catalyst was found to promote the tar reforming reaction and resulted in improvement of heating value, carbon conversion efficiency, and gas yield due to increases in H_2, CO, and CH_4. The presence of steam and dolomite had an effect on the increasing of tar conversion.

1. Introduction

Energy demand has been growing for the past several decades due to rapid industrial and urban development in industry, but fossil fuel reserves have been in decline [1]. Renewable energy has been very popular as an obvious candidate to substitute fossil fuels. Biomass is one of the renewable fuel sources that can claim to have significant environmental benefits with regards to neutral carbon emissions and reduction in global warming [2, 3]. There are many biomass materials that can be utilized for energy [4]. Fast growing plants, which do not compete with food crops, may be used as sustainable energy resources [5, 6] for developed and developing countries. Biomass can be converted to biofuels via several pathways such as biochemical or thermochemical conversion. Gasification process is one of the promising technologies to produce syngas from solid feedstock [2, 7–9]. Producer gas containing simple molecular gas can be used, instead of fossil fuels, in combustion engines.

Gas production is dependent on input streams, operating conditions, and gas output conditioning. Input of gasification process is referred to by type and components of feedstock materials and type and flow of gasifying agent. Gas output conditioning is a process involved in cooling and disposing particulate matter and tar in the gas product. Gasification reactions are controlled by operation conditions such as temperature, pressure, and residence time. Reaction temperature is one of the most influential parameters for the gasification operation. Gasification temperature is normally classified into three ranges; low (400–600°C), medium (600–900°C), and high (>900°C). Increasing temperature tends to result in increasing H_2, CO, gas heating value, carbon conversion efficiency, and gas yields. The advantage of gasification at low temperatures is due to reduced energy input, low tar yield [10], and low cost by partial oxidation, but heating value of fuel gas may be low. To increase the heating value of the product gas, steam may be added to the gasifying medium but additional energy input would be needed. This way, the H_2 content in the producer gas can be improved. H_2 has beneficial properties as a clean energy carrier for heat supply and transportation purposes [2, 8]. Steam gasification takes place at high temperatures because the steam reforming reaction is an endothermic process, but catalytic steam gasification at low temperatures was more useful than high temperature with high content of H_2 [10]. Chang et al. [11] found that maximum H_2 content occurred at steam to biomass ratio of 1.

Normally, producer gas contains a high content of tar which can cause operational problems by blocking gas cooler, filter elements, and engine components. Most producer gas applications also require the removal of dust and tar before the gas can be used [12]. Tar can be effectively minimized in the producer gas by catalytic cracking. Many researchers have extensively studied and shown that a cheap additive such as calcined dolomite (MgO-CaO) was useful in reducing tar, improving gas quality and heating value for biomass gasification [13, 14]. The destruction of tar is more effective at high temperatures, but increasing temperature may lead to higher tar yield [10]. At low temperature of 550°C, Asadullah et al. [15] reported that, with the presence of dolomite, tar conversion was around 63%. At medium temperature, Yu et al. [13] found that tar conversion of around 65–75% could be achieved at 700°C, with the presence of dolomite. Increasing from 700 to 800°C resulted in a decrease in tar conversion. This was contributed to that fact that more stable compounds of tar were formed, so it was harder to crack. Chiang et al. [16] found that increasing content of CaO and temperature (600–900°C) resulted in an increase of gas heating value and carbon conversion rate. At high temperatures, Akay and Jordan [17] used CaO as an in-bed catalyst in a fixed-bed gasifier at 1,040°C and obtained minimum tar yields of less than 0.8 g/kg and maximum gas yield of $4\,Nm^3/kg$. In addition, the gas produced can be applied into an internal combustion engine and gas burners fixed in the combustion chamber with the downstream process similar to the diesel burner [18].

Many reports exist in the open literature on gasification of various types of biomass, but studies on bamboo are rather limited. Bamboo is one of the fast growing and widely cultivated plants in many countries. It can be harvested within a few years [19]. So far, it has been used mainly as structural material for construction and household furniture [20–23]. Its utilization as renewable energy has not yet been investigated extensively. There have been several studies on thermal conversion of bamboo, but most work was about production of activated carbon [24–29], or bio-oil via pyrolysis [30–35]. Chiang et al. [16] investigated fixed-bed gasification of waste bamboo chopsticks. The heating value was found to be in the range of $2.0–10.6\,MJ/m^3$, obtained between 600–900°C using CaO as a catalyst. Kantarelis et al. [5] used steam as a medium in the gasification of bamboo powder and reported that 10–20% H_2 and 15–20% CO can be obtained between 797 and 865 K. Increased steam to biomass ratio resulted in increased H_2, CO_2, and gas yield, but with reduction in CO, CH_4, and heating value. Gasification in a fluidized bed reactor was conducted by Xiao et al. [36] at 400–700°C. The heating value was found to be $7.2\,MJ/Nm^3$ at 700°C with excess air ratio of 0.2. Kannang et al. [37] found that the heating value was $5.26\,MJ/Nm^3$ at 700 K.

There appears to be a lack of information with respect to the gasification of bamboo in a fluidization reactor with the presence of a catalyst, especially at low temperature range. In this study, an experimental study in a fluidized bed reactor was conducted for the gasification of bamboo. The objective of this study was to investigate the effects of reactor temperature, gasifying medium, and catalyst to biomass ratio on composition of product, gas yield, heating value, and carbon conversion efficiency.

TABLE 1: Analysis of bamboo.

Property	Unit	Method	Quantity
Proximate analysis			
Moisture	(% w/w)	ASTM D 3173	5.73
Volatile	(% w/w)	ASTM D 3175	74.68
Fixed carbon	(% w/w)	ASTM D 3172	14.04
Ash	(% w/w)	ASTM D 3177	5.55
Ultimate analysis			
Carbon	(%)	ASTM D 3174	45.66
Hydrogen	(%)	ASTM D 3174	4.32
Nitrogen	(%)	ASTM D 3174	0.24
Oxygen	(%)	By difference	49.78
LHV	(MJ/kg)	ASTM 5865	17.80
C/O			1.22
H/O			1.39

2. Materials and Methods

2.1. Biomass Materials and Catalysts. Samples of bamboo collected in Chiang Mai, Thailand, were used. The collected samples were dried, crushed, and ground in a high-speed rotary mill and sieved to provide a feed sample in the size range between 0.10 and 0.25 mm. The moisture, volatile, fixed carbon, and ash were determined, following the ASTM standards. The carbon, hydrogen, and nitrogen contents were determined using a Thermo Scientific Instrument CHN elemental analyzer. The oxygen content was calculated by difference. The heating value of the dried bamboo was determined with a Parr bomb calorimeter. The analysis results of the bamboo samples are shown in Table 1 [37].

Silica sand with a particle size of 45 micron was used as inert bed material in the fluidized bed gasifier, while calcined dolomite was used as a catalyst. The dolomite was sieved to obtain a fraction with a particle size of 45 micron and then calcined in the oven at 900°C for 4 h. The calcined dolomite was kept in a desiccator for later use.

2.2. Experimental Apparatus and Procedure. Figure 1 shows a schematic diagram of a fluidized bed gasification system used in this study. There are six main components. Air (1) and steam (2) supply units were composed of air and water supplies, control valves, flow meters, and preheaters. An external heating system (3) supplied heat at the bottom of the fluidized bed reactor. The fluidized bed reactor unit (4) comprised a biomass feeder and a reactor. The reactor was made from a stainless steel cylinder and was externally covered with a thick insulator. The total height of the reactor was 2 m, with an internal diameter of 50 mm. The reactor was installed with a series of thermocouples along the length in

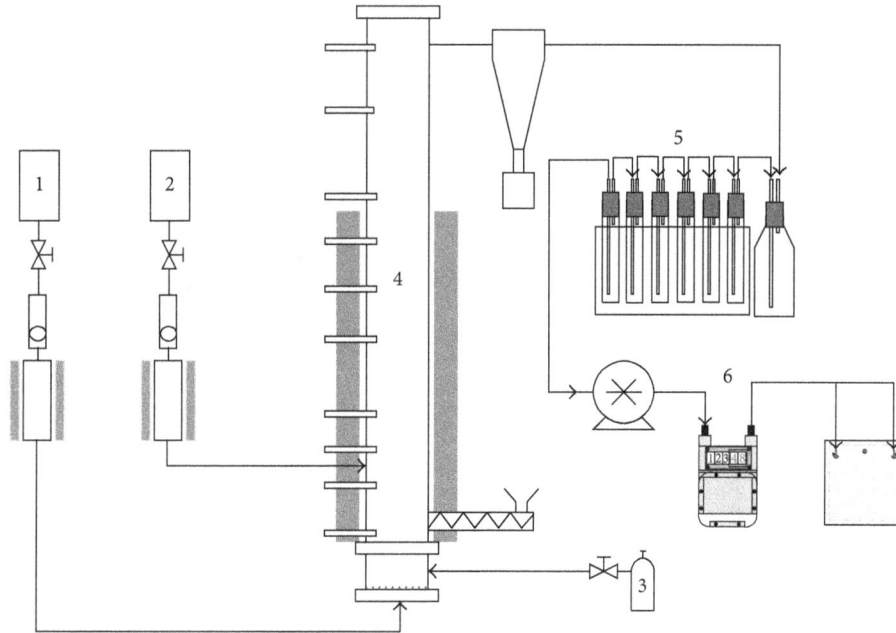

FIGURE 1: Schematic diagram of gasification reactor: (1) air supply, (2) steam supply, (3) external heating, (4) fluidized bed reactor, (5) gas conditioning, and (6) gas sampling unit.

order to measure the temperature distribution inside. An air distributor was installed to ensure uniform air distribution. A gas cleaning unit (5) consisted of a series of glass tubes containing isopropanol for gas cleaning and the collection of tar. The gas collecting unit (6) was constituted of a pump, a gas meter, and gas sampling bag. Cold fluidization tests in a transparent tube of the same dimension to the reactor were carried out for these biomass, dolomite, and silica sand particles between 5 and 30 l/min of air flows. It was visually confirmed that proper mixing was achieved at 15 l/min of air flow or higher.

The biomass was fed into the reactor. Air was used as the fluidizing agent and supplied from an air compressor at a constant flow rate of 15 l/min with the equivalent ratio (ER) of about 0.4. Before the air entered the reactor, it was preheated to 300–500°C. When steam was used, it was produced by a steam generator which fed steam to biomass (S/B) ratio of 1 : 1 at 150°C into the reactor. Biomass was continuously fed into the reactor by the screw feeder from the hopper at a constant rate of 10 g/min.

Each run was started by filling the bed with silica sand, with or without dolomite up to the desired catalyst to biomass (C/B) ratio between 0 : 1, 1 : 1, and 1.5 : 1 w/w. The burner was then turned on to provide heat externally. The start-up period was necessary to preheat the bed. After the bed temperature reached the desired level and became steady, the air compressor was turned on to drive the air through the preheater, air distributor, and into the reactor before the commencement of the biomass feeding. The product gas exited the reactor through a cyclone, via gas conditioning unit, and gas sampling system. Char was separated from the producer gas in the cyclone. The produced gas passed through an ice trap for cooling and cleaning. The dry and

clean gas was sampled using gas bags and analyzed by gas chromatography. The Shimadzu GC model GC-8A fitted with a Shin-carbon column, and TCD detector was used to detect H_2, O_2, N_2, CH_4, CO, and CO. The gas chromatograph was calibrated using standard gases. Helium was used as a carrier gas. Tar was condensed and collected in a series of glass tubes containing isopropanol and it was measured by gravimetric method. The collected tar contained primary and condensed tertiary that dilute in a solvent [38]. Each experimental test's conditions were repeated at least three times. Average values were subsequently presented with error bar.

From the data collected, lower heating value (LHV) and carbon conversion efficiency (CCE) were calculated from the following equations [39]:

$$\text{LHV} = \left(30\text{CO}\% + 25.7\text{H}_2\% + 85.4\text{CH}_4\%\right) \times 4.2 \left(\frac{\text{kJ}}{\text{Nm}^3}\right),$$

$$\text{CCE} = \frac{V_g/22.4 \times \left[\text{CH}_4\% + \text{CO}\% + \text{CO}_2\%\right]}{W\left(1 - \text{ash}\right) \times \text{C}\%/12} \times 100\%,$$

(1)

where CO, CO_2, H_2, and CH_4 are the concentrations of the gas products, V_g is the gas yield measured from the gas meter, and W, C%, and ash are referred to as the biomass feeding rate, carbon, and ash contents in the biomass, respectively.

3. Results and Discussion

3.1. Effects of Temperature and Gasifying Medium. The effects of temperature and gasifying medium on the gas content with nitrogen-free are shown in Figure 2. The reactor temperature was varied between 400 and 600°C. Precision of temperature was about ±20°C for all experiments. Increasing temperature

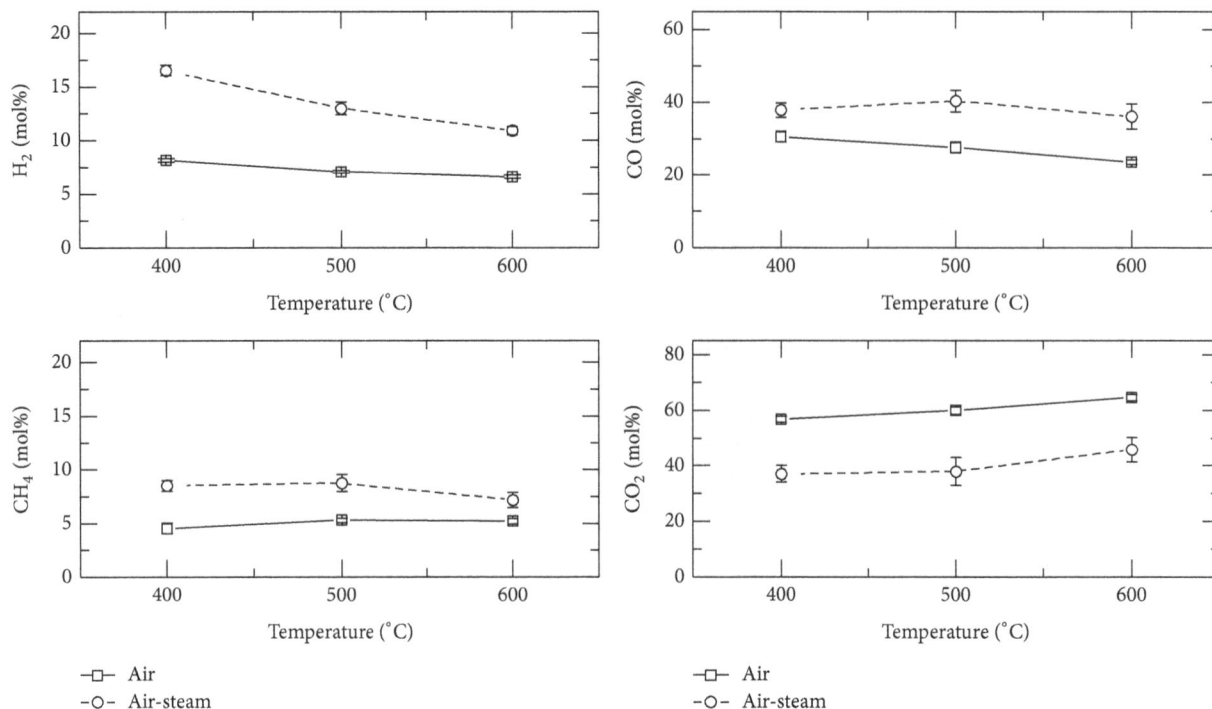

FIGURE 2: Effect of gasifying medium on gas composition at different temperatures.

from 400 to 600°C was found to decrease H_2 and CO contents by around 1.5–5.6% and 4.2–6.9%, respectively. CO content decreased with increasing temperature due to the endothermic reaction [37], while CO_2 increased by around 7.8–8.7% with increasing temperature from 400 to 600°C. CH_4 decreased with increasing temperature from 400 to 600°C due to the reaction of steam and CH_4 [10].

H_2, CO, and CH_4 contents in the product gas were higher when an air/steam mixture was used as the gasifying medium. But CO_2 content from air gasification was higher than that from air-steam gasification [40]. With air gasification, an increased degree of combustion rate may occur to release more CO_2 due to the higher ER (0.4) used in the gasification process. Combustion reactions hardly produce H_2 [41]. It was reported that H_2 content in product gas was obtained from the dehydrogenation of biomass components and secondary decomposition of their pyrolyzed products (reactions (2) or (3)) [37, 41]:

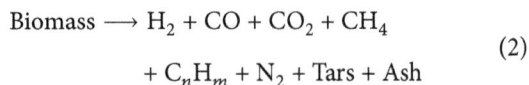

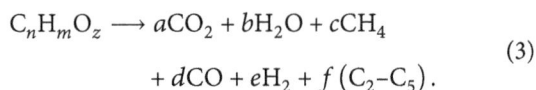

$$\text{Biomass} \longrightarrow H_2 + CO + CO_2 + CH_4 \\ + C_nH_m + N_2 + \text{Tars} + \text{Ash} \qquad (2)$$

$$C_nH_mO_z \longrightarrow aCO_2 + bH_2O + cCH_4 \\ + dCO + eH_2 + f\left(C_2\text{-}C_5\right). \qquad (3)$$

H_2 and CO were decreased at higher temperatures [4] due to the high reactivity of char with air at a higher ER. Increasing temperature favors the endothermic reaction; hence CO content decreased due to a more complete oxidation reaction (reaction (4)) and partial oxidation reaction (reaction (5)) which resulted in increasing CO_2 [4]. Skoulou et al. [42]

suggested that the reaction of char combustion was composed of absorption and desorption. O_2 was absorbed by C which represented carbon active site and released C(O) which was a carbon-oxygen complex in reaction (6). Subsequently, CO_2 was released by desorption (reactions (7) and (8)). Oxygen from CO_2 was absorbed by char to form C(O) and released CO (reaction (9)). CO was desorbed from surface oxygen complex from reaction (10) and absorbed and reacted with C(O) in reaction (11) and released CO_2 [42, 43]:

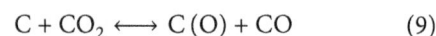

$$C + O_2 \longleftrightarrow CO_2 \qquad (4)$$

$$C + \frac{1}{2}O_2 \longleftrightarrow CO \qquad (5)$$

$$2C + O_2 \longleftrightarrow 2C\left(O\right) \qquad (6)$$

$$2C\left(O\right) \longleftrightarrow CO_2 \qquad (7)$$

$$C\left(O\right) + 2C\left(O\right) \longleftrightarrow CO_2 + C \qquad (8)$$

$$C + CO_2 \longleftrightarrow C\left(O\right) + CO \qquad (9)$$

$$C\left(O\right) \longleftrightarrow C + CO \qquad (10)$$

$$C\left(O\right) + CO \longleftrightarrow C + CO_2 \qquad (11)$$

Figure 3 shows that increasing temperature affected LHV due to the reduction in H_2 and CO content [44]. Carbon conversion efficiency was found to increase due to an increase in CO_2 content while gas yields remained constant. LHV, carbon conversion efficiency, and gas yield were found to increase when air/steam was used as agent in the gasification

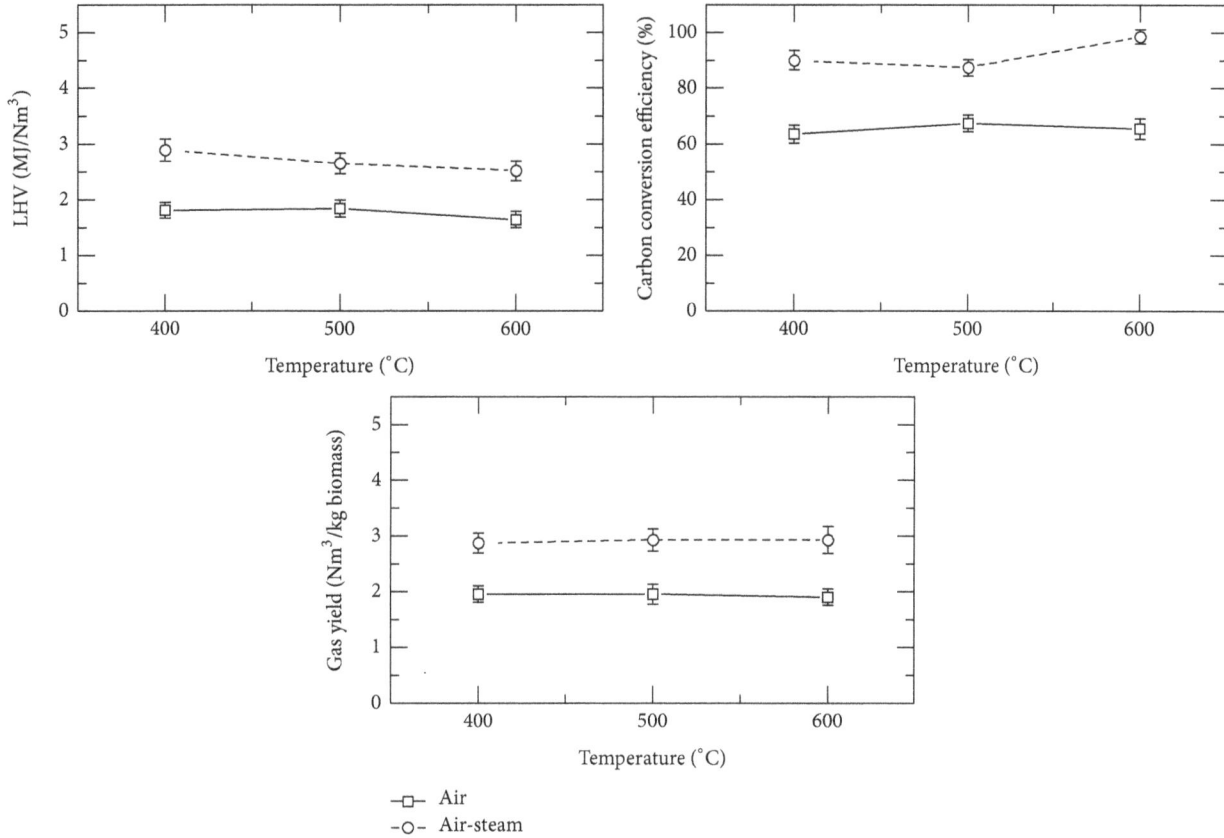

FIGURE 3: Effect of gasifying medium on LHV, carbon conversion efficiency, and gas yield at different temperatures.

process. This was so because higher contents of H_2, CO, and CH_4 were expected from the water-gas reaction [39].

3.2. Effect of Catalyst to Biomass Ratio. The effects of catalyst on gas compositions with nitrogen-free are shown in Figure 4. It was found that the content of H_2 and CO increased, while the content of CH_4 and CO_2 slightly decreased with increasing temperature and catalyst to biomass ratio [10]. It was reported that tar cracking and tar reforming reactions resulted in increased content of H_2 at higher temperatures, according to reactions (12)–(14) [9, 48–50]. The content of CO_2 increased initially with increasing temperature and then decreased when temperatures were higher than 500°C. With increased catalyst to biomass ratio from 0 to 1.5, higher content of CO_2 was obtained due to the release of CO_2 from dolomite. At higher temperatures, reforming of tar with CO_2 with the presence of dolomite led to a decrease of CO_2. The reforming reaction of tar on a dolomite surface occurred by capturing carbon to produce more H_2 and CO, according to reaction (14) [50]:

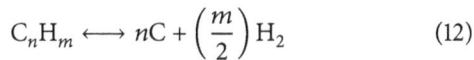

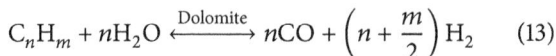

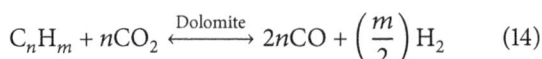

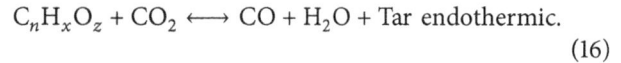

$$C_nH_m \longleftrightarrow nC + \left(\frac{m}{2}\right)H_2 \qquad (12)$$

$$C_nH_m + nH_2O \xrightarrow{\text{Dolomite}} nCO + \left(n + \frac{m}{2}\right)H_2 \qquad (13)$$

$$C_nH_m + nCO_2 \xrightarrow{\text{Dolomite}} 2nCO + \left(\frac{m}{2}\right)H_2 \qquad (14)$$

$$C + CO_2 \longleftrightarrow 2CO \qquad (15)$$

$$C_nH_xO_z + CO_2 \longleftrightarrow CO + H_2O + \text{Tar endothermic.} \qquad (16)$$

At high temperatures, increase in CO and decrease in CO_2 contents were due to reactions (15) and (16). CO_2 reacted with excess carbon in char, producing CO (reaction (15)) similar to those previously reported [41, 46]. The reforming of CO_2 with tar (reaction (16)) produced CO and H_2O. The presence of catalyst promoted the tar reforming reaction, leading to a decrease in tar yield. From Figure 5, it was clear that increasing temperature and catalyst amount resulted in improvement of LHV, carbon conversion efficiency, and gas yield due to the increase in H_2, CO, and CH_4 contents.

From Figure 6, it was found that tar conversion increased with increasing temperature for all cases. With air-steam gasification, tar conversion was increased from 84 to 92% and 77 to 90% when catalyst to biomass ratio was increased from 1 : 1 to 1.5 : 1, respectively. The trend was similar to those previously reported [13, 15].

3.3. Comparison with the Literature. Table 2 shows producer gas composition, gas ratio and yields, carbon conversion efficiency from different biomass materials, and reactor configurations found from the reported literature. In the comparison of bamboo with other agricultural residues [45, 47], it was found that the H_2 content obtained was in

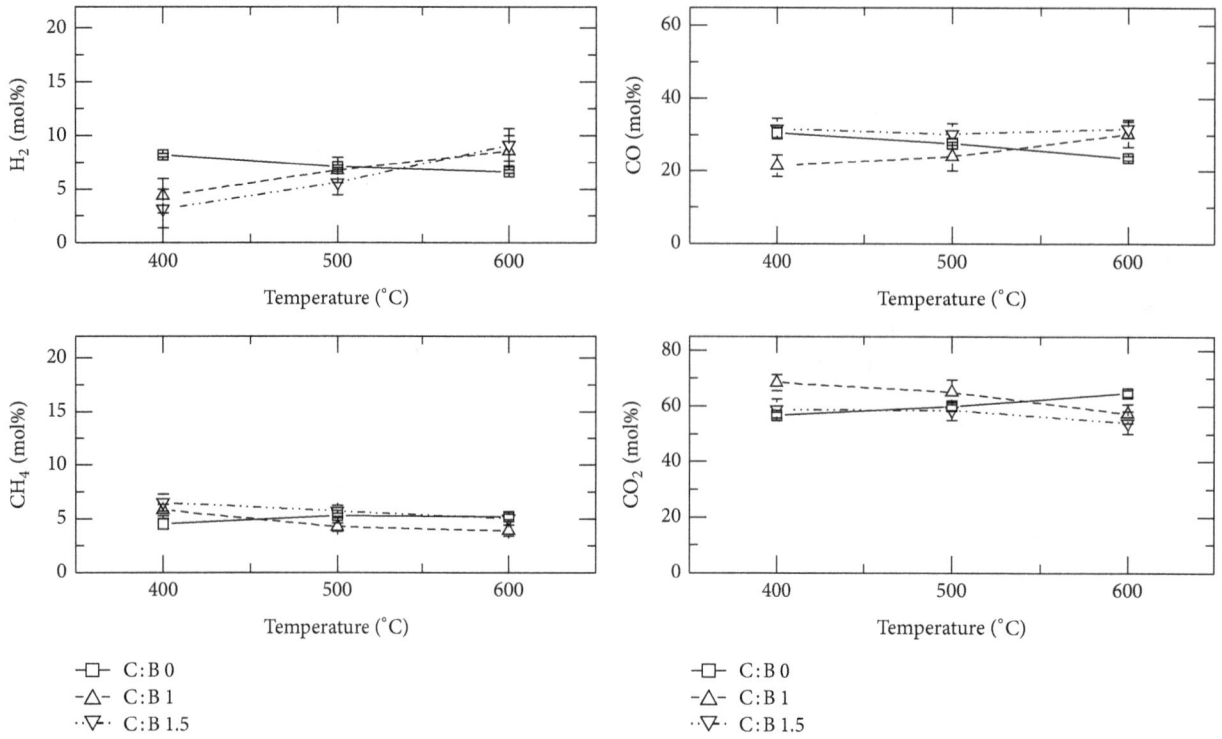

FIGURE 4: Effect of catalyst ratio on gas composition at different temperatures.

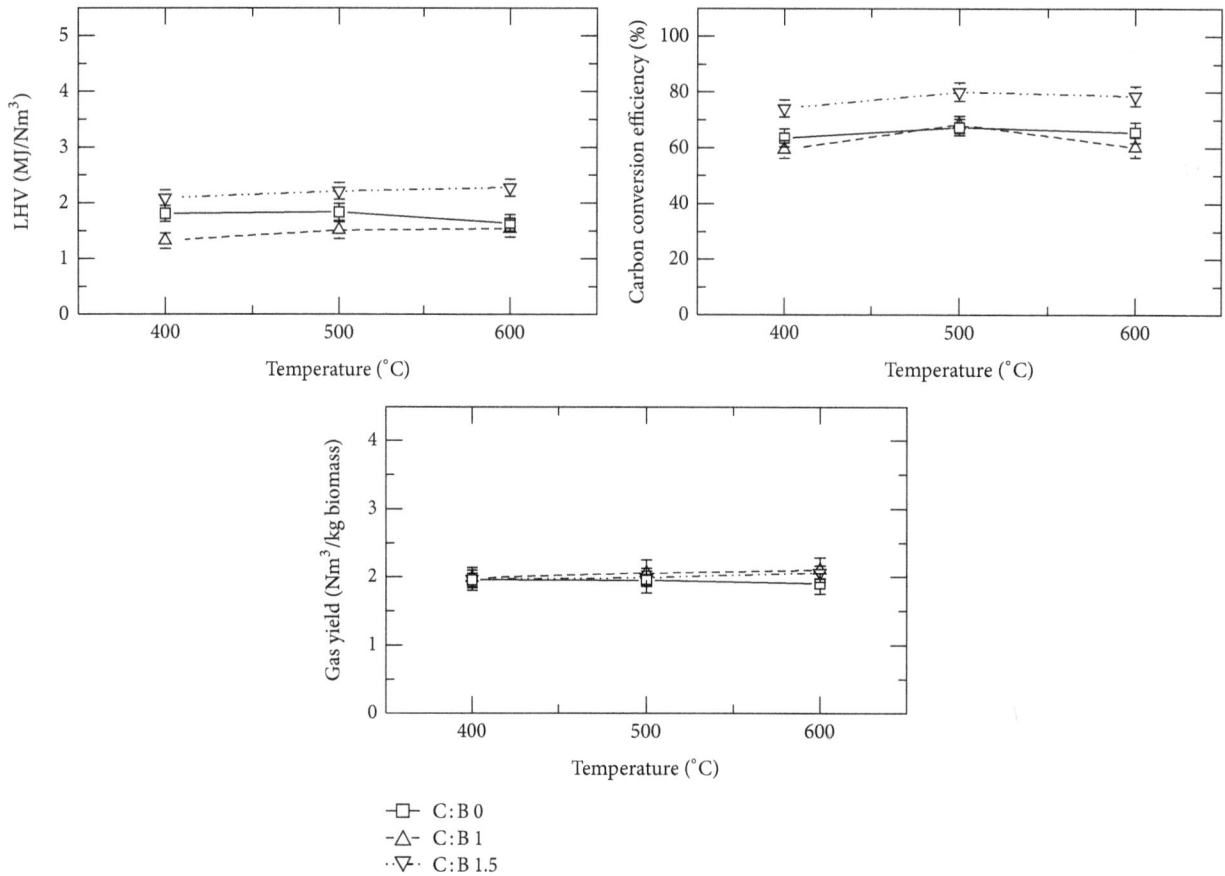

FIGURE 5: Effect of catalyst ratio on LHV, carbon conversion efficiency, and gas yield at different temperatures.

TABLE 2: Comparison with the existing literature.

| Ref | Biomass | C : B | S/B | Agent | ER | T (°C) | (% mol) | | H₂/CO | Gas yield (Nm³/kg) | CCE (%) |
							H₂	CO			
This work	Bamboo	—	—	Air	0.4	400–600	6.6–8.16	23.5–30.6	0.25–0.28	1.9–2.0	63.6–67.4
		1:1–1:1.5	—				3.2–9.1	21.4–31.7	0.1–0.3	1.9–2.1	59.5–80.1
		—	1:1	Air-steam			10.9–16.5	36.1–40.3	0.3–0.4	2.8–2.9	87.3–98.5
[1]	Pine sawdust	3%	0.75–0.9	Air-steam	0.22–0.3	800	35.87–40.67	22.71–29.88	1.2–1.79	1.5–2.3	—
[45]	Rice husk	—	—	Air	0.25–0.35	665–1103	4.0–3.3	19.9–12.3	0.20–0.26	1.3–1.9	55–81
[46]	Sewage sludge	10, 15%	—	Air	0.3	750–850	9.9–15.2	6.1–14.4	1.05–1.62	2.5–2.8	69.9–88.6
		10%	0.5, 1	Air-steam		800	15.7–16.8	7.5–9.1	1.7–2.2	2.7–2.8	82.5–83.0
[39]	Pine sawdust	—	2.81:1	Air-steam	0.22	700–900	21.0–38.0	33.0–43.0	0.5–1.2	1.4–2.5	78.2–92.6
[47]	Coir pith						5.6–10.6	6.5–12.7	0.7–1.0	1.9–3.2	—
	Rice husk	—	—	Air	0.3–0.5	400–900	5.3–8.6	9.3–19.6	0.4–0.67	1.8–2.8	—
	Saw dust						5.2–10.1	10.78–16.84	0.41–0.58	2.2–3.7	—
[11]	Cellulose	—	—	Air	0.2	600–1000	5.29–9.5	12.13–23.6	0.44–1.46	0.3–1.2	12.13–23.6
			0–1.5	Air-steam	0.27	800	13.5–18.6	6.4–11.2	1.2–2.2	0.8–1.0	—

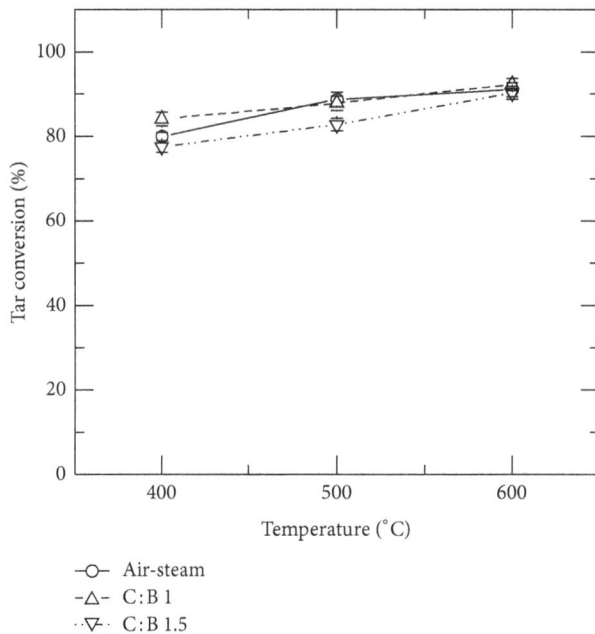

FIGURE 6: Effect of steam and catalyst ratio on tar conversion at different temperatures.

similar magnitude, but CO appeared to be higher than other biomass materials. The H₂ content obtained from air/steam gasification of bamboo was lower than that from pine sawdust but was in a similar range to those from cellulose and sewage sludge. Regarding the CO content, bamboo was similar to pine sawdust but higher than cellulose and sewage sludge [1, 11, 39, 46]. As far as catalytic gasification was concerned, H₂ from bamboo showed a similar magnitude to that from sewage sludge but was lower than pine sawdust [1, 46]. H₂/CO was found to be lower than those from the literature. Gas yields and carbon conversion efficiency obtained from

bamboo gasification were similar to other biomass materials but higher than cellulose.

4. Conclusions

In this study, the investigation of low-temperature gasification of bamboo in a fluidized bed reactor has been carried out. The effects of temperature, gasifying medium, and catalyst to biomass ratio on gas composition, LHV, and carbon conversion efficiency of fuel gas were evaluated. The results showed that the content of H₂ and CO in the fuel gas decreased with increasing temperature while the content of CO₂ increased. Added steam was found to enhance the quality of fuel gas, showing higher contents of H₂, CO, and LHV. The presence of a catalyst was found to increase the amount of H₂ and CO in fuel gas at higher temperature, whereas the opposite trends were true for CO₂.

Acknowledgments

The authors would like to acknowledge the support from the Thailand Research Fund (Contract no. RSA5080010) and the National Research Council of Thailand. Dr. Suphatz Sukamolson was thanked for assistance in preparing the manuscript.

References

[1] P. Lv, J. Chang, T. Wang, Y. Fu, Y. Chen, and J. Zhu, "Hydrogen-rich gas production from biomass catalytic gasification," *Energy and Fuels*, vol. 18, no. 1, pp. 228–233, 2004.

[2] L. Wei, S. Xu, L. Zhang, C. Liu, H. Zhu, and S. Liu, "Steam gasification of biomass for hydrogen-rich gas in a free-fall reactor," *International Journal of Hydrogen Energy*, vol. 32, no. 1, pp. 24–31, 2007.

[3] S. Panigrahi, A. K. Dalai, S. T. Chaudhari, and N. N. Bakhshi, "Synthesis gas production from steam gasification of biomass-derived oil," *Energy and Fuels*, vol. 17, no. 3, pp. 637–642, 2003.

[4] M. K. Ko, W.-Y. Lee, S.-B. Kim, K.-W. Lee, and H.-S. Chun, "Gasification of food waste with steam in fluidized bed," *The Korean Journal of Chemical Engineering*, vol. 18, no. 6, pp. 961–964, 2001.

[5] E. Kantarelis, J. Liu, W. Yang, and W. Blasiak, "Sustainable valorization of bamboo via high-temperature steam pyrolysis for energy production and added value materials," *Energy and Fuels*, vol. 24, no. 11, pp. 6142–6150, 2010.

[6] T. Wongsiriamnuay and N. Tippayawong, "Thermogravimetric analysis of giant sensitive plants under air atmosphere," *Bioresource Technology*, vol. 101, no. 23, pp. 9314–9320, 2010.

[7] B. Acharya, A. Dutta, and P. Basu, "An investigation into steam gasification of biomass for hydrogen enriched gas production in presence of CaO," *International Journal of Hydrogen Energy*, vol. 35, no. 4, pp. 1582–1589, 2010.

[8] J. Li, Y. Yin, X. Zhang, J. Liu, and R. Yan, "Hydrogen-rich gas production by steam gasification of palm oil wastes over supported tri-metallic catalyst," *International Journal of Hydrogen Energy*, vol. 34, no. 22, pp. 9108–9115, 2009.

[9] T. Wongsiriamnuay and N. Tippayawong, "Product gas distribution and composition from catalyzed gasification of mimosa," *International Journal of Renewable Energy Research*, vol. 2, no. 3, pp. 363–368, 2012.

[10] B. Moghtaderi, "Effects of controlling parameters on production of hydrogen by catalytic steam gasification of biomass at low temperatures," *Fuel*, vol. 86, no. 15, pp. 2422–2430, 2007.

[11] A. C. C. Chang, H.-F. Chang, F.-J. Lin, K.-H. Lin, and C.-H. Chen, "Biomass gasification for hydrogen production," *International Journal of Hydrogen Energy*, vol. 36, no. 21, pp. 14252–14260, 2011.

[12] C. Li and K. Suzuki, "Tar property, analysis, reforming mechanism and model for biomass gasification-An overview," *Renewable and Sustainable Energy Reviews*, vol. 13, no. 3, pp. 594–604, 2009.

[13] Q.-Z. Yu, C. Brage, T. Nordgreen, and K. Sjöström, "Effects of Chinese dolomites on tar cracking in gasification of birch," *Fuel*, vol. 88, no. 10, pp. 1922–1926, 2009.

[14] A. K. Dalai, E. Sasaoka, H. Hikita, and D. Ferdoust, "Catalytic gasification of sawdust derived from various biomass," *Energy and Fuels*, vol. 17, no. 6, pp. 1456–1463, 2003.

[15] M. Asadullah, S.-I. Ito, K. Kunimori, M. Yamada, and K. Tomishige, "Biomass gasification to hydrogen and syngas at low temperature: novel catalytic system using fluidized-bed reactor," *Journal of Catalysis*, vol. 208, no. 2, pp. 255–259, 2002.

[16] K.-Y. Chiang, Y.-S. Chen, W.-S. Tsai, C.-H. Lu, and K.-L. Chien, "Effect of calcium based catalyst on production of synthesis gas in gasification of waste bamboo chopsticks," *International Journal of Hydrogen Energy*, vol. 37, no. 18, pp. 13737–13745, 2012.

[17] G. Akay and C. A. Jordan, "Gasification of fuel cane bagasse in a downdraft gasifier: influence of lignocellulosic composition and fuel particle size on syngas composition and yield," *Energy and Fuels*, vol. 25, no. 5, pp. 2274–2283, 2011.

[18] S. Dasappa, H. V. Sridhar, G. Sridhar, P. J. Paul, and H. S. Mukunda, "Biomass gasification—a substitute to fossil fuel for heat application," *Biomass and Bioenergy*, vol. 25, no. 6, pp. 637–649, 2003.

[19] G. Xiao, M.-J. Ni, H. Huang et al., "Fluidized-bed pyrolysis of waste bamboo," *Journal of Zhejiang University A*, vol. 8, no. 9, pp. 1495–1499, 2007.

[20] W. K. Yu, K. F. Chung, and S. L. Chan, "Axial buckling of bamboo columns in bamboo scaffolds," *Engineering Structures*, vol. 27, no. 1, pp. 61–73, 2005.

[21] Y. Li, H. Shen, W. Shan, and T. Han, "Flexural behavior of lightweight bamboosteel composite slabs," *Thin-Walled Structures*, vol. 53, pp. 83–90, 2012.

[22] T. Y. Lo, H. Z. Cui, P. W. C. Tang, and H. C. Leung, "Strength analysis of bamboo by microscopic investigation of bamboo fibre," *Construction and Building Materials*, vol. 22, no. 7, pp. 1532–1535, 2008.

[23] K. Ghavami, "Bamboo as reinforcement in structural concrete elements," *Cement and Concrete Composites*, vol. 27, no. 6, pp. 637–649, 2005.

[24] W. H. Cheung, S. S. Y. Lau, S. Y. Leung, A. W. M. Ip, and G. McKay, "Characteristics of chemical modified activated carbons from bamboo scaffolding," *The Chinese Journal of Chemical Engineering*, vol. 20, no. 3, pp. 515–523, 2012.

[25] K. K. H. Choy, J. P. Barford, and G. McKay, "Production of activated carbon from bamboo scaffolding waste—process design, evaluation and sensitivity analysis," *Chemical Engineering Journal*, vol. 109, no. 1, pp. 147–165, 2005.

[26] A. W. M. Ip, J. P. Barford, and G. McKay, "Production and comparison of high surface area bamboo derived active carbons," *Bioresource Technology*, vol. 99, no. 18, pp. 8909–8916, 2008.

[27] Q.-S. Liu, T. Zheng, P. Wang, and L. Guo, "Preparation and characterization of activated carbon from bamboo by microwave-induced phosphoric acid activation," *Industrial Crops and Products*, vol. 31, no. 2, pp. 233–238, 2010.

[28] S.-F. Lo, S.-Y. Wang, M.-J. Tsai, and L.-D. Lin, "Adsorption capacity and removal efficiency of heavy metal ions by Moso and Ma bamboo activated carbons," *Chemical Engineering Research and Design*, vol. 90, no. 9, pp. 1397–1406, 2011.

[29] E. L. K. Mui, W. H. Cheung, M. Valix, and G. Mckay, "Activated carbons from bamboo scaffolding using acid activation," *Separation and Purification Technology*, vol. 74, no. 2, pp. 213–218, 2010.

[30] Z. Jiang, Z. Liu, B. Fei, Z. Cai, Y. Yu, and X. Liu, "The pyrolysis characteristics of moso bamboo," *Journal of Analytical and Applied Pyrolysis*, vol. 94, pp. 48–52, 2012.

[31] S.-H. Jung, B.-S. Kang, and J.-S. Kim, "Production of bio-oil from rice straw and bamboo sawdust under various reaction conditions in a fast pyrolysis plant equipped with a fluidized bed and a char separation system," *Journal of Analytical and Applied Pyrolysis*, vol. 82, no. 2, pp. 240–247, 2008.

[32] M. Krzesińska and J. Zachariasz, "The effect of pyrolysis temperature on the physical properties of monolithic carbons derived from solid iron bamboo," *Journal of Analytical and Applied Pyrolysis*, vol. 80, no. 1, pp. 209–215, 2007.

[33] R. Lou, S.-B. Wu, and G.-J. Lv, "Effect of conditions on fast pyrolysis of bamboo lignin," *Journal of Analytical and Applied Pyrolysis*, vol. 89, no. 2, pp. 191–196, 2010.

[34] E. L. K. Mui, W. H. Cheung, V. K. C. Lee, and G. McKay, "Compensation effect during the pyrolysis of tyres and bamboo," *Waste Management*, vol. 30, no. 5, pp. 821–830, 2010.

[35] K. Umeki, T. Namioka, and K. Yoshikawa, "The effect of steam on pyrolysis and char reactions behavior during rice straw gasification," *Fuel Processing Technology*, vol. 94, no. 1, pp. 53–60, 2012.

[36] G. Xiao, Y. Chi, M. Ni, and K. Cen, "Study on fluidized-bed gasification of bamboo," *Acta Energiae Solaris Sinica*, vol. 28, no. 7, pp. 814–818, 2007.

[37] N. Kannang, T. Wongsiriamnuay, and N. Tippayawong, "Fuel gas production from low temperature gasification of bamboo in

fluidized bed reactor," in *Proceedings of the International conference of the Thai Society of Agricultural Engineering*, Chiangmai, Thailand., 2012.

[38] G. Gautam, S. Adhikari, S. Thangalazhy-Gopakumar, C. Brodbeck, S. Bhavnani, and S. Taylor, "Tar analysis in syngas derived from pelletized biomass in a commercial stratified downdraft gasifier," *BioResources*, vol. 6, no. 4, pp. 4652–4661, 2011.

[39] P. M. Lv, Z. H. Xiong, J. Chang, C. Z. Wu, Y. Chen, and J. X. Zhu, "An experimental study on biomass air-steam gasification in a fluidized bed," *Bioresource Technology*, vol. 95, no. 1, pp. 95–101, 2004.

[40] S. H. Lee, K. B. Choi, J. G. Lee, and J. H. Kim, "Gasification characteristics of combustible wastes in a 5 ton/day fixed bed gasifier," *The Korean Journal of Chemical Engineering*, vol. 23, no. 4, pp. 576–580, 2006.

[41] Y. Zhang, S. Kajitani, M. Ashizawa, and Y. Oki, "Tar destruction and coke formation during rapid pyrolysis and gasification of biomass in a drop-tube furnace," *Fuel*, vol. 89, no. 2, pp. 302–309, 2010.

[42] V. Skoulou, G. Koufodimos, Z. Samaras, and A. Zabaniotou, "Low temperature gasification of olive kernels in a 5-kW fluidized bed reactor for H_2-rich producer gas," *International Journal of Hydrogen Energy*, vol. 33, no. 22, pp. 6515–6524, 2008.

[43] A. Montoya, T.-T. T. Truong, F. Mondragón, and T. N. Truong, "CO desorption from oxygen species on carbonaceous surface: 1. Effects of the local structure of the active site and the surface coverage," *Journal of Physical Chemistry A*, vol. 105, no. 27, pp. 6757–6764, 2001.

[44] H.-K. Seo, S. Park, J. Lee et al., "Effects of operating factors in the coal gasification reaction," *The Korean Journal of Chemical Engineering*, vol. 28, no. 9, pp. 1851–1858, 2011.

[45] K. G. Mansaray, A. E. Ghaly, A. M. Al-Taweel, F. Hamdullahpur, and V. I. Ugursal, "Air gasification of rice husk in a dual distributor type fluidized bed gasifier," *Biomass and Bioenergy*, vol. 17, no. 4, pp. 315–332, 1999.

[46] J. M. de Andrés, A. Narros, and M. E. Rodríguez, "Behaviour of dolomite, olivine and alumina as primary catalysts in air-steam gasification of sewage sludge," *Fuel*, vol. 90, no. 2, pp. 521–527, 2011.

[47] P. Subramanian, A. Sampathrajan, and P. Venkatachalam, "Fluidized bed gasification of select granular biomaterials," *Bioresource Technology*, vol. 102, no. 2, pp. 1914–1920, 2011.

[48] I. Narváez, A. Orío, M. P. Aznar, and J. Corella, "Biomass gasification with air in an atmospheric bubbling fluidized bed. Effect of six operational variables on the quality of the produced raw gas," *Industrial and Engineering Chemistry Research*, vol. 35, no. 7, pp. 2110–2120, 1996.

[49] T.-Y. Mun, P.-G. Seon, and J.-S. Kim, "Production of a producer gas from woody waste via air gasification using activated carbon and a two-stage gasifier and characterization of tar," *Fuel*, vol. 89, no. 11, pp. 3226–3234, 2010.

[50] P. Basu, *Biomass Gasification and Pyrolysis: Practical Design and Theory*, Academic Press, Boston, Mass, USA, 2010.

20

Canonical Analysis Technique as an Approach to Determine Optimal Conditions for Lactic Acid Production by *Lactobacillus helveticus* ATCC 15009

Marcelo Teixeira Leite,[1] Marcos Antonio de Souza Barrozo,[2] and Eloizio Júlio Ribeiro[2]

[1] Center of Technology and Regional Development, Federal University of Paraíba, 58051-900 João Pessoa, PB, Brazil
[2] Chemical Engineering College, Federal University of Uberlândia, 38408-100 Uberlândia, Brazil

Correspondence should be addressed to Marcelo Teixeira Leite, leitemarcelo@terra.com.br

Academic Editor: Iftekhar A. Karimi

The response surface methodology and canonical analysis were employed to find the most suitable conditions for *Lactobacillus helveticus* to produce lactic acid from cheese whey in batch fermentation. The analyzed variables were temperature, pH, and the concentrations of lactose and yeast extract. The experiments were carried out according to a central composite design with three center points. An empiric equation that correlated the concentration of lactic acid with the independent variables was proposed. The optimal conditions determined by the canonical analysis of the fitted model were $40°C$, pH 6.8, 82 g/L of lactose, and 23.36 g/L of yeast extract. At this point, the lactic acid concentration reached 59.38 g/L. A subsequent fermentation, carried out under optimal conditions, confirmed the product concentration predicted by the adjusted model. This concentration of lactic acid is the highest ever reported for *Lactobacillus helveticus* ATCC 15009 in batch process using cheese whey as substrate.

1. Introduction

1.1. Cheese Whey as a Pollutant. Cheese whey is the liquid remaining after the precipitation and removal of milk casein during cheese making. This byproduct represents 85–90% of the milk volume and retains 55% of milk nutrients. Among the most abundant of these nutrients are lactose (4.5–5.0% w/v), soluble proteins (0.6–0.8% w/v), lipids, and mineral salts [1]. It is produced in large amounts by the dairy industry. To produce 1 kg of cheese are produced, on average, 9 kg of whey.

Because it is a byproduct of low economic value, formerly the cheese whey was simply dumped in watercourses without any previous treatment. This caused the whey to become a serious environmental problem due to its high content of organic matter, with BOD = 30,000–50,000 mg/L and COD = 60,000–80,000 mg/L [2].

1.2. Cheese Whey as a Resource. The whey produced in the dairy industry has two destinations: disposal or utilization. Disposal here means launching over the field, pumping in

water courses, and treatment in effluent systems. In the first case, the use of whey as a fertilizer can over time unduly increase soil salinity. The pumping of untreated whey in watercourses is an environmentally incorrect solution and it is banned in several countries.

Although it solves the legal problems associated with the launch in watercourses, treating whey as an effluent increases production costs and brings no economic return to the producer [3, 4].

The recovery of nutrients from whey and its transformation into higher value-added compounds are better alternatives than the disposal. Thus, it is possible to have a parallel economic activity to reduce the polluting effect. The organic component available in the whey can be used in various ways.

The proteins can be separated by ultrafiltration and used as food supplement and raw material to manufacture nutritional products [5]. However, the recovery of the proteins little contributes to the decrease of the polluting load of the whey, consisted mainly of the lactose present in the permeated. Therefore, the study of the several possibilities to

Canonical Analysis Technique as an Approach to Determine Optimal Conditions for Lactic Acid Production by
Lactobacillus helveticus ATCC 15009

175

use this sugar is of great interest. One of the most promising alternatives is to use it as a source of low cost carbon, for the production of organic acids by fermentation. Most of the researches developed in the last years seek the production of lactic acid of high-added value, through the fermentation of the lactose present in the whey [6].

1.3. The Lactic Acid. Lactic acid is a versatile product that finds applications in several areas. It is used as acidulant in the food, cosmetics, and pharmaceutical industries [7]. Its isomers $L(+)$ and $D(-)$ can be polymerized to obtain compounds with different properties, depending on their intended application. In the medical area, the lactic acid has been used in the production of biodegradable polymers, used as scaffold in tissue transplants [8, 9]. The lactic acid polymers are also used in the production of biodegradable packages [10].

From the amount of lactic acid produced annually all over the world, about 90% is obtained by fermentation. The rest is synthesized starting from the hydrolysis of the lactonitrile. The production of lactic acid by fermentation presents advantages. One of them is the possibility to use renewable substrate, such as starch and cellulose. Another great advantage is that some strains can produce pure forms $L(+)$ or $D(-)$, while the chemical course always takes to the formation of a racemic mixture. The most commonly used substrates for the fermentative production of lactic acid are cheese whey, molasses, and starch. Before its use in fermentative processes, the whey must be deproteinized and demineralized. The separation of the proteins is usually made by ultrafiltration. Yeast extract, peptone, powder milk, and soybean meal are used as supplements in the fermentation of the whey. The microorganisms most frequently used are *Lactobacillus bulgaricus*, *Lactobacillus helveticus*, *Lactobacillus delbrueckii*, and *Lactobacillus casei* [11].

1.4. Canonical Analysis. Nowadays, most of the studies over the fermentative production of lactic acid have the process optimisation as a goal. The experiments are designed to find better media for growth and fermentation, besides optimum values for process variables such as pH and temperature. Several researchers have been using the response surface methodology (RSM) to achieve these goals. Hujanen et al. [12] studied the optimisation of the production of lactic acid using *Lactobacillus casei*. The optimum operating conditions were obtained at 35°C and pH 6.3. Téllez-Luis et al. [13] and Bustos et al. [10] studied the optimisation of a low cost fermentation medium, made up of corn steep liquor supplemented with peptone and yeast extract. Naveena et al. [14] achieved the optimisation of a solid medium made with wheat bran to produce lactic acid using *Lactobacillus amylophilus*. However, the authors of all these works did not use the canonical analysis to find the point of maximum response.

Canonical correlation analysis is a multivariate statistical model that facilitates the study of interrelationships among sets of multiple dependent variables and multiple independent variables. Whereas multiple regression predicts a single dependent variable from a set of multiple independent

TABLE 1: Coded and actual values of the central composite design.

	X_i				
	−1.55	−1	0	1	1.55
LC (g/L)	46.8	55	70	85	93.2
YE (g/L)	0	4.37	12.37	20.37	24.74
T (°C)	33.8	36	40	44	46.2
pH	4.5	5	6	7	7.5

variables; canonical correlation simultaneously predicts multiple dependent variables from multiple independent variables. Canonical correlation places the fewest restrictions on the types of data on which it operates. Because the other techniques impose more rigid restrictions, it is generally believed that the information obtained from them is of higher quality and may be presented in a more interpretable manner. For this reason, many researchers view canonical correlation as a last-ditch effort to be used when all other higher-level techniques have been exhausted. But in situations with multiple dependent and independent variables, canonical correlation is the most appropriate and powerful multivariate technique [15]. This technique has been used successfully in several studies on process optimization [16, 17]. It has gained acceptance in many fields and represents a useful tool for multivariate analysis, particularly as interest has spread to considering multiple dependent variables.

The objective of the present work was to use the response surface methodology and canonical analysis to find the most suitable conditions for *Lactobacillus helveticus* ATCC 15009 to produce lactic acid from cheese whey in batch fermentation. The influences of four process variables have been studied: temperature, pH, and the concentrations of lactose and yeast extract. *Lactobacillus helveticus* (ATCC 15009) has been chosen because it appears to be the most productive bacteria for lactic acid production from lactose [18]. The experiments were carried out according to a central composite design with three center points. The point of maximum response was obtained through a canonical analysis of the adjusted response surface [19].

2. Material and Methods

2.1. Inoculum and Fermentation Medium. *Lactobacillus helveticus* ATCC 15009 was supplied by the André Tosello Foundation (Campinas, SP, Brazil). Stock cultures were maintened on MRS broth [20] and deep frozen at −18°C. As required, these cultures were thawed and reactivated by two transfers in MRS broth (24 h, 37°C, 120 rpm). The cellular concentration of the inoculum was of 1.0×10^7 cells/mL.

The fermentation medium was made up of deproteinized reconstituted cheese whey, supplemented with yeast extract. The concentrations of lactose (LC) and yeast extract (YE) used are shown in Table 1.

2.2. Fermentations and Analyses. Batch fermentations were carried out at 150 rpm for 32 hours under anaerobic conditions in a 3.0 L fermentor NBS Bioflo 110 (New Brunswick Scientific, USA), in which 1800 mL of sterile culture medium was inoculated with 200 mL of seed culture at 1.0×10^7 cells/mL. The pH was maintained at desired values by automatic addition of 6.0 N NaOH.

Lactic acid concentration was determined by a lactate analyzer, Accutrend Lactate (Roche, Germany). In this equipment, the lactic acid is measured by reflectance photometry, after a reaction between lactate and lactate-oxidase. The lactose was quantified using the 3.5-dinitrosalicylic acid method [21]. Biomass measurements were estimated from OD readings at 650 nm. The samples were centrifuged, washed twice with deionized water, and then rediluted, before being introduced in the spectrophotometer. By means of a calibration curve (X = OD$_{650}$/2.35), OD readings were converted to grams of dry cell per mL.

2.3. Experimental Design. The experiments were carried out according to a central composite design with three center points, leading to a set of 27 experiments. The independent variables were lactose concentration (LC), yeast extract concentration (YE), temperature (T), and pH. The chosen value for the extreme level of the design was $\alpha = 1.55$. This value was selected in order to obtain an orthogonal design, where the variance and covariance matrix is diagonal and the estimated parameters are not correlated amongst themselves. The coded (dimensionless) variables were defined as

$$x_1 = \frac{LC(g/L) - 70\,g/L}{15\,g/L},$$
$$x_2 = \frac{YE(g/L) - 12.37\,g/L}{8\,g/L}, \tag{1}$$

$$x_3 = \frac{T(^\circ C) - 40\,^\circ C}{4\,^\circ C}, \tag{2}$$

$$x_4 = pH - 6, \tag{3}$$

where x_1, x_2, x_3, and x_4 are coded variables related to lactose concentration, yeast extract concentration, temperature, and pH, respectively. The levels used to code these variables are in Table 1. The variables range was selected on the basis of previous works [11].

2.4. Canonical Analysis. To quantify the effects of independent variables and their interactions, an empirical equation was fitted to the experimental data. This second order equation can be represented by matricial notation as

$$\hat{y} = b_0 + x'b + x'Bx, \tag{4}$$

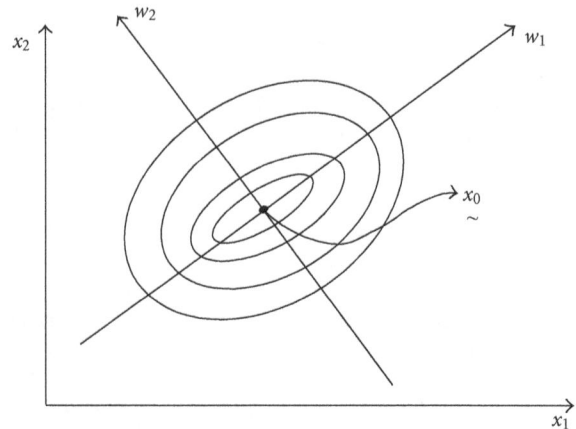

FIGURE 1: Graphical representation of axis translation.

where $\hat{y}$ is the predicted response, $\mathbf{x} = \begin{bmatrix} x_1 \\ x_2 \\ \vdots \\ x_k \end{bmatrix}$ is the independent variables vector, $\mathbf{b} = \begin{bmatrix} b_1 \\ b_2 \\ \vdots \\ b_k \end{bmatrix}$ is the parameters vector of the first order terms, and

$$B = \begin{bmatrix} b_{11} & \dfrac{b_{12}}{2} & \cdots & \dfrac{b_{1k}}{2} \\ \dfrac{b_{21}}{2} & b_{22} & \cdots & \dfrac{b_{2k}}{2} \\ \cdots & & \cdots & \cdots \\ & & \cdots & \dfrac{b_{k-1,k}}{2} \\ & & & b_{kk} \end{bmatrix} \tag{5}$$

is the parameters matrix of the quadratic terms.

The expected maximum for the response (lactic acid concentration in this case), if it exists, will be a set of conditions ($x_1, x_2, \ldots, x_k$), such that the derivatives $\partial \hat{y}/\partial x_1$, $\partial \hat{y}/\partial x_2, \ldots, \partial \hat{y}/\partial x_k$ are simultaneously zero. This value, say $\mathbf{x}_0' = [\mathbf{x}_{1,0}, \mathbf{x}_{2,0}, \ldots, \mathbf{x}_{k,0}]$, is the stationary point of the fitted surface. So, the derivative of $\hat{y}$ with respect to the vector $\mathbf{x}$, equated to zero, is

$$\frac{\partial \hat{y}}{\partial x} = \frac{\partial}{\partial x}\left[b_0 + x'b + x'Bx\right] = b + 2Bx = 0. \tag{6}$$

Thus the stationary point is

$$\mathbf{x}_0 = -B^{-1}\frac{\mathbf{b}}{2}. \tag{7}$$

Here, the stationary point $\mathbf{x}_0$ (7) can be a point at which the fitted surface attains a maximum or a minimum, or a saddle point. The analysis of the nature of the stationary point begins with a translation of the response function from the origin ($x_1 = 0, x_2 = 0, \ldots, x_k = 0$) to the $\mathbf{x}_0$. Figure 1 shows a schematic representation of axis translation for two variables.

Canonical Analysis Technique as an Approach to Determine Optimal Conditions for Lactic Acid Production by
Lactobacillus helveticus ATCC 15009

177

Due to the translation of the axes to stationary point $\mathbf{x_0}$, (4) should be written in terms of the vector, $\mathbf{z}$, given by $\mathbf{z} = \mathbf{x} - \mathbf{x_0}$:

$$\hat{y} = b_0 + (\mathbf{z'} + \mathbf{x'_0})\mathbf{b} + (\mathbf{z'} + \mathbf{x'_0})B(\mathbf{z} + \mathbf{x_0})$$
$$= b_0 + \mathbf{x'_0}\mathbf{b} + \mathbf{x'_0}B\mathbf{x_0} + \mathbf{z'}\mathbf{b} + \mathbf{z'}B\mathbf{x_0} + \mathbf{x'_0}B\mathbf{z} + \mathbf{z'}B\mathbf{z}. \quad (8)$$

Since $\mathbf{z'}B\mathbf{x_0} = \mathbf{x'_0}B\mathbf{z}$, and since the first three terms represent the response function evaluated at the stationary point, (8) is written as

$$\hat{y} = \hat{y}_0 + \mathbf{z'}(\mathbf{b} + 2B\mathbf{x_0}) + \mathbf{z'}B\mathbf{z} = y_0 + \mathbf{z'}B\mathbf{z}. \quad (9)$$

Equation (9) represents the response surface, translated to the new origin. Thus, there is an orthogonal transformation, $\mathbf{z} = M\mathbf{w}$, such that

$$\mathbf{z'}B\mathbf{z} = \mathbf{w'}M'BM\mathbf{w} = \lambda_1 w_1^2 + \lambda_2 w_2^2 + \cdots + \lambda_k w_k^2, \quad (10)$$

where M is the orthogonal matrix $k \times k$ ($M'M = I_k$) and $\lambda_1, \lambda_2, \ldots, \lambda_k$ are characteristic roots of the matrix B. The determination of matrix M is important because the transformation $\mathbf{w} = M'\mathbf{z}$ enables one to obtain the expression relating the original variables $\mathbf{z}_i$ (as a result $\mathbf{x}_i$, since $\mathbf{z} = \mathbf{x} - \mathbf{x_0}$) to the canonical one, $\mathbf{w}_i$. M is the matrix of eigenvector associated with the characteristic roots λ_i.

Therefore, based on (9) and (10), the response function can be expressed in terms of new variables $w_1, w_2, \ldots w_k$, whose axes correspond to the principal axes of the contour system. The form of the function in terms of these variables is called the *canonical form* and is given by

$$\hat{y} = \hat{y}_0 + \lambda_1 w_1^2 + \lambda_2 w_2^2 + \cdots + \lambda_k w_k^2. \quad (11)$$

The sign and magnitude of the characteristic roots λ_i can determine the nature of the stationary point. If $\lambda i < 0$, a move in any direction from the stationary point results in a decrease in $\hat{y}$ (see (11)). Therefore, the stationary point represents a point of maximum response for the fitted surface. On the other hand, $\lambda_i > 0$, $\mathbf{x_0}$ will be a minimum for the fitted surface. Finally, in the case where the λ's differ in sign, the stationary point is a saddle point.

In this work, the canonical analysis was accomplished through a routine implemented using the software *Maple 7*. The regression parameters of the model (4) were estimated using the software *Statistica 7*.

3. Results and Discussion

3.1. Experimental Results and Fitted Equation. Table 2 shows the central composite experimental design and the results obtained for lactic acid production. The lactic acid concentration varied in a wide range, from 0.45 g/L in experiment 26 to 56.88 g/L in experiment 12. The results obtained at the central level show good reproducibility.

To obtain the equation that describes the response (lactic acid concentration) as a function of the independent variables, hypothesis tests were performed using the t-Student statistics to identify the significant parameters. The parameters, estimated by the least-square method, with significance level greater than 10% ($P > 0.1$) were neglected.

TABLE 2: First central composite design and experimental results.

Run	LC (g/L)	YE (g/L)	T (°C)	pH	Lactic acid (g/L)
1	55.0	4.37	36.0	5.0	1.67
2	85.0	4.37	36.0	5.0	7.20
3	55.0	20.37	36.0	5.0	1.28
4	85.0	20.37	36.0	5.0	13.50
5	55.0	4.37	44.0	5.0	1.12
6	85.0	4.37	44.0	5.0	12.87
7	55.0	20.37	44.0	5.0	5.67
8	85.0	20.37	44.0	5.0	8.46
9	55.0	4.37	36.0	7.0	9.90
10	85.0	4.37	36.0	7.0	16.38
11	55.0	20.37	36.0	7.0	37.08
12	85.0	20.37	36.0	7.0	56.88
13	55.0	4.37	44.0	7.0	7.20
14	85.0	4.37	44.0	7.0	19.58
15	55.0	20.37	44.0	7.0	37.80
16	85.0	20.37	44.0	7.0	43.65
17	70.0	12.37	40.0	6.0	37.22
18	70.0	12.37	40.0	6.0	36.94
19	70.0	12.37	40.0	6.0	37.15
20	46.8	12.37	40.0	6.0	12.87
21	93.2	12.37	40.0	6.0	42.13
22	70.0	0	40.0	6.0	23.94
23	70.0	24.74	40.0	6.0	27.81
24	70.0	12.37	33.8	6.0	26.19
25	70.0	12.37	46.2	6.0	9.18
26	70.0	12.37	40.0	4.5	0.45
27	70.0	12.37	40.0	7.5	48.15

The fitted equation, in matricial notation, is given by ($R^2 = 0.92$)

$$\hat{y} = b_0 + \mathbf{x'}\mathbf{b} + \mathbf{x'}B\mathbf{x}, \quad (12)$$

where

$$\mathbf{x} = \begin{bmatrix} x_1 \\ x_2 \\ x_3 \\ x_4 \end{bmatrix}, \quad \mathbf{b} = \begin{bmatrix} 5.57 \\ 6.56 \\ 0 \\ 12.35 \end{bmatrix},$$

$$B = \begin{bmatrix} -3.41 & 0 & 0 & 0 \\ 0 & -2.99 & 0 & 3.44 \\ 0 & 0 & -7.51 & 0 \\ 0 & 3.44 & 0 & -4.75 \end{bmatrix}. \quad (13)$$

Vector $\mathbf{b}$ shows that the linear effect of temperature was not significant ($b_3 = 0$). The only significant interaction occurred between yeast extract concentration and pH ($b_{42}/2 = b_{24}/2 = 3.44$, matrix B). All quadratic

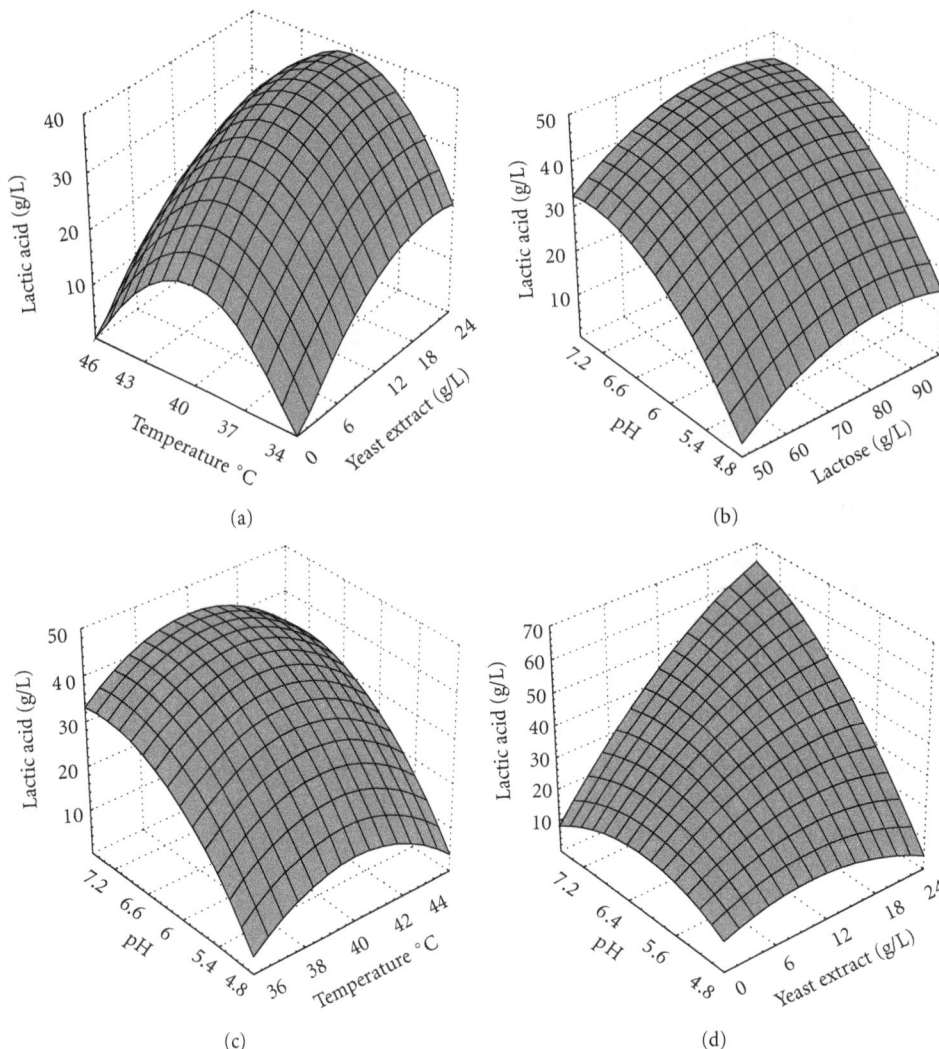

FIGURE 2: Response surfaces for the production of lactic acid as a function of the variables: (a) temperature and yeast extract concentration; (b) lactose concentration and pH; (c) temperature and pH; (d) yeast extract concentration and pH. For each surface, the other variables are in the center point.

effects, including the temperature, were significant ($b_{ii} \neq 0$, matrix B).

The influences of the independent variables and their interactions on the lactic acid concentration can be also analyzed through the response surfaces (Figure 2) obtained from (11). The production of lactic acid was maximum when the temperature was around 40°C ($x_3 = 0$), as it can be seen in Figures 2(a) and 2(c). The effect of temperature on the production of lactic acid by Lactobacilli has only been studied in a few reports [11]. Lactobacilli are mesophilic, which makes plausible the result obtained in this work, although some strains of *Lactobacillus helveticus* grow at 50–52°C [22]. The fact that the optimum temperature is close to 40°C can be considered acceptable when compared with the works available in the literature using cheese whey as substrate and strains of *Lactobacillus helveticus*. Tango and Ghaly [18], while studying the effect of temperature on product formation, performed experiments at 23°C, 37°C, and

42°C. The highest concentration of lactic acid was obtained at 42°C. Kulozik and Wilde [23] performed experiments at five levels of temperature: 35°C, 38°C, 42°C, 45°C, and 49°C. The product formation was slightly higher at 45°C than at 42°C. Norton et al. [24], Amrane and Prigent [25], and Schepers et al. [26] used a temperature of 42°C in their work, which corroborates the results obtained here, although they do not justify their choices.

Figure 2(b) shows that the lactic acid concentration was maximum when the lactose concentration was in the range of 70–85 g/L ($0 < x_1 < 1$). The response decreased for larger concentrations, probably due to the inhibition by the substrate. Several studies showed that cellular growth and lactic acid production by bacteria of the *Lactobacillus* genus are inhibited by the substrate [27, 28]. Tango and Ghaly [29] report that the growth of *Lactobacillus helveticus* and the production of lactic acid were inhibited for lactose concentrations greater than 80 g/L, which confirms

Canonical Analysis Technique as an Approach to Determine Optimal Conditions for Lactic Acid Production by
Lactobacillus helveticus ATCC 15009

179

the results of this work. However, Aeschlimann and von Stockar [30] published a similar study where there was no inhibition by the substrate for initial lactose concentrations smaller than 132 g/L.

The strong influence of the pH on the process can be seen in Figures 2(b) and 2(c). The amount of lactic acid produced when the pH was smaller than 6 ($x_4 < 0$) was very little. The major influence of pH over lactic acid production in fermentation processes is due to the fact that the catalytic activity of the enzymes and metabolic activity of the microorganism depend on extracellular pH. According to Hofvendahl and Hahn-Hägerdal [11] and Wood and Holzapfel [22], optimal pH for lactic acid production by lactic acid bacteria varies between 5.0 and 7.0 and is dependent on the strain.

For values of pH equal or larger than 6 ($x_4 > 0$), there was higher production of lactic acid, except when the yeast extract concentration was smaller than 12.37 g/L ($x_2 < 0$). That interaction between pH and yeast extract is shown in Figure 2(d). Even in favorable conditions of pH, low concentrations of yeast extract led to low product concentrations. This happened due to the complex nutritional demands of the genus *Lactobacillus* [22]. The source of nitrogen is the main factor influencing the growth of these microorganisms. Because the fermentative production of lactic acid is associated to the cellular growth, there is no product yield if the medium does not have an appropriate concentration of nitrogen to promote growth [31]. According to Figure 2(a), the optimum concentration of yeast extract seems to be in the range from 20.37 g/L to 24.74 g/L ($1 < x_2 < 1.55$). This result is in accordance with Amrane and Prigent [25]. They carried out a study on the influence of the yeast extract concentration on the homolactic fermentation of cheese whey, using *Lactobacillus helveticus*. The experiments used yeast extract concentrations in the range 2–30 g/L. According to the authors, larger concentrations of this substance led to the same final concentration of lactic acid; however, the fermentation time decreased significantly. In this work, the production of lactic acid was very low at concentrations of yeast extract smaller than 12.37 g/L, for the fermentation time chosen for that study. Several authors have been looking for low cost nitrogen sources substitutes for yeast extract. Corn steep liquor seems to be a viable alternative [32–34]. Fungi and yeasts [8, 35] and adapted strains of *Lactobacillus* [29] have also been used with the same purpose.

3.2. Canonical Analysis. The point of maximum lactic acid concentration was determined through a canonical analysis of the fitted model (12). The stationary point ($\mathbf{x_0}$) obtained from (6) is given by

$$\mathbf{x_0} = \begin{bmatrix} 0.82 \\ 15.54 \\ 0 \\ 12.56 \end{bmatrix}. \tag{14}$$

TABLE 3: Coded and actual values of variables in the second central composite design.

| | X_i | | | | |
	-1.40	-1	0	1	1.40
YE (yeast extract concentration, g/L)	18.36	20	24	28	29.64
pH	5.6	6	7	8	8.4

The characteristic roots (λ_i) of matrix B (11) are all negative: $\lambda_1 = -7.51$, $\lambda_2 = -7.42$, $\lambda_3 = -3.41$, $\lambda_4 = -0.32$. Therefore, $\mathbf{x_0}$ is a point of maximum response.

From stationary point (12), it can be seen that the optimum coded value obtained for lactose concentration was $x_1 = 0.82$; this value corresponds to 82.3 g/L (see (1)). There are no reports about the ideal concentration of this substrate in the fermentative production of lactic acid using *Lactobacillus helveticus*. In works published on this subject, initial concentrations were in the range 50–150 g/L. Increases of the initial concentration of lactose to values greater than 100 g/L implied in greater product concentrations [26]. However, they caused a decrease of conversion from substrate to product $Y_{P/S}$ and, consequently, an increase in the concentration of residual lactose. Therefore, the value found in the present work can be considered plausible. The optimum coded value obtained for temperature was $x_3 = 0$, corresponding to 40°C (2), confirming the value indicated by the response surface analysis (Figures 2(a) and 2(c)).

The optimum values of yeast extract concentration ($x_2 = 15.54$) and of pH ($x_4 = 12.56$) extrapolated the limits of the experimental design (-1.55 to $+1.55$). Hence, a new experimental design had to be performed to find the best conditions for these two variables. In this new experimental design, the lactose concentration and temperature were set at the best conditions, that is, 82 g/L and 40°C, respectively. The coded variables for this second central composite design were defined as

$$x_2 = \frac{\text{YE(g/L)} - 24\,\text{g/L}}{4\,\text{g/L}}, \tag{15}$$

$$x_4 = \text{pH} - 7.$$

The coded and actual levels of these variables are in Table 3. The results of the new experimental design are showed in Table 4.

As before, an empirical equation was fitted to this new set of experimental data. The fitted equation ($R^2 = 0.95$) in matricial notation is given by

$$\mathbf{x} = \begin{bmatrix} x_2 \\ x_4 \end{bmatrix}, \quad \mathbf{b} = \begin{bmatrix} 0 \\ -6.77 \end{bmatrix}, \quad B = \begin{bmatrix} -0.16 & -0.05 \\ -0.05 & -0.06 \end{bmatrix}. \tag{16}$$

The response surface corresponding to this new equation is shown in Figure 3. The production of lactic acid was very small both for pH greater than 8 ($x_4 > 1$) and for yeast extract concentrations larger than 28 g/L ($x_2 > 1$). On the other hand, the product concentration was maximum for yeast

TABLE 4: New central composite design and experimental results.

YE (g/L)	pH	Lactic acid (g/L)
20	6	42.15
28	6	26.32
20	8	6.80
28	8	39.82
24	7	58.23
24	7	59.14
18.36	7	48.73
29.64	7	38.04
24	5.6	28.02
24	8.4	5.20

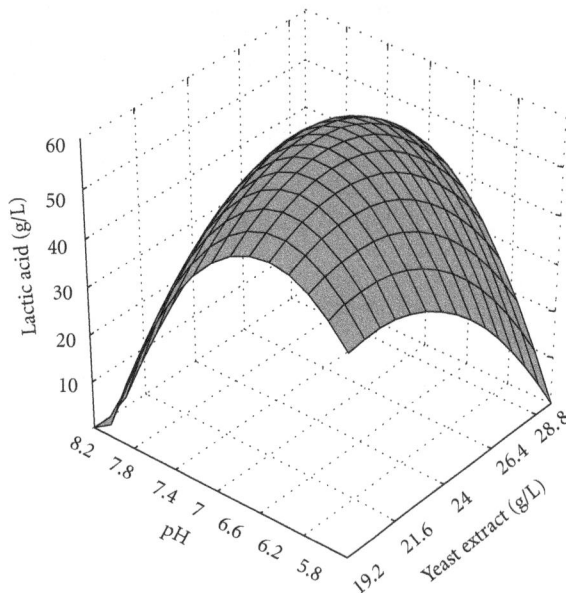

- ○ Biomass
- ◻ Lactose
- ▲ Lactic acid

FIGURE 4: Concentrations of product, substrate, and biomass as a function of the time, under the best conditions (lactose 82 g/L; yeast extract 23.36 g/L; temperature 40°C; pH 6.8).

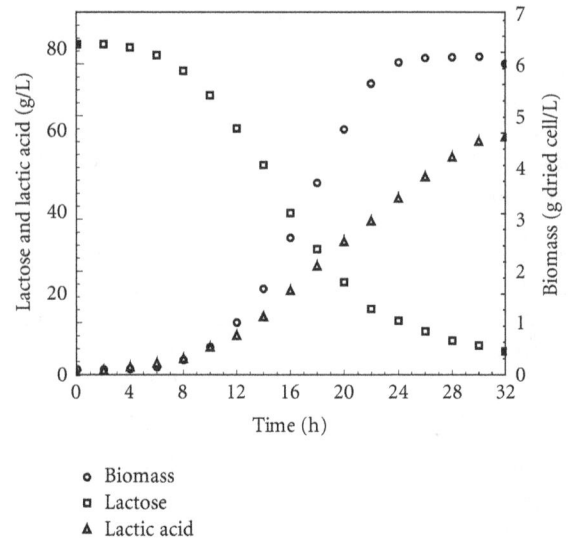

FIGURE 3: Response surface for lactic acid production as a function of the pH and of the yeast extract concentration ($T = 40°C$, lactose 82 g/L).

extract concentrations in the range 20–24 g/L ($0 < X_2 < 1$) and pH around 7 ($x_4 = 0$).

The point of maximum response was determined through a canonical analysis of the fitted model. The stationary point is given by

$$\mathbf{x_0} = \begin{bmatrix} -0.16 \\ -0.20 \end{bmatrix}. \tag{17}$$

The actual values of the coded independent variables were obtained from (15). The coded variables $x_2 = -0.16$ and $x_4 = -0.20$ correspond to 23.36 g/L of yeast extract concentration and pH 6.8, respectively. With the new stationary point, the canonical form of the fitted model is given by

$$Y = 59.38 - 23.79w_2^2 - 5.61w_4^2. \tag{18}$$

This equation shows that both characteristic roots (λ_i) have negative sign. This means that any displacement from

the stationary point causes a decrease of the response and, therefore, the stationary point is a maximum point. As a result, the optimum conditions determined by this study are temperature 40°C, pH 6.8, concentration of lactose 82 g/L, and concentration of yeast extract 23.36 g/L. Under these conditions, the canonical analysis predicts a production of acid lactic of 59.38 g/L.

A new fermentation was accomplished using the optimum operational conditions listed above. The result is shown in Figure 4. The concentration of lactic acid reached 59.15 g/L, confirming the result predicted by the canonical analysis. This value is the biggest ever reported in the literature for works that used cheese whey as substrate and strains of *Lactobacillus helveticus* ATCC 15009. The low residual concentration of the lactose, 5.75 g/L, shows that the substrate was almost totally consumed after 32 hours of fermentation. The cellular concentration reached the maximum value of 6.14 g/L. The factors of conversion from substrate into product $Y_{P/S}$ and from substrate into biomass $Y_{X/S}$ were 0.77 and 0.08, respectively.

4. Conclusions

This study demonstrated that the response surface methodology associated with the canonical analysis was appropriate to find the conditions that maximized the production of lactic acid from cheese whey fermentation. The optimal conditions determined by the canonical analysis were 40°C, pH 6.8, 82 g/L of lactose, and 23.36 g/L of yeast extract. At this point, the lactic acid concentration reached 59.15 g/L, the biggest value ever reported in studies using cheese whey as substrate and strains of *Lactobacillus helveticus* ATCC 15009 in batch fermentation.

The production of lactic acid was strongly linked to the pH and to the yeast extract concentration. The decrease of

Canonical Analysis Technique as an Approach to Determine Optimal Conditions for Lactic Acid Production by
Lactobacillus helveticus ATCC 15009

181

the response for higher lactose concentrations indicates an inhibition by the substrate.

The high concentration of yeast extract at the point of maximum response justifies the search for low cost nitrogen sources.

References

[1] S. T. Yang, H. Zhu, Y. Li, and G. Hong, "Continuous propionate production from whey permeate using a novel fibrous bed bioreactor," *Biotechnology and Bioengineering*, vol. 43, no. 11, pp. 1124–1130, 1994.

[2] A. J. Mawson, "Bioconversions for whey utilization and waste abatement," *Bioresource Technology*, vol. 47, no. 3, pp. 195–203, 1994.

[3] F. V. Kosikowski, "Whey utilization and whey products," *Journal of Dairy Science*, vol. 62, pp. 1149–1160, 1979.

[4] T. Sienkiewicz and C. L. Riedel, *Whey and Whey Utilization*, Verlag Th. Mann, Germany, 1990.

[5] M. I. González Siso, "The biotechnological utilization of cheese whey: a review," *Bioresource Technology*, vol. 57, no. 1, pp. 1–11, 1996.

[6] W. Fu and A. P. Mathews, "Lactic acid production from lactose by Lactobacillus plantarum: kinetic model and effects of pH, substrate, and oxygen," *Biochemical Engineering Journal*, vol. 3, no. 3, pp. 163–170, 1999.

[7] R. Datta, S. P. Tsai, P. Bonsignore, S. H. Moon, and J. R. Frank, "Technological and economic potential of poly(lactic acid) and lactic acid derivatives," *FEMS Microbiology Reviews*, vol. 16, no. 2-3, pp. 221–231, 1995.

[8] M. T. Gao, T. Shimamura, N. Ishida, and H. Takahashi, "Fermentative lactic acid production with a metabolically engineered yeast immobilized in photo-crosslinkable resins," *Biochemical Engineering Journal*, vol. 47, no. 1–3, pp. 66–70, 2009.

[9] X. Liu, L. Smith, G. Wei, Y. Won, and P. X. Ma, "Surface engineering of nano-fibrous poly(L-lactic acid) scaffolds via self-assembly technique for bone tissue engineering," *Journal of Biomedical Nanotechnology*, vol. 1, no. 1, pp. 54–60, 2005.

[10] G. Bustos, A. B. Moldes, J. L. Alonso, and M. Vázquez, "Optimization of D-lactic acid production by Lactobacillus coryniformis using response surface methodology," *Food Microbiology*, vol. 21, no. 2, pp. 143–148, 2004.

[11] K. Hofvendahl and B. Hahn-Hägerdal, "Factors affecting the fermentative lactic acid production from renewable resources," *Enzyme and Microbial Technology*, vol. 26, no. 2–4, pp. 87–107, 2000.

[12] M. Hujanen, S. Linko, Y. Y. Linko, and M. Leisola, "Optimisation of media and cultivation conditions for L(+)(S)-lactic acid production by *Lactobacillus casei* NRRL B-441," *Applied Microbiology and Biotechnology*, vol. 56, no. 1-2, pp. 126–130, 2001.

[13] S. J. Téllez-Luis, A. B. Moldes, J. L. Alonso, and M. Vázquez, "Optimization of lactic acid production by Lactobacillus delbrueckii through response surface methodology," *Journal of Food Science*, vol. 68, no. 4, pp. 1454–1458, 2003.

[14] B. J. Naveena, M. Altaf, K. Bhadrayya, S. S. Madhavendra, and G. Reddy, "Direct fermentation of starch to L(+) lactic acid in SSF by Lactobacillus amylophilus GV6 using wheat bran as support and substrate: medium optimization using RSM," *Process Biochemistry*, vol. 40, no. 2, pp. 681–690, 2005.

[15] J. F. Hair, R. L. Tatham, R. E. Anderson, and W. Black, *Multivariate Data Analysis*, Prentice Hall, New York, NY, USA, 5th edition, 1998.

[16] M. C. B. Fortes, A. A. M. Silva, R. C. Guimarães, C. H. Ataíde, and M. A. S. Barrozo, "Pre-separation of siliceous gangue in apatite flotation," *Industrial and Engineering Chemistry Research*, vol. 46, no. 21, pp. 7027–7029, 2007.

[17] R. C. Santana, A. C. C. Farnese, M. C. B. Fortes, C. H. Ataíde, and M. A. S. Barrozo, "Influence of particle size and reagent dosage on the performance of apatite flotation," *Separation and Purification Technology*, vol. 64, no. 1, pp. 8–15, 2008.

[18] M. S. A. Tango and A. E. Ghaly, "Effect of temperature on lactic acid production from cheese whey using Lactobacillus helveticus under batch conditions," *Biomass and Bioenergy*, vol. 16, no. 1, pp. 61–78, 1999.

[19] C. F. J. Wu and M. Hamada, *Experiments: Planning, Analysis, and Parameter Design Optimization*, John Wiley & Sons, New York, NY, USA, 2000.

[20] J. de Man, M. Rogosa, and M. Sharpe, "A medium for the cultivation of Lactobacilli," *Journal of Applied Bacteriology*, vol. 23, no. 1, pp. 130–135, 1960.

[21] G. L. Miller, "Use of dinitrosalicylic acid reagent for determination of reducing sugar," *Analytical Chemistry*, vol. 31, no. 3, pp. 426–428, 1959.

[22] B. J. B. Wood and W. H. Holzapfel, *The Genera of Lactic Acid Bacteria*, Blackie Academic & Professional, Glasgow, UK, 1995.

[23] U. Kulozik and J. Wilde, "Rapid lactic acid production at high cell concentrations in whey ultrafiltrate by Lactobacillus helveticus," *Enzyme and Microbial Technology*, vol. 24, no. 5-6, pp. 297–302, 1999.

[24] S. Norton, C. Lacroix, and J. C. Vuillemard, "Kinetic study of continuous whey permeate fermentation by immobilized Lactobacillus helveticus for lactic acid production," *Enzyme and Microbial Technology*, vol. 16, no. 6, pp. 457–466, 1994.

[25] A. Amrane and Y. Prigent, "Influence of yeast extract concentration on batch cultures of *Lactobacillus helveticus*: growth and production coupling," *World Journal of Microbiology and Biotechnology*, vol. 14, no. 4, pp. 529–534, 1998.

[26] A. W. Schepers, J. Thibault, and C. Lacroix, "*Lactobacillus helveticus* growth and lactic acid production during pH-controlled batch cultures in whey permeate/yeast extract medium. Part I. Multiple factor kinetic analysis," *Enzyme and Microbial Technology*, vol. 30, no. 2, pp. 176–186, 2002.

[27] C. N. Burgos-Rubio, M. R. Okos, and P. C. Wankat, "Kinetic study of the conversion of different substrates to lactic acid using *Lactobacillus bulgaricus*," *Biotechnology Progress*, vol. 16, no. 3, pp. 305–314, 2000.

[28] H. Mi-Young, S. Kim, Y. Lee, M. Kim, and S. Kim, "Kinetics analysis of growth and lactic acid production in pH-controlled batch cultures of Lactobacillus casei KH-1 using yeast extract/corn steep liquor/glucose medium," *Journal of Bioscience and Bioengineering*, vol. 96, no. 2, pp. 134–140, 2003.

[29] M. S. A. Tango and A. E. Ghaly, "Kinetic modeling of lactic acid production from batch submerged fermentation of cheese whey," *Transactions of the American Society of Agricultural Engineers*, vol. 42, no. 6, pp. 1791–1800, 1999.

[30] A. Aeschlimann and U. von Stockar, "The production of lactic acid from whey permeate by *Lactobacillus helveticus*," *Biotechnology Letters*, vol. 11, no. 3, pp. 195–200, 1989.

[31] G. G. Pritchard and T. Coolbear, "The physiology and biochemistry of the proteolytic system in lactic acid bacteria," *FEMS Microbiology Reviews*, vol. 12, no. 1–3, pp. 179–206, 1993.

[32] M. Altaf, B. J. Naveena, and G. Reddy, "Use of inexpensive nitrogen sources and starch for l(+) lactic acid production in anaerobic submerged fermentation," *Bioresource Technology*, vol. 98, no. 3, pp. 498–503, 2007.

[33] B. Gullón, J. L. Alonso, and J. C. Parajó, "Experimental evaluation of alternative fermentation media for L-lactic acid production from apple pomace," *Journal of Chemical Technology
and Biotechnology*, vol. 83, no. 5, pp. 609–617, 2008.

[34] L. Yu, T. Lei, X. Ren, X. Pei, and Y. Feng, "Response surface
optimization of l-(+)-lactic acid production using corn steep
liquor as an alternative nitrogen source by Lactobacillus rham-
nosus CGMCC 1466," *Biochemical Engineering Journal*, vol.
39, no. 3, pp. 496–502, 2008.

[35] V. Kitpreechavanich, T. Maneeboon, Y. Kayano, and K. Sakai,
"Comparative characterization of l-Lactic acid-producing
thermotolerant *Rhizopus Fungi*," *Journal of Bioscience and Bioengineering*, vol. 106, no. 6, pp. 541–546, 2008.

21

Biosorption Potential of *Trichoderma gamsii* Biomass for Removal of Cr(VI) from Electroplating Industrial Effluent

B. Kavita and Haresh Keharia

BRD School of Biosciences, Sardar Patel University, Vallabh Vidyanagar, Gujarat 388120, India

Correspondence should be addressed to Haresh Keharia, haresh970@gmail.com

Academic Editor: Jerzy Bałdyga

The potential use of acid-treated biomass of *Trichoderma gamsii* to remove hexavalent chromium ions from electroplating industrial effluent was evaluated. Electroplating industrial effluent contaminated with 5000 mg/L of Cr(VI) ions, collected from industrial estate of Gujarat, India, was mixed with acid-treated biomass of *T. gamsii* at biomass dose of 10 mg/mL. Effect of contact time and initial Cr(VI) ions was studied. The biosorption of Cr(VI) ions attained equilibrium at time interval of 240 minutes with maximum removal of 87% at preadjusted initial Cr(VI) concentration of 100 mg/L. The biosorption of Cr(VI) ions by biomass of *T. gamsii* increased as the initial Cr(VI) ion concentration of the effluent was adjusted in increasing range of 100–500 mg/L. At 500 mg/L, initial Cr(VI) concentration, acid-treated biomass of *T. gamsii* showed maximum biosorption capacity of 44.8 mg/g biomass from electroplating effluent. The Cr(VI) biosorption data were analysed using adsorption isotherms, that is, Freundlich and Langmuir isotherm. The correlation regression coefficients (R^2) and isotherm constant values show that the biosorption process follows Freundlich isotherm ($R^2 > 0.9$, $n > 1$, and $K_f = 8.3$). The kinetic study shows that biosorption of Cr(VI) ions by acid-treated biomass of *T. gamsii* follows pseudo-second-order rate of reaction at increasing concentration of Cr(VI). In conclusion, acid-treated biomass of *T. gamsii* can be used as biosorbent for Cr(VI) ions removal from Cr(VI)-contaminated wastewater generated by industries.

1. Introduction

Variety of anthropogenic sources including leather tanning, electroplating, wood preservation, metal finishing, pigment, and dye industries contribute towards hexavalent chromium in the environment [1–3]. The hexavalent chromium is classified in group A of human carcinogens by United State Environmental Agency (USEPA). Therefore, USEPA has regulated/limited the industrial discharge of Cr(VI) to surface water up to <0.05 mg/L.

Many conventional methods including chemical precipitation, chemical coagulation, ion exchange, electrochemical methods, adsorption using activated carbon and natural zeolite, membrane process, and ultrafiltration have been employed by several industries to remove Cr(VI) from their effluent [4–6]. However, these methods suffer from several disadvantages which include high operating cost, excess production of sludge, decrease in removal efficiency in presence of other metals, and large consumption of chemicals [7]. Hence, remediation of Cr(VI) demands some cost effective, economic, efficient, and eco-friendly methods.

In this context, biosorption is an emerging and attractive technology which is being worked out by many researchers since last two decades [8–10]. In general fungal cell walls are mainly 80–90% polysaccharides, with proteins, lipids, polyphosphates, and inorganic ions. Chitin is a common constituent of fungal cell walls. Chitin is a strong but flexible nitrogen containing polysaccharide consisting of N-acetyl-glucosamine residues. All these biopolymers offer many functional groups such as carboxyl, hydroxyl, sulphate, phosphate, and amino groups that can bind with several metal ions [11]. The ongoing research on Cr(VI) biosorption suggests that fungal biomass can passively bind metal ions via various physicochemical mechanisms or combination of several phenomena, namely, ion exchange, complexation, coordination, adsorption, electrostatic interaction, and

TABLE 1: Comparison of various fungal biomasses for Cr(VI) biosorption.

Fungal biosorbents	Maximum biosorption capacity (mg/g biomass)	Reference
Aspergillus sp.	1.56	[18]
Rhizopus arrhizus	8.40	[19]
Termitomyces clypeatus	11.1	[20]
Aspergillus niger	11.6	[21]
Rhizopus nigricans	12.70	[22]
Neurospora crassa	15.85	[23]
Rhizopus arrhizus	23.2	[24]
Mucor hiemalis	30.5	[25]

chelation [12, 13]. For example, biosorption of Cr(VI) by using *Rhizopus arrhizus* was found to be 23 mg/g biomass [14]. Similarly several other biomasses of fungal origin have been reported extensively for Cr(VI) biosorption (Table 1). The process of biosorption has gained importance over conventional methods due to several advantages like reusability of biomaterial, removal of heavy metal from effluent irrespective of toxicity, short operation time, and no secondary compound production [15]. However, reports on practical application of any of these biosorbents to Cr(VI) containing wastewater are sparse [13, 16].

Electroplating is one of the metal finishing process which contributes in discharging toxic level of Cr(VI) in environment. Apart from Cr(VI), electroplating industrial effluent contains many other metal ions which may affect the Cr(VI) biosorption efficiency of biomass. Therefore, it is of key importance to investigate the biosorption efficiency of biomass in contaminated effluent and effect of other parameters of effluent on biosorption process.

We have isolated a hexavalent chromium tolerant fungal culture, identified as *T. gamsii*, from a chromium contaminated soil. This isolate exhibited a very efficient biosorption capacity for hexavalent chromium from pure solutions (50.6 mg/g biomass). In the present study we have investigated the utility of *T. gamsii* biomass for removal of Cr(VI) from electroplating industrial effluent.

2. Materials and Methods

2.1. Chemicals and Fungal Strain. All the chemicals used were of analytical grade (AR) and purchased from either Qualigens Fine Chemicals, India, or Hi-Media Laboratories, India or Ranbaxy Fine Chemicals Limited, India.

A laboratory Cr(VI) tolerant fungal strain designated as FCR16 was used for present studies. FCR16 was grown and maintained on Potato Dextrose liquid/solidified medium as per the requirement.

FCR16 was identified by 18 s rDNA sequencing. The analysis of the nucleotide sequencing was carried out using Blast-n tool at NCBI (http://blast.ncbi.nlm.nih.gov/Blast.cgi). The phylogenetic tree was constructed by neighbour joining method using MEGA version 4.0 [17].

2.2. Preparation of Fungal Biomass for Cr(VI) Ion Biosorption. FCR16 was inoculated by transferring a block of fungal

TABLE 2: Physical and chemical characteristics waste water collected from effluent of electroplating industry.

Parameters	mg/L
COD	32
Ammonium nitrate	61.6
Total solids	10.30
Total dissolved salts	10.1
Chlorides	350
Sulphate	ND
Sulphides	ND
Nickel	32.8
Iron	10.31
Oil and grease	10.66
Hexavalent Cr(VI)	5,000
pH	1.5
Color	Yellow to brown

growth (16 mm diameter) on Potato Dextrose agar plates (grown for three days at 30°C) using sterile cup borer to 500 mL Erlenmeyer flasks filled with 200 mL of culture medium composed of the following (g/L): Potatoes infusion forms, 200; and Dextrose, 20 and incubated on to a rotary shaker at 150 rpm for five days at 30°C. Upon incubation, the biomass produced was separated by filtration and the resulting biomass was washed thoroughly for several times with distilled water. The biomass was treated with hydrochloric acid (6N), washed with distilled water to bring the pH of the biomass in neutral range, and then was used directly for Cr(VI) adsorption studies from electroplating industrial wastewater, considering the higher Cr(VI) biosorption capacity of acid-treated biomass of FCR16 in pure Cr(VI) solution.

2.3. Characteristics of Effluent Sample. Electroplating effluent was collected from an electroplating unit located in Industrial Estate of Vadodara, Makarpura, Gujarat, India. The characteristics of electroplating effluent are listed in Table 2. The major contaminants of wastewater were Cr(VI), Ni, and Fe. In addition, pH of the waste water was highly acidic that is, 1.5.

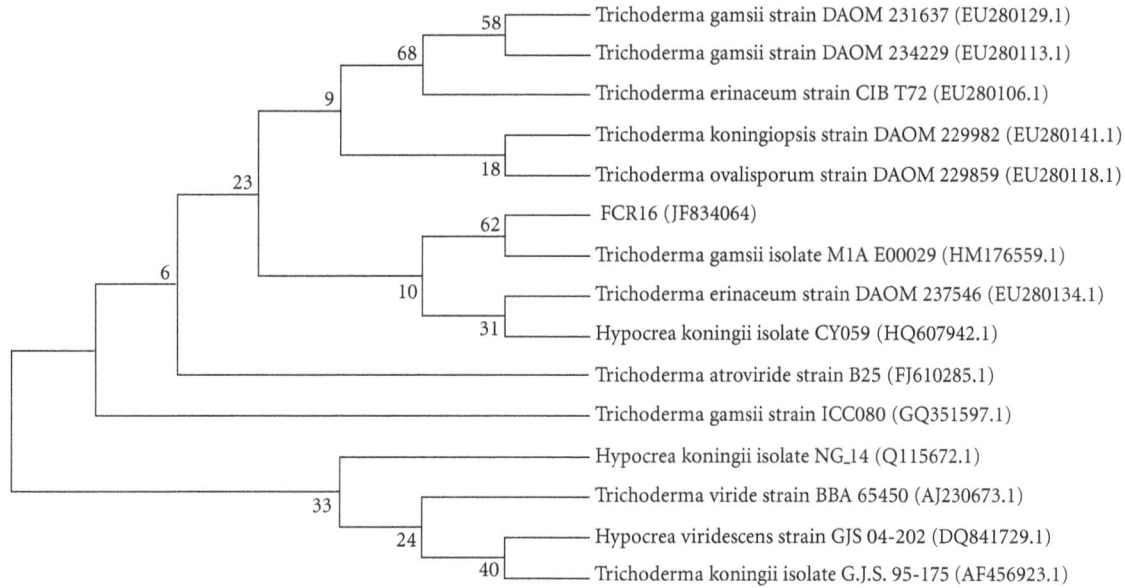

FIGURE 1: Phylogenetic affiliation based on 18 s rRNA gene sequence comparison over 561 nucleotides showing the relationship of FCR16 with other fungal strains. The value at node represents percentage of 1000 bootstrap replicates. Number in bracket represents GenBank accession numbers.

2.4. Batch Experiment. The biomass (at biomass dose of 10 mg/mL) of FCR16 was mixed with 100 mL of diluted effluent containing Cr(VI) in the range from 100 to 500 mg/L. Apart from Cr(VI) concentration adjustments, no other pretreatment was given to the effluent. After mixing, the experimental set was kept on shaker (150 rpm) at 30°C. 1 mL sample was withdrawn at regular time interval and residual Cr(VI) was measured.

2.5. Kinetics of Cr(VI) Biosorption. Pseudo-first-order and pseudo-second-order rate equations have been used for modelling the kinetics of Cr(VI) ion biosorption [26]. Linear form of pseudo-first-order rate equation is expressed as follows:

$$\log\left(q_{eq} - q_t\right) = \log q_{eq} - \frac{k_1 t}{2.303}, \qquad (1)$$

where q_t and q_{eq} are sorption capacity at time t and at equilibrium, respectively, and k_1 is pseudo-first-order rate constant.

In case the biosorption follows pseudo-first-order rate equation, a plot of $\log(q_{eq} - q_t)$ versus t should generate straight line with intercept of $\log q_{eq}$ and slope of $-k_1/2.303$.

Similarly, linear form of pseudo-second-order rate equation is expressed as

$$\frac{t}{q_t} = \frac{1}{k_2\left(q_{eq}\right)^2} + \frac{t}{q_{eq}}, \qquad (2)$$

where k_2 is pseudo-second-order rate constant.

In case the biosorption follows pseudo-second-order rate equation, a plot of t/q_t versus t should generate a straight line with intercept of $1/k_2 q_{eq}^2$ and slope of $1/q_{eq}$.

The shape (linearity) of graph and comparison of experimental and calculated q_{eq} values can help in deciding which kinetic model is followed by biosorption process. Another important factor which influences the kinetic model is the value of coefficient of determination: R^2. A value of $R^2 > 0.9$ shows the suitability of model for describing the kinetics.

2.6. Equilibrium Model for Cr(VI) Biosorption by FCR16 Biomass from Electroplating Wastewater. Biosorption data were analyzed using Langmuir and Freundlich equilibrium isotherms to determine the feasibility of Cr(VI) ion biosorption. The Freundlich isotherm equation is an empirical equation based on the biosorption on a heterogeneous surface suggesting that the binding sites are not equivalent or dependent [27]. Langmuir isotherm equation is based on monolayer sorption onto a surface with finite number of identical sites, which are homogeneously distributed over the sorbent surface [28].

2.7. Analysis of Cr(VI) Ions. The concentration of the Cr(VI) ions was determined spectrophotometrically after complexation of the Cr(VI) with 1, 5-diphenylcarbazide [29]. The absorbance was recorded at 540 nm and concentration was determined from the calibration curve.

Characterization of effluent was done according to standard methods described by APHA [30].

3. Results and Discussion

The Cr(VI) tolerant fungal strain designated as FCR16 was identified as *Trichoderma gamsii* with 99% similarity (accession number: JF834064). The phylogenetic relationship of FCR16 with other related fungal species is presented in Figure 1.

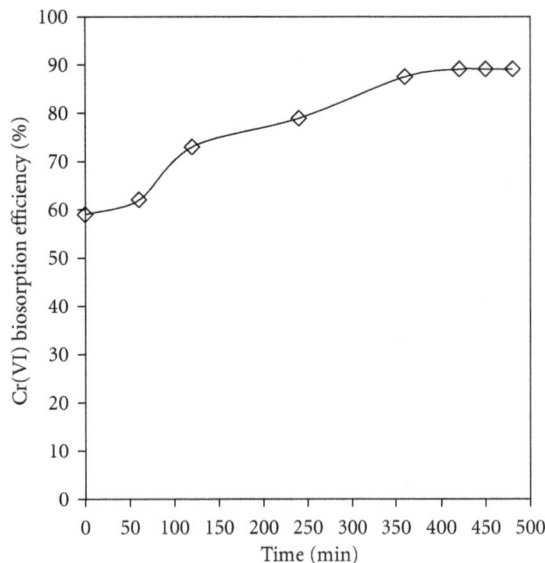

FIGURE 2: Time course for Cr(VI) removal by biosorption using *T. gamsii* biomass from electroplating industrial waste, diluted to a final Cr(VI) concentration of 100 mg/L.

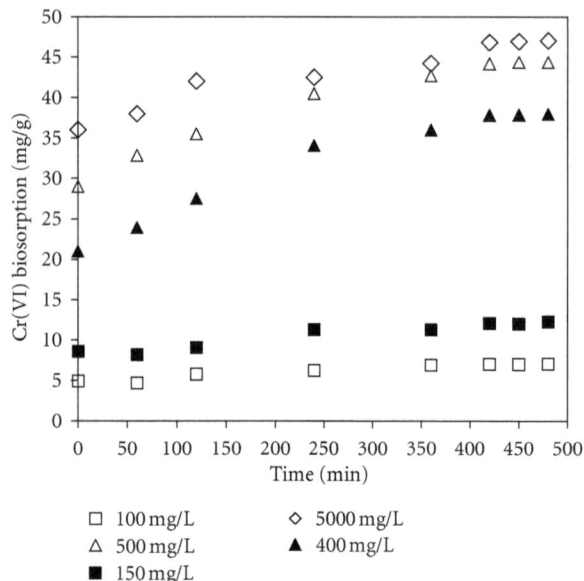

FIGURE 3: Cr(VI) biosorption (mg/g) by *T. gamsii* biomass from electroplating industrial waste diluted to final Cr(VI) concentration in the range from 100 to 500 mg/L.

3.1. Effect of Contact Time on Cr(VI) Ion Biosorption from Electroplating Industrial Effluent by Acid-Treated Biomass of *T. gamsii*.

Electroplating industrial effluent (pH: 1.5) containing 5000 mg/L of Cr(VI) ions was diluted (without any pretreatment of effluent) 50 times with distilled water to get the final concentration of 100 mg Cr(VI)/L.

Figure 2 shows the role of contact time on Cr(VI) biosorption using acid-treated biomass of *T. gamsii* at biomass dose of 10 mg/mL under shaking condition of 150 rpm. It was found that biosorption increased from 50 to 89% as the contact time was increased from 0 to 420 minutes. As illustrated, one gram of *T. gamsii* biomass could remove 89% of Cr(VI) ions at equilibrium. Metal biosorption is reported to be biphasic process, with rapid sorption of metal ions to the surface groups of the biomass constituting the first phase followed by a second phase during which diffusion of metal to internal binding sites on the biomass limits the sorption rate [31, 32]. Furthermore, the Cr(VI) biosorption depends on protonation and deprotonation of the cell wall polymer functional group relative to their pKa. At low pH, the protonation of functional group gives an overall positive charge to the fungal biomass, thereby leading to enhanced Cr(VI) biosorption. In the present study the acidic nature of the electroplating effluent and acid pretreatment of biomass together led to the significant Cr(VI) biosorption as demonstrated by higher Cr(VI) removal (%) in Figure 2. This biosorption efficiency was slightly lower than Cr(VI) ion removal efficiency of *T. gamsii* (50.6 mg/g biomass) from pure solution of Cr(VI) (data not shown). This may be attributed to competition between Cr(VI) and other metal ions present in electroplating effluent for the functional groups on the surface of biomass. Similar reduced Cr(VI) biosorption efficiency from electroplating industrial waste by *A. niger* has been reported by Kumar et al. [33].

3.2. Effect of Initial Cr(VI) Ion Concentration of Effluent on Biosorption.

The biosorption of Cr(VI) ions from electroplating effluent was carried out for 420 minutes at 150 rpm using acid-treated biomass (10 mg/mL) of *T. gamsii* with series of dilutions of effluent to get final Cr(VI) concentration in the range of 100 to 500 mg/L. It can be demonstrated from the experimental results that uptake capacity (q_{eq}, mg/g biomass) increased from 7.06 mg to 42.71 mg Cr(VI)/g acid-treated biomass of *T. gamsii* (Figure 3) when initial Cr(VI) concentration was increased from 100 to 500 mg/L, suggesting the increased propelling force provided by higher initial Cr(VI) ion concentration to overcome all mass transfer resistance of metal ions between the aqueous and solid phases, consequently, resulting in higher probability of collision between Cr(VI) ions and biosorbents [25, 34]. The Cr(VI) biosorption capacity of acid-treated *T. gamsii* biomass (42.71 mg/g) was comparable or better than other biosorbents reported for removal of Cr(VI) from electroplating effluent, namely, *Padina boergesenii* (49 mg/g), *Lentinus edodes* (21.5 mg/g), *C. lipolytica* (10 mg/L), and *A. niger* (65% from electroplating effluent contaminated with 47 mg/L Cr(VI)) [34–36].

Kinetic studies based on pseudo-second-order plot of t/q_t versus t (2) indicated that the biosorption of Cr(VI) ion followed pseudo-second-order rate of reaction in the Cr(VI) concentration range of 100 to 500 mg/L (Figure 4). The values of experimental/calculated equilibrium uptake capacities (q_{eqexp} and q_{eqcal}), correlation regression coefficient (R^2), and second-order rate constants (k_2) are presented in Table 3. The values of equilibrium uptake capacity increased (from 7.26 to 44.8 mg/g biomass) whereas second-order rate constant (k_2) was found to decrease (from 0.376 to 0.075) with increasing concentration of Cr(VI) ions (from 100 to

TABLE 3: Second-order kinetic parameters for biosorption of Cr(VI) by *T. gamsii* biomass at various dilutions of electroplating industrial wastewater.

Cr(VI), mg/L	q_{eqexp}, mg/g (experimental)	q_{eqcal}, mg/g (calculated)	k_2	R^2
100	7.064	7.26	0.376	0.992
150	12.09	12.28	0.218	0.991
400	37.85	39.06	0.054	0.988
500	42.71	44.8	0.075	0.995

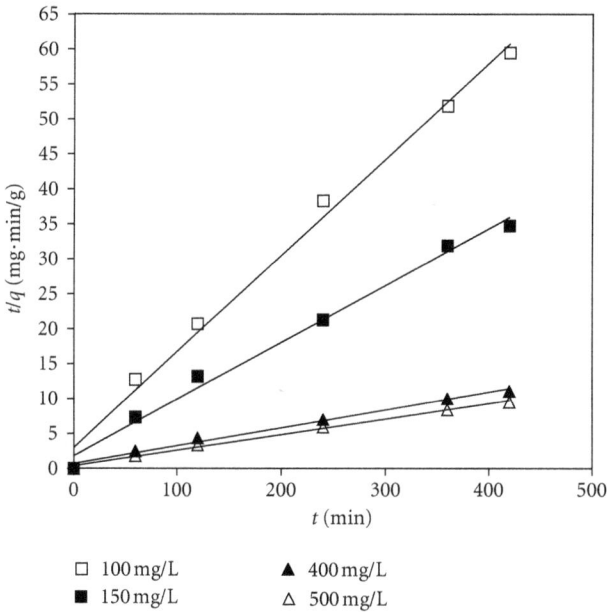

FIGURE 4: Linearized second-order kinetic plot of Cr(VI) biosorption by *T. gamsii* biomass at varying initial concentrations of Cr(VI).

TABLE 4: Isotherm parameters for Cr(VI) biosorption by *T. gamsii* biomass at various dilutions of electroplating industrial wastewater.

Freundlich isotherm constants			Langmuir isotherm constant		
n	K_f, mg/g	R^2	b	Q, mg/g	R^2
1.13	8.3	0.9423	0.068	133.33	0.5046

application to characterize the interaction of metal ions with biomass preparations [13]. The linearized plots of Freundlich and Langmuir isotherm model for biosorption of Cr(VI) ions from electroplating effluent by acid-treated biomass of *T. gamsii* are presented in Figure 5. It can be seen that R^2 value for the Freundlich isotherm is 0.9423 against the Langmuir isotherm R^2 value of 0.5046. Analysis of correlation regression coefficient shows that biosorption process fits better into Freundlich isotherm (Figure 5). The Langmuir and Freundlich adsorption constants calculated from the corresponding isotherms are presented in Table 4. The Freundlich isotherm constants k_f and n were calculated as 8.3 and 1.13, respectively. The high magnitude of k_f and n illustrates high adsorption capacity of biomass.

All these results showed that Freundlich isotherm model fitted the results quite well which are in agreement with the heterogeneity of sorbent (*T. gamsii* biomass) surface. Binding sites are not independent and adsorption energy of a metal binding site depends on whether or not the adjacent sites are already occupied. Thus, the adsorption of Cr(VI) ions by *T. gamsii* seems to be a complex process involving multilayer, interactive, or multiple-site type binding.

4. Conclusion

In conclusion, the present study provides the practical application of the *T. gamsii* biomass. Acid-treated biomass of *T. gamsii* is effective in removing Cr(VI) ions from acidic (pH 1.5) electroplating effluent contaminated with 5000 mg/L of Cr(VI) and other coexisting metal ions. At initial pH of electroplating effluent and biomass dose of 10 mg/mL, 89% of Cr(VI) ions were removed within 420 minutes of contact time. The biosorption of Cr(VI) ions increased with increasing contact time and initial Cr(VI) ion concentration. Kinetic model developed based on the values of equilibrium uptake capacity, correlation regression coefficient, and rate constants illustrated that the biosorption follows second-order rate of reaction. The Freundlich adsorption model was found to better describe the phenomenon of Cr(VI) biosorption onto acid-treated biomass of *T. gamsii*. Thus,

500 mg/L). This shows that the chromium sorption kinetics is strongly dependent on mass transfer phenomenon [25]. The rate of biosorption increases at slower rate compared to the increase in concentration due to sorption site saturation, which thus leads to the decrease in rate constant. The calculated uptake capacity values estimated from second-order kinetic model were in agreement to the experimental values. Additionally, correlation regression coefficients of pseudo-second-order model are quite high ($R^2 > 0.98$), very close to unity. Therefore, Cr(VI) ion biosorption by acid-treated biomass followed pseudo-second-order model. These observations are in agreement with the observations made by Ye et al. [37] where they have used *Candida lipolytica* and dewatered sewage sludge for biosorption of Cr(VI) ions from electroplating wastewater.

3.3. Adsorption Isotherms for Cr(VI) Ion Biosorption. The experimental values of equilibrium uptake capacities of Cr(VI) ions from electroplating effluent (Table 3) by acid-treated biomass of *T. gamsii* were analyzed by Freundlich and Langmuir isotherm models. Langmuir and Freundlich isotherms are single-solute adsorption isotherm models, which are widely used to analyze data for effluent treatment

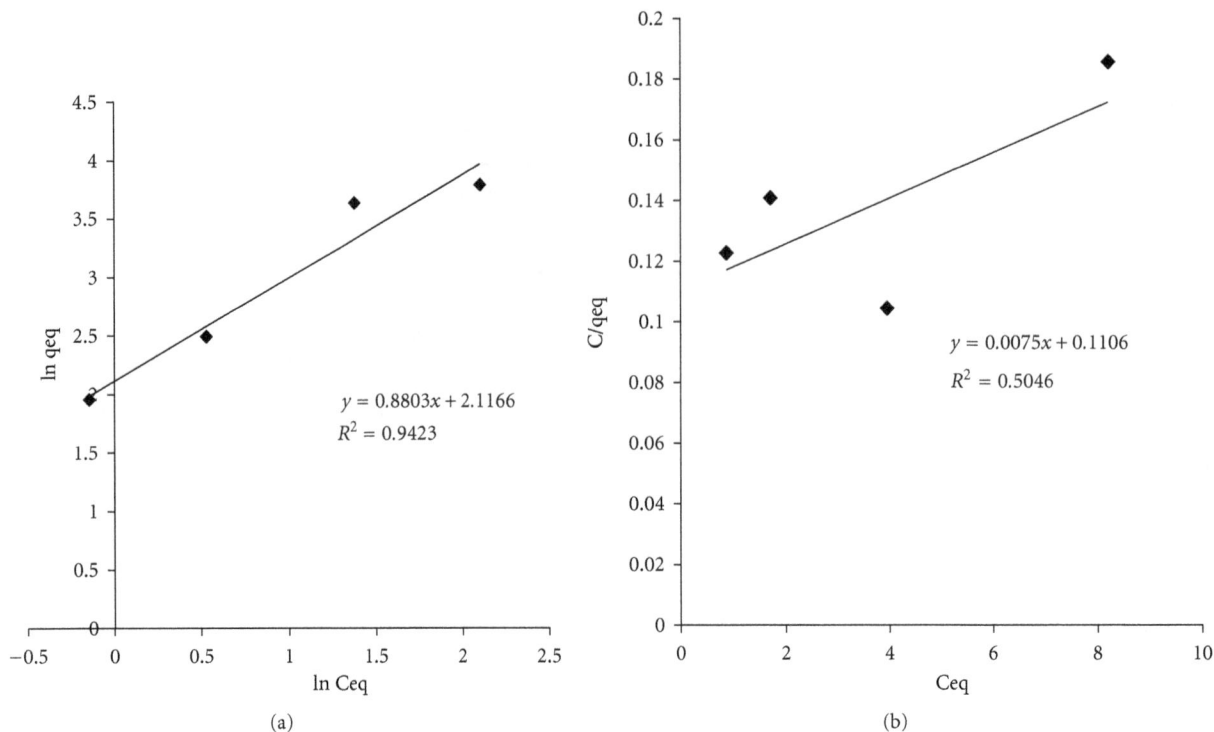

FIGURE 5: Biosorption isotherm: (a) Freundlich and (b) Langmuir isotherm for Cr(VI) biosorption by *T. gamsii* biomass at various dilutions of electroplating industrial waste water.

the results suggest the reasonable potential of acid-treated biomass of *T. gamsii* as sorbent for removal of Cr(VI) from electroplating effluents.

Acknowledgments

H. Keharia gratefully acknowledges Department of Science and Technology (DST), New Delhi, India, for financial assistance and B. Kavita is thankful to University Grant Commission (UGC), New Delhi, India, for meritorious fellowship.

References

[1] R. Aravindhan, B. Madhan, J. R. Rao, B. U. Nair, and T. Ramasami, "Bioaccumulation of chromium from tannery waste water: an approach for chrome recovery and reuse," *Environmental Science and Technology*, vol. 38, no. 1, pp. 300–306, 2004.

[2] C. Quintelas, B. Fonseca, B. Silva, H. Figueiredo, and T. Tavares, "Treatment of chromium(VI) solutions in a pilot-scale bioreactor through a biofilm of *Arthrobacter viscosus* supported on GAC," *Bioresource Technology*, vol. 100, no. 1, pp. 220–226, 2009.

[3] P. Suksabye, P. Thiravetyan, and W. Nakbanpote, "Column study of chromium(VI) adsorption from electroplating industry by coconut coir pith," *Journal of Hazardous Materials*, vol. 160, no. 1, pp. 56–62, 2008.

[4] F. J. Alguacil, M. Alonso, F. Lopez, and A. Lopez-Delgado, "Uphill permeation of Cr(VI) using Hostarex A327 as iono-phore by membrane-solvent extraction processing," *Chemosphere*, vol. 72, no. 4, pp. 684–689, 2008.

[5] C. H. Ko, P. J. Chen, S. H. Chen, F. C. Chang, F. C. Lin, and K. K. Chen, "Extraction of chromium, copper, and arsenic from CCA-treated wood using biodegradable chelating agents," *Bioresource Technology*, vol. 101, no. 5, pp. 1528–1531, 2010.

[6] M. M. Matlock, B. S. Howerton, and D. A. Atwood, "Chemical precipitation of heavy metals from acid mine drainage," *Water Research*, vol. 36, no. 19, pp. 4757–4764, 2002.

[7] R. Gupta, P. Ahuja, S. Khan, R. K. Saxena, and H. Mohapatra, "Microbial biosorbents: meeting challenges of heavy metal pollution in aqueous solutions," *Current Science*, vol. 78, no. 8, pp. 967–973, 2000.

[8] M. Izquierdo, C. Gabaldon, P. Marzal, and F. J. Alvarez-Hornos, "Modeling of copper fixed bed biosorption from waste water by *Posidonia oceanica*," *Bioresource Technology*, vol. 101, no. 2, pp. 510–517, 2010.

[9] T. Srinath, T. Verma, P. W. Ramteke, and S. K. Garg, "Chromium (VI) biosorption and bioaccumulation by chromate resistant bacteria," *Chemosphere*, vol. 48, no. 4, pp. 427–435, 2002.

[10] M. Spinti, H. Zhuang, and E. M. Trujillo, "Evaluation of immobilized biomass beads for removing heavy metals from wastewaters," *Water Environment Research*, vol. 67, no. 6, pp. 943–952, 1995.

[11] B. Volesky, "Biosorption and me," *Water Research*, vol. 41, no. 18, pp. 4017–4029, 2007.

[12] B. Volesky, "Detoxification of metal-bearing effluents: biosorption for the next century," *Hydrometallurgy*, vol. 59, no. 2-3, pp. 203–216, 2001.

[13] J. Wang and C. Chen, "Biosorbents for heavy metals removal and their future," *Biotechnology Advances*, vol. 27, no. 2, pp. 195–226, 2009.

[14] R. S. Prakasham, J. S. Merrie, R. Sheela, N. Saswathi, and S. V. Ramakrishna, "Biosorption of chromium VI by free and immobilized *Rhizopus arrhizus*," *Environmental Pollution*, vol. 104, no. 3, pp. 421–427, 1999.

[15] J. M. Modak and K. A. Natarajan, "Biosorption of metals using nonliving biomass—a review," *Minerals and Metallurgical Processing*, vol. 12, no. 4, pp. 189–196, 1995.

[16] K. Chojnacka, "Biosorption and bioaccumulation—the prospects for practical applications," *Environment International*, vol. 36, no. 3, pp. 299–307, 2010.

[17] K. Tamura, J. Dudley, M. Nei, and S. Kumar, "MEGA4: molecular evolutionary genetics analysis (MEGA) software version 4.0," *Molecular Biology and Evolution*, vol. 24, no. 8, pp. 1596–1599, 2007.

[18] S. Zafar, F. Aqil, and I. Ahmad, "Metal tolerance and biosorption potential of filamentous fungi isolated from metal contaminated agricultural soil," *Bioresource Technology*, vol. 98, no. 13, pp. 2557–2561, 2007.

[19] M. Nourbakhsh, Y. Sağ, D. Özer, Z. Aksu, T. Kutsal, and A. Çağlar, "A comparative study of various biosorbents for removal of chromium(VI) ions from industrial waste waters," *Process Biochemistry*, vol. 29, no. 1, pp. 1–5, 1994.

[20] S. K. Das and A. K. Guha, "Biosorption of hexavalent chromium by *Termitomyces clypeatus* biomass: kinetics and transmission electron microscopic study," *Journal of Hazardous Materials*, vol. 167, no. 1–3, pp. 685–691, 2009.

[21] S. Srivastava and I. S. Thakur, "Evaluation of bioremediation and detoxification potentiality of *Aspergillus niger* for removal of hexavalent chromium in soil microcosm," *Soil Biology and Biochemistry*, vol. 38, no. 7, pp. 1904–1911, 2006.

[22] R. S. Bai and T. E. Abraham, "Studies on chromium(VI) adsorption-desorption using immobilized fungal biomass," *Bioresource Technology*, vol. 87, no. 1, pp. 17–26, 2003.

[23] S. Tunali, I. Kiran, and T. Akar, "Chromium(VI) biosorption characteristics of *Neurospora crassa* fungal biomass," *Minerals Engineering*, vol. 18, no. 7, pp. 681–689, 2005.

[24] Z. Aksu and E. Balibek, "Chromium(VI) biosorption by dried *Rhizopus arrhizus*: effect of salt (NaCl) concentration on equilibrium and kinetic parameters," *Journal of Hazardous Materials*, vol. 145, no. 1-2, pp. 210–220, 2007.

[25] N. Tewari, P. Vasudevan, and B. K. Guha, "Study on biosorption of Cr(VI) by *Mucor hiemalis*," *Biochemical Engineering Journal*, vol. 23, no. 2, pp. 185–192, 2005.

[26] U. Farooq, J. A. Kozinski, M. A. Khan, and M. Athar, "Biosorption of heavy metal ions using wheat based biosorbents—a review of the recent literature," *Bioresource Technology*, vol. 101, no. 14, pp. 5043–5053, 2010.

[27] H. M. F. Freundlich, "Uber die adsorption in lasungen," *The Journal of Physical Chemistry*, vol. 57, pp. 385–470, 1906.

[28] I. Langmuir, "The constitution and fundamental properties of solids and liquids. Part I. Solids," *The Journal of the American Chemical Society*, vol. 38, no. 2, pp. 2221–2295, 1916.

[29] Anon, *"Metals" Standard Methods for Determination of Water and Waste Water*, American Public Health Association, Washington, DC, USA, 20th edition, 1998.

[30] A. D. Eaton, L. S. Clesceri, and A. E. Greenberg, *Standard Methods for the Examination of Water and Waste Water*, American Public Health Association, Washington, DC, USA, 1995.

[31] N. A. Adesola Babarinde, O. O. Oyesiku, and O. F. Dairo, "Isotherm and thermodynamic studies of the biosorption of copper (II) ions by *Erythrodontium barteri*," *International Journal of Physical Sciences*, vol. 2, no. 11, pp. 300–304, 2007.

[32] Y. Liu, X. Chang, Y. Guo, and S. Meng, "Biosorption and preconcentration of lead and cadmium on waste Chinese herb Pang Da Hai," *Journal of Hazardous Materials*, vol. 135, no. 1–3, pp. 389–394, 2006.

[33] R. Kumar, N. R. Bishnoi, Garima, and K. Bishnoi, "Biosorption of chromium(VI) from aqueous solution and electroplating wastewater using fungal biomass," *Chemical Engineering Journal*, vol. 135, no. 3, pp. 202–208, 2008.

[34] Y. Khambhaty, K. Mody, S. Basha, and B. Jha, "Biosorption of Cr(VI) onto marine *Aspergillus niger*: experimental studies and pseudo-second order kinetics," *World Journal of Microbiology and Biotechnology*, vol. 25, no. 8, pp. 1413–1421, 2009.

[35] G. Q. Chen, G. M. Zeng, X. Tu, C. G. Niu, G. H. Huang, and W. Jiang, "Application of a by-product of *Lentinus edodes* to the bioremediation of chromate contaminated water," *Journal of Hazardous Materials*, vol. 135, no. 1–3, pp. 249–255, 2006.

[36] E. Thirunavukkarasu and K. Palanivelu, "Biosorption of Cr(VI) from plating effluent using marine algal mass," *Indian Journal of Biotechnology*, vol. 6, no. 3, pp. 359–364, 2007.

[37] J. Ye, H. Yin, B. Mai et al., "Biosorption of chromium from aqueous solution and electroplating wastewater using mixture of Candida lipolytica and dewatered sewage sludge," *Bioresource Technology*, vol. 101, no. 11, pp. 3893–3902, 2010.

22

Gas-Solid Reaction Properties of Fluorine Compounds and Solid Adsorbents for Off-Gas Treatment from Semiconductor Facility

Shinji Yasui,[1] Tadashi Shojo,[2] Goichi Inoue,[2] Kunihiko Koike,[2] Akihiro Takeuchi,[3] and Yoshio Iwasa[3]

[1] *Nagoya Institute of Technology, Gokiso-cho, Showa-ku, Nagoya 466-8555, Japan*
[2] *Iwatani Corporation, 4-5-1 Katsube, Moriyama-shi, Shiga 524-0041, Japan*
[3] *Chubu Electric Power Co., Inc., 20-1 Kitasekiyama, Odaka-cho, Midori-ku, Nagoya 459-8522, Japan*

Correspondence should be addressed to Shinji Yasui, yasui.shinji@nitech.ac.jp

Academic Editor: Annabelle Couvert

We have been developing a new dry-type off-gas treatment system for recycling fluorine from perfluoro compounds present in off-gases from the semiconductor industry. The feature of this system is to adsorb the fluorine compounds in the exhaust gases from the decomposition furnace by using two types of solid adsorbents: the calcium carbonate in the upper layer adsorbs HF and converts it to CaF_2, and the sodium bicarbonate in the lower layer adsorbs HF and SiF_4 and converts them to Na_2SiF_6. This paper describes the fluorine compound adsorption properties of both the solid adsorbents—calcium carbonate and the sodium compound—for the optimal design of the fixation furnace. An analysis of the gas-solid reaction rate was performed from the experimental results of the breakthrough curve by using a fixed-bed reaction model, and the reaction rate constants and adsorption capacity were obtained for achieving an optimal process design.

1. Introduction

Fluorocarbons and perfluoro compounds (PFCs) contribute to global warming and are used in large quantities in the semiconductor industry, which must reduce the emission of these gases to the atmosphere in order to achieve the requirements of the Kyoto Protocol. In the semiconductor industry, voluntary reduction goals for PFCs were set at the World Semiconductor Council held in April 1999 and ongoing reduction efforts have been made. However, since hydrofluorocarbons (HFCs) and PFCs are used in critical processes, including chamber cleaning and etching during the manufacturing of semiconductors such as LCDs and solar panels, the consumption of these chemicals increases every year. Therefore, in spite of the reduction efforts, the emission of fluorocarbons and PFCs has increased in recent years.

The technology for the treatment of PFCs, which has been supplied in the semiconductor industries as an inexpensive process, is a combination of combustion to decompose PFCs into acid components such as HF and scrubbing to neutralize the acid components [1]. However, because the combustion method has a low decomposition ratio and the treatment of the resultant wastewater has a high energy load, a new dry treatment technology is desirable. Consequently, we are developing a new dry-type treatment technology for PFCs that consists of an electric furnace for decomposition of the PFCs and a dry-fixation furnace for adsorption of the acid components, which enables fluorine recycling. Fluorine is a precious resource for Japan, which imports most of its fluorine from China or Mexico [2]. This paper describes the details of the treatment process, the gas-solid reaction properties of fluorine compounds, and solid adsorbents in the dry-fixation furnace of the fluorine recycling system.

2. Dry-Type Treatment System

The processing flow of the treatment system being developed is shown in Figure 1. The treatment system is divided into three parts. In the semiconductor industry, silane (SiH_4) is used for Si deposition, and NF_3 is mainly used as the chamber-cleaning gas. These gases are used alternately in the

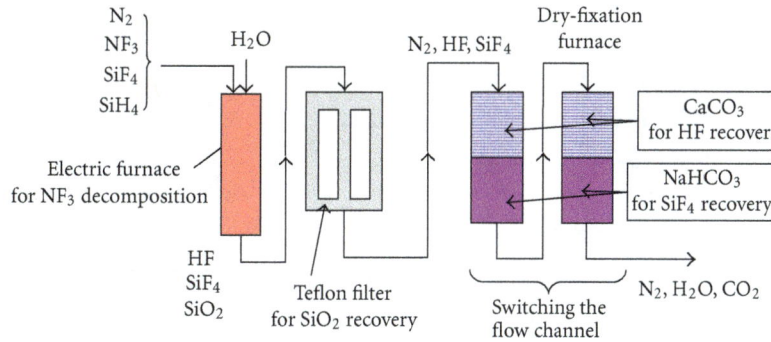

FIGURE 1: Schematic of the dry-type gas treatment system.

chamber; that is, the exhaust gases from the process chamber include SiH_4, NF_3, and SiF_4 diluted with N_2 through the vacuum pump. Initially in the treatment process, NF_3 and SiH_4 are decomposed in the electric furnace to form HF, SiF_4, and SiO_x. The SiO_x powder is then removed using a Teflon filter, and the remaining HF and SiF_4 are adsorbed by solid adsorbents in the subsequent dry-fixation furnace. The dry-fixation furnace is a switching system with two columns, each filled with an upper layer of calcium carbonate and a lower layer of sodium bicarbonate; the calcium carbonate in the upper layer adsorbs HF and converts it to CaF_2 and the sodium bicarbonate in the lower layer adsorbs HF and SiF_4 to convert them to Na_2SiF_6. These chemical reactions are shown as follows:

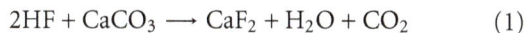

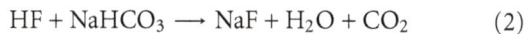

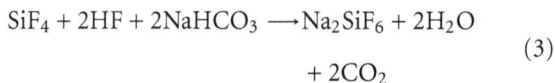

$$2HF + CaCO_3 \longrightarrow CaF_2 + H_2O + CO_2 \quad (1)$$

$$HF + NaHCO_3 \longrightarrow NaF + H_2O + CO_2 \quad (2)$$

$$SiF_4 + 2HF + 2NaHCO_3 \longrightarrow Na_2SiF_6 + 2H_2O$$
$$+ 2CO_2 \quad (3)$$

The vapor and CO_2 byproducts do not require further treatment. Because the Teflon filters have to be processed at low temperatures, less than 200°C, it is desirable to process the gas-solid reaction of each adsorbent at a low temperature of 150°C for negligible reheating system. The treatment flow rate in the developing system is 300 L/min (3% NF_3 and 0.5% SiH_4 in N_2) from emissions from the three chambers of the semiconductor-processing device.

The purpose of the dry-fixation furnace in this system is to generate high-purity calcium fluoride during the off-gas treatment. The reaction properties of each adsorbent are shown in Figure 2. The role of the sodium bicarbonate is to adsorb any HF gas breakthrough from the $CaCO_3$ layer by converting it to high-purity calcium fluoride and also to adsorb SiF_4 gas during the treatment process, because SiF_4 gas does not react with $CaCO_3$ at a low temperature of 150°C.

3. Fixed-Bed Reaction Model

In order to achieve the optimal design of the reaction vessel of the dry-fixation furnace, we investigated the reaction property of HF gas and the solid adsorbent of $CaCO_3$. We used the fixed-bed reaction model for analyzing this reaction property. By assuming that the distribution of the HF gas concentration as the gas moved from inlet to outlet in the $CaCO_3$ layer remained constant, the reaction rate of HF and $CaCO_3$ in the fixed-bed furnace can be estimated from the differential time of the HF concentration in the outlet gas breakthrough from the fixed-bed furnace.

Under the isothermal conditions in the fixed-bed furnace and by negligible volume change of the reactive gases, the mass balance of the HF component at the cross-sectional area in the fixed-bed furnace is given by the following equation using the symbols defined in Figure 3:

$$v\frac{\partial C_{HF}}{\partial x} + \varepsilon\frac{\partial C_{HF}}{\partial t} = -\frac{\partial \eta}{\partial t}. \quad (4)$$

The reaction zone of the HF gas and $CaCO_3$ is formed in the fixed-bed furnace, as shown in Figure 4, and the zone is moved forward to the outlet at the constant velocity of u_S. Under these conditions, the HF gas flowing into the furnace accumulates in the solid adsorbent, and the reaction zone is moved only by the amount of the accumulation. The mass balance at this time is represented by the following equation:

$$v \cdot C_{HF}^0 = u_S \cdot \eta^0. \quad (5)$$

In the steady state, the time change of the molar concentration of HF in the fixed-bed furnace is determined by the moving velocity of the HF concentration curve with a constant shape, and the relationship between them is represented by the following equation:

$$\frac{\partial C_{HF}}{\partial t} = -u_S\frac{\partial C_{HF}}{\partial x}. \quad (6)$$

The following equation is obtained from (4), (5), and (6):

$$\left(\varepsilon - \frac{\eta^0}{C_{HF}^0}\right)\frac{\partial C_{HF}}{\partial t} = -\frac{\partial \eta}{\partial t}. \quad (7)$$

The following relationship is satisfied in our fixed-bed furnace:

$$\varepsilon \ll \frac{\eta^0}{C_{HF}}. \quad (8)$$

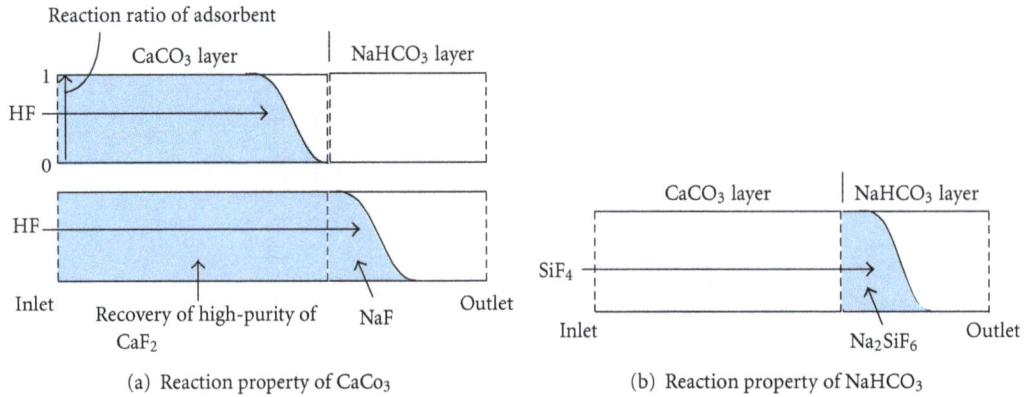

FIGURE 2: Schematic of the dry-type gas treatment system.

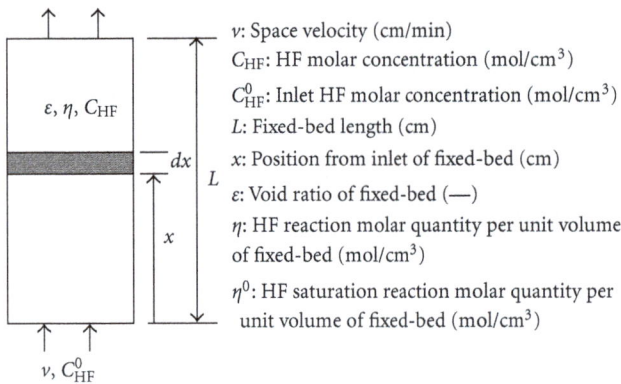

FIGURE 3: Description of material balance of fixed-bed reaction furnace.

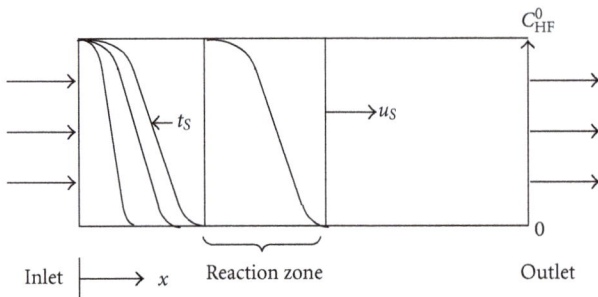

FIGURE 5: Breakthrough curve as compared with inside phenomenon on fixed-bed reactor.

By evaluating this relationship at the outlet part of the fixed-bed furnace, this equation can be expressed as ordinary differential equations:

$$\frac{d\left(C_{HF}/C_{HF}^0\right)}{dt} = \frac{d(\eta/\eta^0)}{dt}. \qquad (10)$$

Therefore, HF molar adsorption quantity per unit volume at the outlet of the fixed-bed furnace and the HF gas concentration in the gas exhausted from the fixed-bed furnace obtained by the experiments show the same changes for the reaction time, as shown in Figure 5. The ratio η/η^0 is the normalized fluoride ratio of $CaCO_3$. The fluoride ratio does not change even if the saturation amount is changed from the per unit volume to the per unit weight, so the following equation is obtained:

$$\frac{\eta}{\eta^0} = \frac{\eta_g}{\eta_g^0}, \qquad (11)$$

where η_g is the HF molar adsorption quantity per unit weight and η_g^0 is the HF adsorption capacity per unit weight of $CaCO_3$. Finally, the following relationship can be obtained from (10) and (11):

$$\frac{d(\eta/\eta^0)}{dt} = \frac{d\left(\eta_g/\eta_g^0\right)}{dt}. \qquad (12)$$

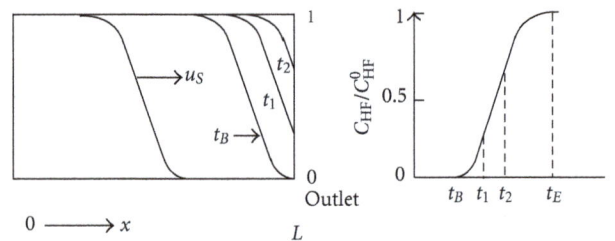

FIGURE 4: Inside phenomenon on the fixed-bed furnace.

From (7) and (8), the relationship between HF molar adsorption quantity per unit volume of fixed-bed furnace and the HF molar concentration can be expressed by the following equation:

$$\frac{\partial\left(C_{HF}/C_{HF}^0\right)}{\partial t} = \frac{\partial(\eta/\eta^0)}{\partial t}. \qquad (9)$$

TABLE 1: Experimental conditions for the gas-solid reaction.

	Practical system	Experimental conditions
Processing flow rate	400 L/min	1.6 L/min
Inner diameter	45 cm	2.8 cm
Superficial velocity	6 cm/s	6 cm/s
L/D ratio	1.2	1.2
$CaCO_3$ weight	120 kg	31 g
Space velocity (SV)	400 h^{-1}	6700 h^{-1}
Adsorbent size	1~2 mm, 2~5 mm	1~2 mm, 2~5 mm
Fixation temperature	150°C	150°C

In this research, the reaction rate of HF gas and the solid adsorbent $CaCO_3$ was evaluated by the differential value of the breakthrough curve of HF concentration in the exhaust gases obtained by the fundamental experiments. This fixed-bed reaction model was used in the analysis of the desulfurization catalyst quality [3, 4].

4. HF Breakthrough Properties of $CaCO_3$

In our previous studies, the reaction rates of HF gas and the solid adsorbent $CaCO_3$ were investigated in the fundamental experiments for the purpose of the recovering fluoride from the waste fluorocarbons [5, 6]. In these studies, the reaction conditions of HF gas concentration were very high, 35%–60%. In a semiconductor factory, the concentration of the cleaning gas NF_3 in the off-gas is less than about 5% at most, so the HF concentration in the exhaust gases from the electric decomposition furnace is assumed to be less than about 10%. Therefore, the reaction rate of HF and $CaCO_3$ was investigated in the low HF concentration conditions.

4.1. Experimental Conditions and Method. The important parameters for the design of reactors for gas-solid reactions are the particle size of the solid adsorbents, reaction temperature, and superficial velocity. The experimental conditions are consistent with the design specifications of these parameters for a practical system. The design specification and the experimental parameters are shown in Table 1.

The experiment flow is shown in Figure 6. The reaction gas, including HF gas, was obtained by thermal decomposition of HFC134a ($C_2H_2F_4$) with water vapor and diluted at the entrance of the fixed-bed reaction furnace. The fixed-bed reactor filled with calcium carbonate is made of stainless steel, which is heated uniformly in a vertical electric furnace. The temperature of the vertical electric furnace was set at 150°C. HF gas concentrations of 1%, 5%, and 10% and $CaCO_3$ grain sizes of 1-2 mm and 2–5 mm were investigated. The breakthrough properties for each experimental condition were investigated by measuring the concentration of F ions in the impinger. The concentration of F ions was measured by using a F^- ion sensor and ion chromatograph. The experimental conditions are summarized in Table 2.

4.2. Experimental Results. The results of the breakthrough properties are shown in Figure 7. The HF concentration in

the off-gas from the fixed-bed reactor increased gradually after the moment of breakthrough and finally reached the initial concentration at the entrance of the fixed-bed reactor. The results shown in the vertical axis in Figure 7 were normalized by the initial HF concentration at the entrance.

The elapsed time from the breakthrough point to the complete breakthrough time ($C_{HF}/C_{HF}^0 = 1$) increased under low HF concentration conditions. For a fixed-bed reactor, the differential of the breakthrough curve ($d(C_{HF}/C_{HF}^0)/dt$) can be regarded as equal to the reaction rate (dX/dt) of the $CaCO_3$ filling the outside edge of the fixed-bed reactor:

$$X = \frac{C_{HF}}{C_{HF}^0}, \quad (13)$$

where X is the reaction ratio of $CaCO_3$. Therefore, the slopes of the breakthrough curves of these results were evaluated and plotted against the HF concentration at the slope points, as shown in Figure 8. From these results, the reaction rate of $CaCO_3$ can be determined using a first-order reaction formula of the HF concentration.

The results of the reaction rates of $CaCO_3$ were investigated using the gas-solid reaction model of a shrinking unreacted core system. The reaction rate of a shrinking unreacted core system is expressed using the following equation [7]:

$$\frac{dX}{dt} = \frac{C_{HF}}{f_S^{-1} \cdot (1-X)^{-2/3} + f_p^{-1} \cdot \left[(1-X)^{-1/3} - 1\right] + f_g^{-1}}. \quad (14)$$

Here, f_S is the rate constant for the particle surface chemical reaction, f_p is the rate constant for diffusion through the product layer of a spherical particle, and f_g is the rate constant for the external mass transport. By evaluating the results of Figure 8 using (14), the overall reaction rate of $CaCO_3$ could be expressed in terms of the two rate constants: the rate constant for diffusion through the product layer (f_p) and the rate constant for external mass transport (f_g) as shown in Figure 9.

Evaluating the slope and the intercept of the straight lines in Figure 9, each rate constant was obtained as shown in Table 3.

By using each obtained rate constant, it was possible to calculate the elapsed time from the initial breakthrough point to the complete breakthrough point by setting the initial concentration of HF (C_{HF}^0) and the reaction ratio of $CaCO_3$ (e.g., 0.95). From this calculated time, the amount of HF gas leaking from the $CaCO_3$ layer until the $CaCO_3$ was converted into high-purity CaF_2 was calculated from the treatment flow rate. This indicates the amount of HF gas that needs to be adsorbed in the subsequent $NaHCO_3$ layer.

5. SiF_4 Adsorption Properties of Sodium-Based Adsorbents

The purpose of the $NaHCO_3$ layer is to adsorb the HF and SiF_4 exhausted from the $CaCO_3$ layer. To determine the amount of $NaHCO_3$ needed, it is necessary to calculate

TABLE 2: Experimental conditions of the gas-solid reaction of HF and $CaCO_3$.

Test no.	Flow rate of the treated gases						Pyrolysis gas conditions		Adsorbent			Fixation conditions			
	$C_2H_2F_4$ [mL/min]	Ar [mL/min]	H_2O [mL/min]	Air [mL/min]	N_2 [mL/min]	Total [mL/min]	HF flow rate [mL/min]	HF concentration [mol/m³]	Type	Weight [g]	Grain size [mm]	Reaction temperature [°C]	Space velocity [cm/s]	SV [h⁻¹]	L/D
1	5	300	50	100	1100	1555	20	0.574	$CaCO_3$	31	1~2	150	6.1	4408.4	1.2
2	20	300	75	150	1000	1595	80	2.239	$CaCO_3$	31	1~2	150	6.2	4295.5	1.2
3	40	300	150	300	700	1590	160	4.492	$CaCO_3$	31	1~2	150	6.2	4394.3	1.2
4	5	300	50	100	1100	1555	20	0.574	$CaCO_3$	31	2~5	150	6.1	4408.4	1.2
5	20	300	75	150	1000	1595	80	2.239	$CaCO_3$	31	2~5	150	6.2	4295.5	1.2
6	40	300	150	300	700	1590	160	4.492	$CaCO_3$	31	2~5	150	6.2	4394.3	1.2

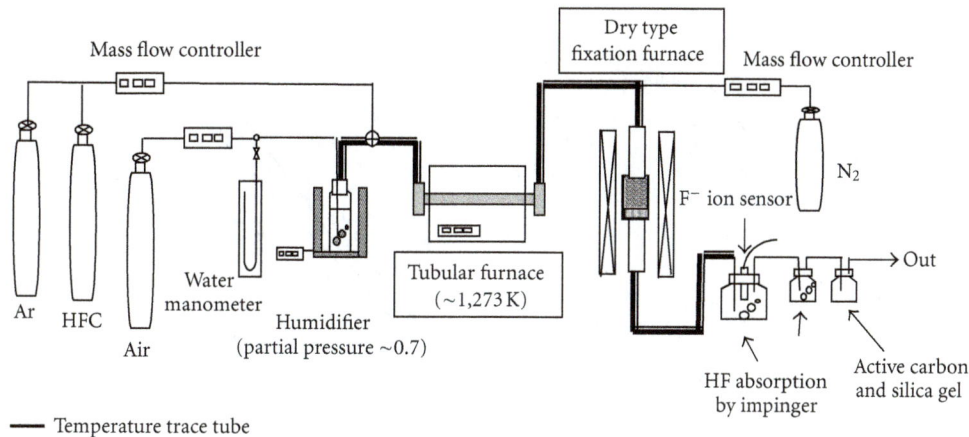

FIGURE 6: Experiment flow of the gas-solid reaction of HF and $CaCO_3$.

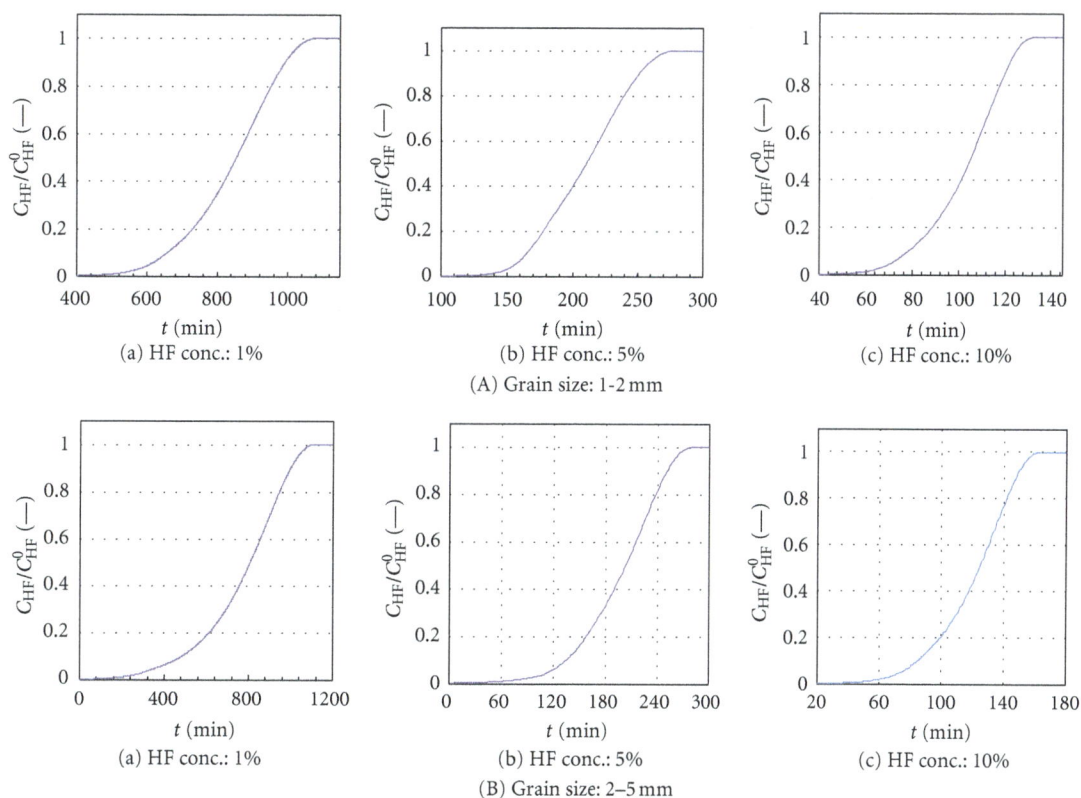

(a) HF conc.: 1%

(b) HF conc.: 5%

(c) HF conc.: 10%

(A) Grain size: 1-2 mm

(a) HF conc.: 1%

(b) HF conc.: 5%

(c) HF conc.: 10%

(B) Grain size: 2–5 mm

FIGURE 7: HF breakthrough properties of $CaCO_3$.

the adsorption capacity of each gas. By the fundamental experiments in our studies, the amount of $NaHCO_3$ needed for HF adsorption was confirmed to supply an equimolar amount of HF leaked from the $CaCO_3$ layer because the reaction of $NaHCO_3$ and HF at 150°C generate not only NaF and $NaHF_2$ but also HF can be completely adsorbed by an equimolar amount of $NaHCO_3$. Therefore, this section describes the results of the experiments for examination of the adsorption capacity of SiF_4.

5.1. Experimental Conditions and Method. The experimental flow is shown in Figure 10. At first, HF was generated by thermal decomposition of HFC134a ($C_2H_2F_4$) with water vapor and air; SiF_4 was generated by the reaction of the HF and pieces of quartz. Then, these reaction gases were introduced into the fixed-bed furnace with N_2 dilution gas to establish the SiF_4 concentration and then reacted with sodium-based adsorbents in the furnace. We used two types of sodium-based adsorbents: $NaHCO_3$ and NaF. NaF was

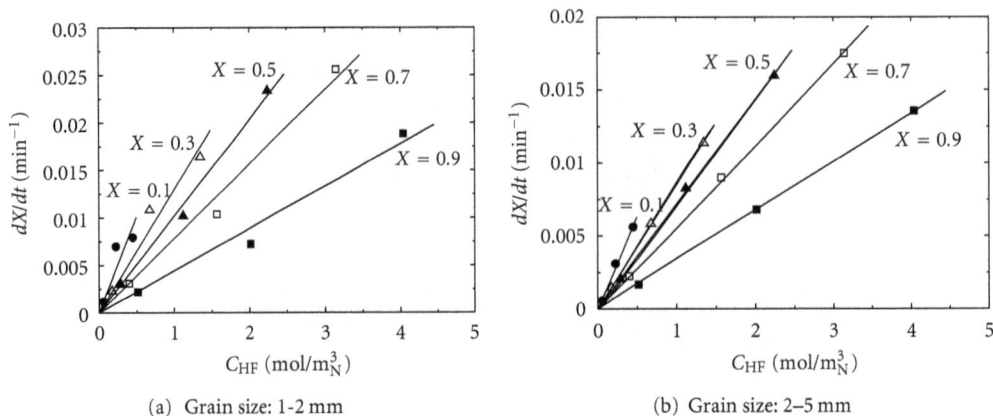

(a) Grain size: 1-2 mm

(b) Grain size: 2–5 mm

FIGURE 8: Reaction rate of $CaCO_3$ plotted against the HF concentration.

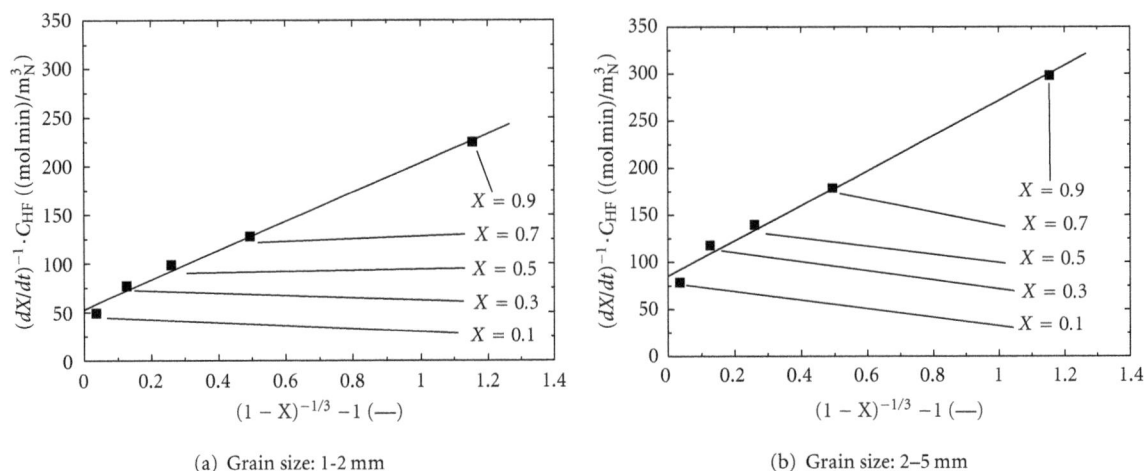

(a) Grain size: 1-2 mm

(b) Grain size: 2–5 mm

FIGURE 9: Plots of $(dX/dt)^{-1} \cdot C_{HF}$ versus $(1-X)^{-1/3}-1$.

TABLE 3: Rate constants of the gas-solid reaction of HF and $CaCO_3$.

| Grain size 1-2 mm | $f_p = 6.67 \times 10^{-3}$ [$m^3/(mol \cdot min)$] | $f_g = 1.88 \times 10^{-2}$ [$m^3/(mol \cdot min)$] |
| Grain size 2–5 mm | $f_p = 5.37 \times 10^{-3}$ [$m^3/(mol \cdot min)$] | $f_g = 1.17 \times 10^{-2}$ [$m^3/(mol \cdot min)$] |

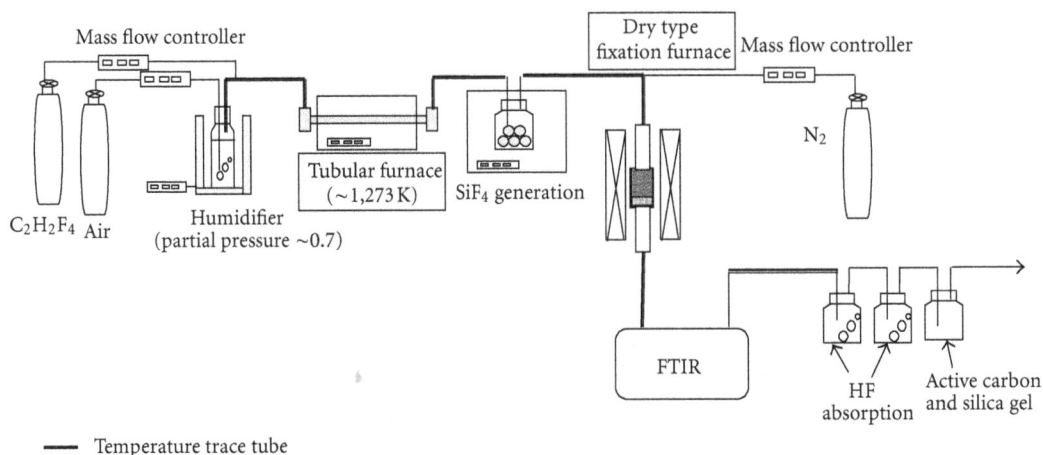

—— Temperature trace tube

FIGURE 10: Experimental flow of the gas-solid reaction of SiF_4 and sodium adsorbents.

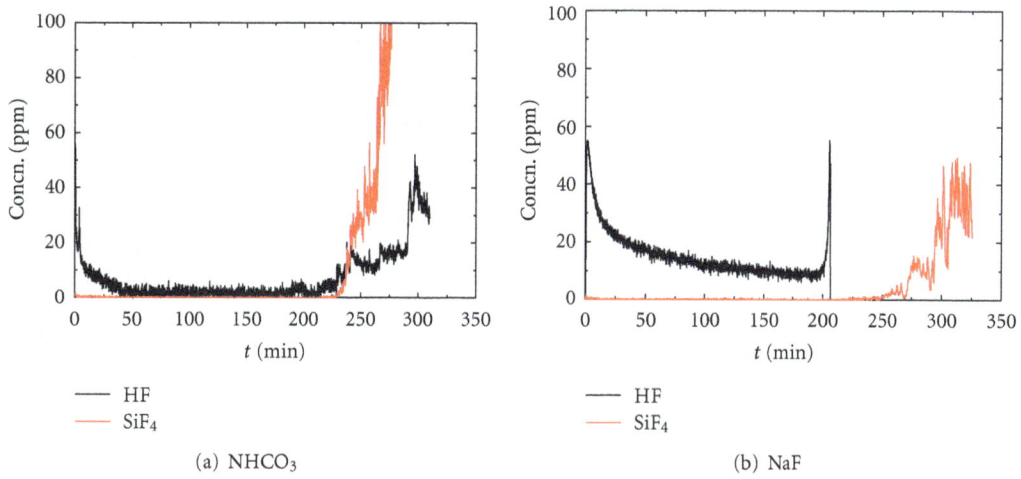

(a) NHCO$_3$ (b) NaF

FIGURE 11: Breakthrough properties of HF and SiF$_4$.

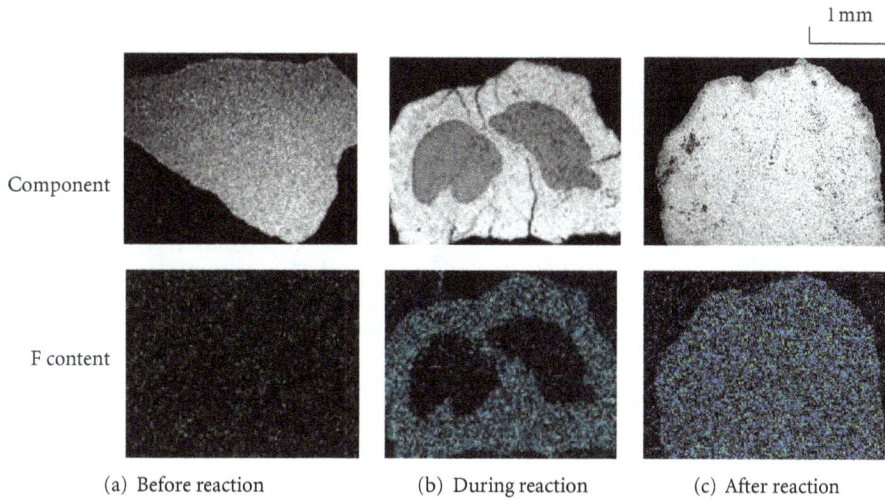

(a) Before reaction   (b) During reaction   (c) After reaction

FIGURE 12: EPMA photographs of the CaCO$_3$ adsorbents before and after experiments.

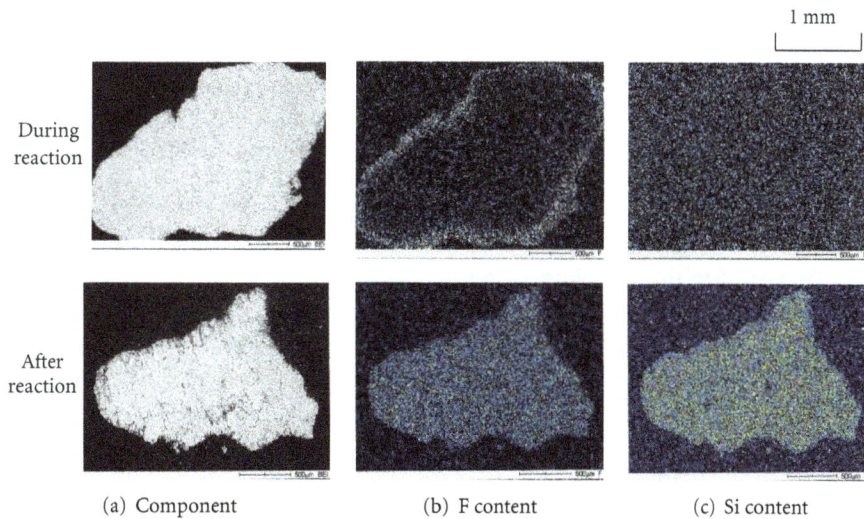

(a) Component   (b) F content   (c) Si content

FIGURE 13: EPMA photographs of the NaHCO$_3$ adsorbents before and after experiments.

TABLE 4: Experimental conditions of the fixed-bed furnace.

	W	T	V	L	L/D	v	SV
	g	°C	cm³	cm	—	cm/s	h⁻¹
NaHCO₃ or NaF	12	150	10.4	1.68	0.60	11.7	25066

used because it is generated by the reaction of HF and NaHCO₃ in the fixed-bed furnace. The concentrations of HF and SiF₄ in the exhaust gas from the fixed-bed furnace were analyzed by Fourier-transform infrared spectroscopy (FT-IR), and the amount of SiF₄ adsorbed by each adsorbent up to the moment of breakthrough was determined. Table 4 shows the gas-solid reaction conditions of these experiments.

5.2. Experimental Results. Figure 11 shows the results of the breakthrough properties of each adsorbent. The breakthrough times of SiF₄ in both adsorbents were more than 200 min from after the start of the gas-solid reaction. Because the concentration of SiF₄ was less than 3 ppm, which is the TLV value of SiF₄, until the breakthrough time, it was confirmed that both sodium-based adsorbents can completely adsorb SiF₄. The breakthrough times for the NaHCO₃ and NaF adsorbents were 230 min and 250 min, respectively. The adsorption capacity of each adsorbent was then calculated using that obtained time to be 0.133 g-SiF₄/g-NaHCO₃ and 0.145 g-SiF₄/g-NaF. NaHCO₃ can adsorb both the HF and SiF₄ that leak from CaCO₃ layer. The amount of NaHCO₃ filling the fix-bed furnace can be calculated using the obtained adsorption capacities for each gas by the amount of CaCO₃ loading in the fixation furnace.

6. Chemical Analysis of the Obtained Fluorides

The major advantage of this dry-treatment system is the ability to recycle the obtained fluorides. This section shows the chemical properties of the obtained fluorides. Figure 12 shows the EPMA photographs of the cross-section of the calcium adsorbents before and after the reactions. It is confirmed that the reaction with fluorine proceeds to the core of the calcium adsorbent. Mg and Si impurities are detected at about several hundred or several thousand ppm in the obtained calcium fluorides. These impurities are originally contained in the raw calcium carbonate materials as dolomite (CaMg(CO₃)₂) and silica. However, the purity of the obtained CaF₂ is over 97%, which is sufficient for it to be recycled as fluorine resources.

The EPMA photographs of the cross-section of the NaHCO₃ adsorbents before and after the experiments are shown in Figure 13. Similar to the CaCO₃ adsorbent, the fluorine and silicon reaction proceeded to the core of the adsorbents. Silicon was detected as Na₂SiF₆ by X-ray diffraction analysis. This sodium-based material can be also reused as a special material by the fluorine industry.

7. Conclusions

In order to develop a new dry-type gas treatment system for semiconductor facilities, the gas-solid reaction properties of HF, SiF₄ gas, and solid adsorbents in a fixation furnace were investigated. From the experimental results of the HF breakthrough properties of the CaCO₃ adsorbent, the apparent gas-solid reaction rate constants were obtained by using the rate formula of the unreacted core model. In addition, the HF and SiF₄ breakthrough properties of sodium base adsorbents of NaHCO₃ and NaF were investigated, and the adsorption capacities of each adsorbent were obtained. From the experimental results of the gas-solid reaction rate constants and the adsorption capacities, the optimum dry-type fixation furnace can be designed. Furthermore, by examining the chemical compositions of the obtained fluorides, it was confirmed that the calcium fluoride could be recycled as a fluorine resource.

Acknowledgments

This work was supported by a Grant-in-Aid for Scientific Research (C), the Iwatani Naoji Foundation's Research Grant.

References

[1] Ministry of Economy, Trade and Industry, "About the Future State of the Cure Against Discharge Control of Chlorofluorocarbon," Ministry of Economy, Trade and Industry Report, 2006.

[2] Mineral commodity summarizes, 2010, http://minerals.usgs.gov/minerals/pubs/mcs/.

[3] H. Shirai, M. Kobayashi, and M. Nunokawa, "Modeling of desulfurization reaction for fixed bed system using honeycomb type iron oxide sorbent and desulfurization characteristics in coal gas," *Kagaku Kogaku Ronbunshu*, vol. 27, no. 6, pp. 771–778, 2001.

[4] C. B. Shumaker and R. Schuhmann Jr., "Reaction rates for sulfur fixation with iron at 1100 to 1275 K," *Metallurgical Transactions B*, vol. 14, no. 2, pp. 291–300, 1983.

[5] S. Yasui, K. Ikeda, and H. Shirai, "Research on dry-processing technology for reconverted resources fluorine from waste chlorofluorocarbons," in *Proceedings of the AIChE Annual Meeting*, Salt Lake City, Utah, USA, November 2007, Paper no.173d.

[6] S. Yasui, S. Nakai, A. Ueno, T. Utsumi, T. Imai, and T. Murakami:, "Low temperature dry processing technology for exhaust gases containing fluorine using calcium absorbents," in *Proceedings of the 18th International Congress of Chemical and Process Engineering (CHISA '08)*, Prague, Czech Republic, 2008.

[7] J. Szekely, J. W. Evans, and H. Y. Sohn, *Gas-Solid Reaction*, Academic Press, New York, NY, USA, 1976.

Permissions

The contributors of this book come from diverse backgrounds, making this book a truly international effort. This book will bring forth new frontiers with its revolutionizing research information and detailed analysis of the nascent developments around the world.

We would like to thank all the contributing authors for lending their expertise to make the book truly unique. They have played a crucial role in the development of this book. Without their invaluable contributions this book wouldn't have been possible. They have made vital efforts to compile up to date information on the varied aspects of this subject to make this book a valuable addition to the collection of many professionals and students.

This book was conceptualized with the vision of imparting up-to-date information and advanced data in this field. To ensure the same, a matchless editorial board was set up. Every individual on the board went through rigorous rounds of assessment to prove their worth. After which they invested a large part of their time researching and compiling the most relevant data for our readers. Conferences and sessions were held from time to time between the editorial board and the contributing authors to present the data in the most comprehensible form. The editorial team has worked tirelessly to provide valuable and valid information to help people across the globe.

Every chapter published in this book has been scrutinized by our experts. Their significance has been extensively debated. The topics covered herein carry significant findings which will fuel the growth of the discipline. They may even be implemented as practical applications or may be referred to as a beginning point for another development. Chapters in this book were first published by Hindawi Publishing Corporation; hereby published with permission under the Creative Commons Attribution License or equivalent.

The editorial board has been involved in producing this book since its inception. They have spent rigorous hours researching and exploring the diverse topics which have resulted in the successful publishing of this book. They have passed on their knowledge of decades through this book. To expedite this challenging task, the publisher supported the team at every step. A small team of assistant editors was also appointed to further simplify the editing procedure and attain best results for the readers.

Our editorial team has been hand-picked from every corner of the world. Their multi-ethnicity adds dynamic inputs to the discussions which result in innovative outcomes. These outcomes are then further discussed with the researchers and contributors who give their valuable feedback and opinion regarding the same. The feedback is then collaborated with the researches and they are edited in a comprehensive manner to aid the understanding of the subject.

Apart from the editorial board, the designing team has also invested a significant amount of their time in understanding the subject and creating the most relevant covers. They scrutinized every image to scout for the most suitable representation of the subject and create an appropriate cover for the book.

The publishing team has been involved in this book since its early stages. They were actively engaged in every process, be it collecting the data, connecting with the contributors or procuring relevant information. The team has been an ardent support to the editorial, designing and production team. Their endless efforts to recruit the best for this project, has resulted in the accomplishment of this book. They are a veteran in the field of academics and their pool of knowledge is as vast as their experience in printing. Their expertise and guidance has proved useful at every step. Their uncompromising quality standards have made this book an exceptional effort. Their encouragement from time to time has been an inspiration for everyone.

The publisher and the editorial board hope that this book will prove to be a valuable piece of knowledge for researchers, students, practitioners and scholars across the globe.

List of Contributors

Mohit Katragadda and Nilanjan Chakraborty
School of Mechanical and Systems Engineering, Newcastle University, Claremont Road, Newcastle-upon-Tyne NE1 7RU, UK

Haruki Furukawa, Yoshihito Kato, Tomoho Kato and Yutaka Tada
Department of Life and Materials Engineering, Nagoya Institute of Technology, Gokiso-cho, Showa-ku, Nagoya-shi, Aichi 466-8555, Japan

Shunsuke Hashimoto and Yoshiro Inoue
Division of Chemical Engineering, Graduate School of Engineering Science, Osaka University, 1-3 Machikaneyama-cho, Toyonaka-shi, Osaka 560-8531, Japan

Ahmad Mousa
Department of Materials Engineering, Faculty of Engineering, Al Balqa Applied University, Salt 19117, Jordan

Gert Heinrich, Bernd Kretzschmar, Udo Wagenknecht and Amit Das
Leibniz-Institut fur Polymerforschung Dresden e.V., Hohe Straβe 6, 01069 Dresden, Germany

Wadood T. Mohammed and Sarmad A. Rashid
Chemical Engineering Department, College of Engineering, University of Baghdad, Baghdad, Iraq

Kent E. Wardle
Chemical Sciences and Engineering Division, Argonne National Laboratory, Argonne, IL 60439, USA

Henry G. Weller
OpenCFD Limited Bracknell, Berkshire RG12 1BW, UK

Yoshinobu Tanaka
IEM Research, 1-46-3 Kamiya, Ushiku-shi, Ibaraki 300-1216, Japan

Julian Fasano
Mixer Engineering Co., 2673 Stonebridge Drive, Troy, OH 45373, USA

Eric E. Janz
Chemineer, Inc., 5870 Poe Avenue, Dayton, OH 45414, USA

Kevin Myers
Department of Chemical & Materials Engineering, University of Dayton, 300 College Park, Dayton, OH 45469-0246, USA

Ruan C. A. Moura and Franco D. R. Amado
PROCIMM, State University of Santa Cruz, Road Ilheus-Itabuna km 16, 45662-000 Ilheus, BA, Brazil

Daniel A. Bertuol
DEQ, Federal University of Santa Maria, 97105-900 Santa Maria, RS, Brazil

Carlos A. Ferreira
PPGEM, Federal University of Rio Grande do Sul, 91501-970 Porto Alegre, RS, Brazil

Ziaul Huque, Ghizlane Zemmouri and Donald Harby
Department of Mechanical Engineering, Prairie View A&M University, P.O. Box 519, Mail Stop 2525, Prairie View, TX 77446, USA
Center for Energy and Environmental Sustainability, Prairie View A&M University, P.O. Box 519, Mail Stop 2500, Prairie View, TX 77446, USA

Raghava Kommalapati
Center for Energy and Environmental Sustainability, Prairie View A&M University, P.O. Box 519, Mail Stop 2500, Prairie View, TX 77446, USA
Department of Civil and Environmental Engineering, Prairie View A&M University, P.O. Box 519, Mail Stop 2510, Prairie View, TX 77446, USA

Gabriela C. Lopes, Leonardo M. Rosa, Jose R. Nunhez and Milton Mori
School of Chemical Engineering, University of Campinas, 500 Albert Einstein Avenue, 13083-970 Campinas, SP, Brazil

Waldir P. Martignoni
PETROBRAS, 65 Republica do Chile Avenue, 20031-912 Rio de Janeiro, RJ, Brazil

Ho Shing Wu and Yeng Shing Fu
Department of Chemical Engineering and Materials Science, Yuan Ze University, Zhongli, Taiwan

M. F. El-Amin
Computational Transport Phenomena Laboratory (CTPL), Division of Physical Sciences and Engineering (PSE), King Abdullah University of Science and Technology (KAUST), Thuwal 23955-6900, Saudi Arabia
Department of Mathematics, Aswan Faculty of Science, South Valley University, Aswan 81528, Egypt

Amgad Salama and Shuyu Sun
Computational Transport Phenomena Laboratory (CTPL), Division of Physical Sciences and Engineering (PSE), King Abdullah University of Science and Technology (KAUST), Thuwal 23955-6900, Saudi Arabia

Jozef Mikulec, Andrea Kleinova, Jan Cvengros, Ludmila Jorıkova and Marek Banic
VURUP a.s., Vlcie hrdlo, 820 03 Bratislava, Slovakia
Faculty of Chemical and Food Technology, Slovak University of Technology, Radlinskeho 9, 812 37 Bratislava, Slovakia

Shahram Moradi, Sanaz Raeis Farshid and Saeed Abedini Khorrami
Department of Chemistry, Tehran North Branch, Islamic Azad University, Tehran 1913674711, Iran

Parviz Aberoomand Azar and Mohammad Hadi Givianrad
Department of Chemistry, Science and Research Branch, Islamic Azad University, Tehran, Iran

Pritam V. Hule, B. N. Murthy and Channamallikarjun S. Mathpati
Department of Chemical Engineering, Institute of Chemical Technology, Matunga, Mumbai 400 019, India

Thanh Hai Ngo and Adrian Schumpe
Institut fur Technische Chemie, Technische Universitat Braunschweig, Hans-Sommer-Straße 10, 38106 Braunschweig, Germany

Nor Hanizah Shahirudin, Alatengtuya, Norihisa Kumagai, Takafumi Horie and Naoto Ohmura
Department of Chemical Science and Engineering, Kobe University, 1-1 Rokkodai, Nada, Hyogo, Kobe 657-8501, Japan

B. R. Rout
Department of Mathematics, Krupajal Engineering College, Prasanti Vihar, Pubasasan, Kausalyaganga, Bhubaneswar, Odisha 751002, India
S. K. Parida
Department of Mathematics, Institute of Technical Education and Research (ITER), SOA University, Bhubaneswar, Odisha 751019, India

S. Panda
Department of Mathematics and Civil Engineering, National Institute of Technology (NIT) Calicut, Calicut 673601, India

Thanasit Wongsiriamnuay
Division of Agricultural Engineering, Faculty of Engineering and Agro-Industry, Maejo University, Chiang Mai 50290, Thailand

Nattakarn Kannang and Nakorn Tippayawong
Department of Mechanical Engineering, Faculty of Engineering, Chiang Mai University, Chiang Mai 50200, Thailand

Marcelo Teixeira Leite
Center of Technology and Regional Development, Federal University of Paraíba, 58051-900 Joao Pessoa, PB, Brazil

Marcos Antonio de Souza Barrozo and Eloizio Julio Ribeiro
Chemical Engineering College, Federal University of Uberlandia, 38408-100 Uberlandia, Brazil

B. Kavita and Haresh Keharia
BRD School of Biosciences, Sardar Patel University, Vallabh Vidyanagar, Gujarat 388120, India

Shinji Yasui
Nagoya Institute of Technology, Gokiso-cho, Showa-ku, Nagoya 466-8555, Japan

Tadashi Shojo, Goichi Inoue and Kunihiko Koike
Iwatani Corporation, 4-5-1 Katsube, Moriyama-shi, Shiga 524-0041, Japan

Akihiro Takeuchi and Yoshio Iwasa
Chubu Electric Power Co., Inc., 20-1 Kitasekiyama, Odaka-cho, Midori-ku, Nagoya 459-8522, Japan